OCR
Geography

OCR and Heinemann are working together to provide better support for you

MAGDALEN COLLEGE SCHOOL
BRACKLEY

MAGDALEN COLLEGE SCHOOL
BRACKLEY

Jane Dove
Paul Guinness
Chris Martin
Garrett Nagle
Michael Witherick

www.heinemann.co.uk
✓ Free online support
✓ Useful weblinks
✓ 24 hour online ordering

01865 888118

Official Publisher Partnership

Heinemann is an imprint of Pearson Education Limited, a company incorporated in England and Wales, having its registered office at Edinburgh Gate, Harlow, Essex, CM20 2JE. Registered company number: 872828

www.heinemann.co.uk

Heinemann is the registered trademark of Pearson Education Limited

Text © Pearson Education Limited 2009

First published 2009

13 12 11 10 09

10 9 8 7 6 5 4 3 2 1

British Library Cataloguing in Publication Data is available from the British Library on request.

ISBN 978 0 435 357 627

Copyright notice
All rights reserved. No part of this publication may be reproduced in any form or by any means (including photocopying or storing it in any medium by electronic means and whether or not transiently or incidentally to some other use of this publication) without the written permission of the copyright owner, except in accordance with the provisions of the Copyright, Designs and Patents Act 1988 or under the terms of a licence issued by the Copyright Licensing Agency, Saffron House, 6–10 Kirby Street, London EC1N 8TS (www.cla.co.uk). Applications for the copyright owner's written permission should be addressed to the publisher.

Edited by Lucy Tritton and Caroline Low, Virgo Editorial
Designed by Wooden Ark Studio
Typeset by Phoenix Photosetting, Chatham, Kent
Original illustrations © Pearson Education Limited 2009
Illustrated by Phoenix Photosetting
Cover design by Pearson Education
Picture research by Helen Reilly
Printed in Italy by Rotolito Lombarda

Websites
There are links to relevant websites in this book. In order to ensure that the links are up to date, that the links work, and that the sites are not inadvertently linked to sites that could be considered offensive, we have made the links available on the Heinemann website at www.heinemann.co.uk/hotlinks. When you access the site, the express code is 7627P.

Acknowledgements

Garrett Nagle would like to thank Angela, Rosie, Patrick and Bethany for their support and help.

The author and publisher would like to thank the following individuals and organisations for permission to reproduce copyright material.

Maps, diagrams and extracts

Figures 1.14 and 1.17: from Yang, S. et al 'Trends in annual discharge from the Yangtze River to the sea (1865–2004)', *Hydrological Sciences Journal* vol. 50(5) pp.825–836 reprinted by permission of the International Association of Hydrological Sciences.

Figure 1.32: Risk map for Montserrat, September 1997, reprinted by permission of the Montserrat Volcano Observatory.

Figure 3.4: Figures from *Weather Systems* by Leslie F. Musk, 1988 Cambridge University Press by permission of Cambridge University Press.

Figure 3.13: Figures from *AS Geography: concepts and cases* by Paul Guinness and Garrett Nagle, Hodder Murray 2000 reprinted by permission of John Murray (Publishers) Ltd.

Figure 3.14: Figure 8.5: 'The Great Gale of 16 October 1987', *Geography Review* vol. 1 Issue 3 1987 reproduced with permission of Philip Allan Updates.

Figure 3.17: Extract from www.ussartf.org/blizzards.htm used by permission of the United States Search and Rescue Task Force.

Figure 4.3: Figure from Population Reference Bureau reproduced by permission of Population Reference Bureau.

Figures 4.4 and 4.9: Figure from *World Population Data Sheet 2005* reproduced by permission of Population Reference Bureau.

Figure 4.6: Figure from *Population Bulletin* vol. 60(4) 2005 reproduced by permission of Population Reference Bureau.

Figures 4.7 and 4.8: Figure from *World Population Prospects: the 2004 Revision medium scenario* reproduced by permission of Population Reference Bureau.

Figures 4.18 and 4.19: Figures from *BP Statistical Review of World Energy 2008* with permission of BP p.l.c.

Figure 4.20: Figure 25.2 from *New patterns: process and change in Human Geography* by Michael Carr, 1997, ISBN 0-17-438681-8 reprinted by permission of Nelson Thornes Ltd.

Figures 4.23 and 4.24: Figures from *Advanced Geography: concepts and cases* by Paul Guinness and Garrett Nagle, revised edition 2002 reproduced by permission of Hodder & Stoughton Ltd.

Figures 4.30 and 4.36: Figure used by permission of WRI EarthTrends.

Figure 4.31: Figure 'Humanity's ecological footprint 1961–2001' from *Eco-footprints* by David Holmes, GeoFactsheet no. 183, April 2005, fig 3 © Curriculum Press reprinted by permission of the Press.

Figures 4.32 and 4.33: Figures used by permission of the World Wildlife Fund.

Figure 4.44: Graph 'Decline in North Sea cod' from 'How do we balance conservation with the interests of the fishing industry?' by Martin Hickman, *The Independent*, 21 November 2007 reprinted by permission of The Independent.

Figures 5.2 and 5.26: Two figures from *Global Shift* by Peter Dicken, PCP (3rd edn, 1998) reproduced by permission of SAGE Publications, London, Los Angeles, New Delhi and Singapore © Peter Dicken 1998.

Figures 5.7, 5.8 and 5.9: Figure from *The CSGR Globalisation Index: an Introductory Guide* by Ben Lockwood, Michela Redoano (2005) Centre for the Study of Globalisation and Regionalisation Working Paper 155/04, reproduced by permission of Michela Redoano.

Figure 5.18: Figure from 'China vs India: the world's biggest two-horse race', *Business Review* vol. 14(1) 2007 reprinted by permission of Philip Allan Updates and Hodder Education.

Figure 5.20: Figure 4a from *World Development Indicators 2007*, p.185, reprinted by permission of The International Bank for Reconstruction and Development/World Bank.

Figures 5.21 and 5.41: Figure from Human Development Report 2007/08 reproduced by permission of Palgrave Macmillan.

Figure 5.22: Figure from *UN World Economic and Social Survey 2006*, Maddison (2001) and UN/DESA, reproduced by permission of the UN.

Figure 5.27: Figure 'The Commodity Circuitry of the Nike Shoe' from *Nike Culture* by Robert Goldman & Stephen Papson, 1998, p.8, reproduced by permission of SAGE Publications, London, Los Angeles, New Delhi and Singapore

Figure 5.32: Figure from http://trade.wtosh.com/english/res_e/statis_e/statis_e.htm reproduced with permission of the World Trade Organisation.

Figures 5.34 and 5.35: Reproduced with permission of the World Trade Organisation.

Figures 5.37, 5.38 and 5.40: UK, China and Bolivia Trade statistics from http://stat.wto.org/Country profile used with permission of the World Trade Organisation.

Figures 5.42 and 5.44: Figures *DFID provides emergency aid towards Bangladesh cyclone* and *UK's long-term goal*

for *Bangladesh* © Crown Copyright material is reproduced with the permission of the Controller, Office of Public Sector Information (OPSI) Click Use Licence No. C2008002327.

Figure 5.47: Figure from *The CSGR Globalisation Index: an Introductory Guide* by Ben Lockwood, Michela Redoano (2005) Centre for the Study of Globalisation and Regionalisation Working Paper 155/04, reproduced by permission of Michela Redoano.

Figures 6.1 and 6.20: Figures from *Development, Disparity and Dependence* by Michael Witherick, Nelson Thornes 1998, used by permission of the author.

Figure 6.5: Map 'The global distribution of four income groupings (2003)', reproduced with permission of the World Bank.

Figure 6.9: Table adapted from Philip's *Modern School Atlas*, 95th edition, 2006.

Figure 6.14: Figure from *The Changing Face of Japan* by Michael Witherick and Michael Carr, Hodder Arnold 1993, reproduced by permission of Edward Arnold (Publishers) Ltd.

Figure 6.19: Figure from Human Development Report 2007/08 reproduced by permission of Palgrave Macmillan.

Figure 6.21: Figure from Robertson Economic Information Services www.economic.co.zw with permission of John Robertson.

Figure 6.29: Table 'Quality of Life Index from The World in 2005' reprinted by permission of The Economist Newspaper Limited, London 2005.

Figure 6.30: From *The Little Green Data Book 2006*, reproduced by permission of the World Bank and the Copyright Clearance Center.

Figure 6.36: Reproduced with the permission of Nelson Thornes Ltd from EPICS: *Development, Globalisation and sustainability*, John Morgan, 978-0748758227, first published in 2001.

Figures 7.2 and 7.5: Reproduced by permission of Ordnance Survey on behalf of HMSO. © Crown copyright 2008. All rights reserved. Ordnance Survey licence number 100030901.

Figure 7.15: Figure from Getting started with GIS from www.rgs.org/OurWork/Schools/Resources/GIS/Getting+started+with+GIS.htm reproduced by permission of the Royal Geographical Society.

Figure 7.16: Image of Southend-on-Sea Bathing Waters from 'What's in Your Back Yard' section of the Environment Agency website © Environment Agency reproduced by permission of the Environment Agency.

CD-Rom only:

Figure 6.37: Map: 'The distribution of multiple deprivation in the UK (2006)'. Reproduced by permission of Ordnance Survey on behalf of HMSO © Crown Copyright 2003. All rights reserved. Ordnance Survey number 100000230 and by permission of the Office of National Statistics.

Figure 6.38: Map: 'The distribution of multiple deprivation in London by borough (2004)'. © Crown Copyright material is reproduced with the permission of the Controller, Office of Public Sector Information (OPSI) Click Use Licence No. C2008002327.

Figure 6.41: Figure from www.carbontrust.co.uk by permission of the Carbon Trust.

Figure 6.42: Figure 'The Distribution of SO_2 emissions in the UK' © Crown Copyright material is reproduced with the permission of the Controller, Office of Public Sector Information (OPSI) Click Use Licence No. C2008002327 and the NAEI.

Photographs:

Alamy/Anthony Thorogood: 66; .../Arctic Images: 99; .../Chad Ehlers: 119; .../Colin Palmer Photography: 275 (both); .../Danita Delimont: 148 (t); .../EmmePi Europe: 6–7; .../ephotocorp: 259; .../George Brice: 39 (b); .../Michael Grant: 55 (t); .../Robert Harding Picture Library Ltd: 248–9

Artville: 156

Corbis/A.J. Sisco: 108 (t); .../Andrew Brown/Ecoscene: 36 (t); .../Andrew Holbrooke: 223; .../Bettmann: 35; .../Justin Guariglia: 187; .../Lester Lefkowitz: 134 (t); .../Paulo Fridman: 173; .../Reuters/Richard Clement: 184; .../Stephanie Maze: 134 (b); .../Sygma: 11 (b)

Fotolia/Jean François Lefevre: 107; ...Jon R. Peters: 139; .../Jim Parkin: 75

Fotolia.com/bev: 149

Garret Nagle: 11 (t), 33 (t), 39 (t), 40, 103 (both); 108 (bl, br) 109

Getty Images: 150; 162–3; .../AFP: 193; .../Alan Staats/Stringer: 141; .../Christopher Pillitz: 36 (b); .../Doug Armand: 167; .../Eitan Abramovich/Stringer: 124–5; .../Keystone/Stringer: 21; .../Time & Life Pictures: 144

iStockphoto/George Clerk: 71 (b); .../Graham Bedingfield: 274; .../Martin Lovatt: 52; Martin Strmko: 74; .../Ricardo De Mattos: 221; .../Thomas Lammeyer: 76

Jane Dove: 53, 54, 55 (b), 59, 68, 77

Magnum Photos/Susan Meiselas: 33 (b)

Oxford Mail: 9

PA Photos/AP: 206–7, 228; .../Greg Baker/AP: 17; .../Robert Keilman/Ap: 197; .../Rui Vieira/PA Archive: 185

Panos Pictures/Justin Jin: 177; Mark Henley: 210; .../Qilai Shen: 148 (b); .../Rob Huibers: 31

Photodisc: 143, 155

Rex Features: 216; .../AS/TS/Keystone USA: 168; .../Kommer Keystone USA/Rex Features: 190; .../Phil Yeomans: 104; .../Ray Roberts: 116; .../Sipa Press: 15, 28, 166

Science Photo Library/Bernhard Edmaier: 19; .../CC Studio: 100; .../David C. Clegg: 172; .../David R. Frazier: 115; .../Gregory Dimijian: 4–5; .../NASA: 222, 246–7; .../NOAA: 84–85; .../Spot Image: 106; .../Steve Vowles: 46–47; Wayne Lawler: 71 (t)

Still Pictures/Mark Edwards: 151, 179; Reinhard Dirscherl/Water Frame: 83

CD-Rom only:

Corbis/Mike Theiss/Ultimate Chase: Figure 3.38
Science Photo Library/Planetary Visions Ltd: Figure 6.40

Every effort has been made to contact copyright holders of material reproduced in this book. Any omissions will be rectified in subsequent printings if notice is given to the publishers.

Contents

Introduction ... 1

Unit 3 Global issues

SECTION A: ENVIRONMENTAL ISSUES

Chapter 1 Earth hazards

1.1 What are the hazards associated with mass movement and slope failure? ... 8
1.2 What are the hazards associated with flooding? ... 16
1.3 What are the hazards associated with earthquake and volcanic activity? ... 24
1.4 Why do the impacts on human activity of such hazards vary over time and location? ... 34
1.5 How can hazards be managed to reduce their impacts? ... 37
Exam café ... 42

Chapter 2 Ecosystems and environments under threat

2.1 What are the main components of ecosystems and environments and how do they change over time? ... 48
2.2 What factors give the chosen ecosystem or environment its unique characteristics? ... 55
2.3 In what ways are physical environments under threat from human activity? ... 62
2.4 Why does the impact of human activity on the physical environment vary over time and location? ... 67
2.5 How can physical environments be managed to ensure sustainability? ... 72
Exam café ... 80

Chapter 3 Climatic hazards

3.1 What conditions lead to tropical storms and tornadoes and in what ways do they represent a hazard to people? ... 86
3.2 How do atmospheric systems cause heavy snowfall, intense cold spells, heatwaves and drought and in what ways do they represent a hazard to people? ... 96
3.3 Why do the impacts of climatic hazards vary over time and location? ... 105
3.4 What can humans do to reduce the impact of climatic hazards? ... 108
3.5 In what ways do human activities create climatic hazards? ... 113
Exam café ... 120

SECTION B: ECONOMIC ISSUES

Chapter 4 Population and resources

4.1	How and why does the number and rate of growth of population vary over time and space?	126
4.2	How can resources be defined and classified?	133
4.3	What factors affect the supply and use of resources?	137
4.4	Why does the demand for resources vary with time and location?	144
4.5	In what ways does human activity attempt to manage the demand and supply of resources and development?	149
Exam café		158

Chapter 5 Globalisation

5.1	What is meant by the term 'globalisation' and why is it occurring?	164
5.2	What are the issues associated with globalisation?	172
5.3	What are transnational corporations (TNCs) and what is their contribution to the countries in which they operate?	180
5.4	How far do international trade and aid influence global patterns of production?	186
5.5	How can governments evaluate and manage the impact of globalisation?	198
Exam café		202

Chapter 6 Development and inequalities

6.1	In what ways do countries vary in their levels of economic development and quality of life?	208
6.2	Why do levels of economic development vary and how can they lead to inequalities?	216
6.3	To what extent is the development gap increasing or decreasing?	225
6.4	In what ways do economic inequalities influence social and environmental issues?	232
6.5	To what extent can social and economic inequalities be reduced?	235
Exam café		242

Unit 4 Geographical skills

Chapter 7 Geographical skills

7.1	Identify a suitable geographical question or hypothesis for investigation	250
7.2	Develop a plan and strategy for conducting the investigation	253
7.3	Collect and record data appropriate to the geographical question or hypothesis	258
7.4	Present the data collected in appropriate forms	265
7.5	Analyse and interpret the data	276
7.6	Present a summary of the findings and an evaluation of the investigation	293
Exam café		296

Index	301

Your A2 Geography CD-ROM

Introduction to the Student Book and CD-ROM

Student Book

Units 3 and 4

Your A2 Geography course is divided into two units: Global issues and Geographical skills. This Student Book provides an exact match to the OCR specification. In addition to teaching and learning materials it also includes activities, exam support (through the Exam Café sections) and extension opportunities.

Features of the book

- **Key terms**: through the text you will find definitions of important key terms that may be new to you. These are a useful reference source.

- **Activities**: there is a range of motivating activities to help you practise what you are learning. These include '**discussion point**' activities that provide opportunities for small-group discussions and '**theory into practice**' activities which encourage the application of concepts to real-life contexts.

- **Case studies**: new and up-to-date case studies provide real-world examples of the topics you are studying. Questions on the case studies will enable you to explore the topic further, understand the key issues and deepen your understanding.

- **Knowledge check**: questions at the end of each chapter will check that you have taken on board all the concepts within the chapter and that you can use your knowledge synoptically.

Additional material contained on the CD-ROM is clearly signposted in the text, for example:

'Take it further' activity 1.3 on CD-ROM

In the unique Exam Café you'll find lots of ideas to help you prepare for the Unit 3 and Unit 4 exams. You'll find the Exam Café at the end of each chapter. You can **Relax** because there's handy advice on getting started on your A2 Geography course, **Refresh your memory** with summaries and checklists of the key ideas you need to revise, and **Get the result** through practising exam-style questions accompanied by hints and tips from examiners.

Student CD-ROM

LiveText

On the CD-ROM you will find an electronic version of the Student Book powered by LiveText. In addition to the Student Book and LiveText tools there are:

- additional case studies to support the main chapter content
- 'Take it further' extension activities which provide opportunities for you to undertake further work on a topic
- a glossary of all the key terms in the Student Book.

Within the electronic version of the Student Book you will also find the interactive Exam Café.

Immerse yourself in the contemporary interactive Exam Café environment! With a click of your mouse you can visit three separate areas in the café to **Relax**, **Refresh your memory** or **Get the result**. You'll find a wealth of material including:

- Revision timetable – a blank template for you to complete
- Revision tips from students, Bytesize concepts, Common mistakes and Examiner's hints
- Language of the exam (an interactive activity)
- Revision help in the form of checklists
- Multiple choice quizzes
- A 'Thinking and planning' tool to help you create your own mind maps
- Sample exam questions (which you can try) with student answers and examiner comments
- Sample exam papers
- Up-to-date weblinks that direct you to extra exciting resources and encourage further research on topics.

UNIT 3

Global issues

Environmental issues: Chapters 1–3
Economic issues: Chapters 4–6

The aims of this unit are to develop:

- the ability to identify and quantify issues of global concern
- an understanding that such issues are dynamic, that they change over time and place
- an appreciation of place and the diverse nature of its interdependent connections
- an understanding of the interdependence of environments and the dynamic interaction between people and the environment
- a knowledge of the use of modern technologies, such as GIS, remote sensing, etc., to understand the nature and impact of global issues
- a knowledge and understanding of the potential of ICT and its relevance to global issues
- the ability to select and use appropriate GIS skills and techniques to explore global issues
- an understanding of the diverse and dynamic factors responsible for global issues
- an understanding of how the effects of global issues may vary between countries at different stages on the development continuum (MEDCs, NICs and LEDCs)
- an understanding and evaluation of the diversity of responses to global issues

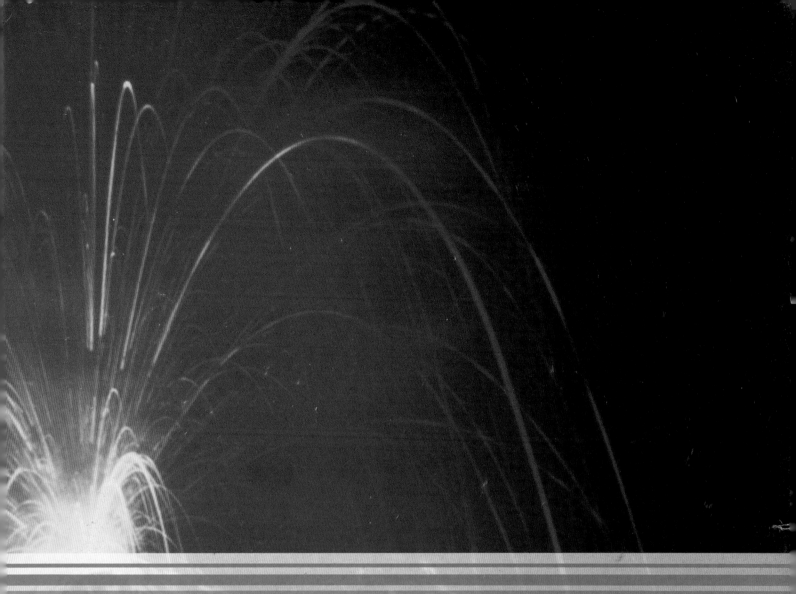

- the ability to synthesise understanding and knowledge from physical and human geography to develop explanation, connections and make evaluative judgements
- an ability to carry out individual research/investigative work, including fieldwork
- an understanding of geographical ideas, concepts and processes
- the skills to identify, analyse and evaluate the connections between the different aspects of geography
- the ability to analyse and synthesise geographical information in a variety of forms and from a range of sources
- an understanding of new ideas and developments about the changing nature of geography in the twenty-first century
- the ability to critically reflect on, and evaluate the potential and limitations of, approaches and methods used both inside and outside the classroom.

Chapter 1
Earth hazards

Since the South Asian tsunami of 2004 there has been an increased awareness of Earth hazards and the devastation they can cause, and more attention has been directed towards dealing with Earth hazards. The impact of the earthquakes in Northridge, Los Angeles (1994) and Kobe, Japan (1995) show that MEDCs as well as LEDCs can be badly affected by hazards in loss of life, disruption to the daily routine and economic losses. Even some of the wealthiest countries in the world are shown to be vulnerable. However, vulnerability to Earth hazards is very uneven – the impact of hazards is socially selective and generally the poor are more at risk than the rich.

Questions for investigation

- What are the hazards associated with mass movement and slope failure?
- What are the hazards associated with flooding?
- What are the hazards associated with earthquake and volcanic activity?
- Why do the impacts on human activity of such hazards vary over time and location?
- How can hazards be managed to reduce their impacts?

Consider this

- Are natural hazards becoming more frequent or are there more people living in hazardous environments?

- Are there ways in which human activities are leading to an increase in the frequency and intensity of hazards?

- In what ways are Earth hazards socially selective?

1.1 What are the hazards associated with mass movement and slope failure?

Mass movements include any large-scale movement of the Earth's surface that is not accompanied by a moving agent such as a river, glacier or ocean wave.

Causes of mass movements

Shear strength and shear resistance

Shear strength and shear resistance are key ideas in the understanding of mass movements and slope failure. Slope failure is caused by:

1. a reduction in the internal resistance, or shear strength, of the slope, and/or
2. an increase in shear stress, that is the forces attempting to pull a mass down slope.

Both can occur at the same time.

> **Key terms**
>
> **Shear strength:** the internal resistance of a body to movement.
>
> **Shear stress:** the force acting on a body that causes movement of the body down slope.

Increases in shear stress can be caused by a number of factors (Figure 1.1). These include: material characteristics; slope angle; weathering processes; water availability and climate; and weather. Weaknesses in rocks include: joints; bedding planes; and faults. Stress may be increased by:

- steepening or undercutting of a slope – this increases the slope angle and reduces slope strength
- the addition of a mass of material, such as the dumping of mining waste – this increases the stress on the slope and may increase slope angle
- weathering which may reduce cohesion and resistance
- vibrational shock and earthquakes – these destabilise the slopes and may trigger landslides even in urban areas (Figure 1.2).

Mass movements may be more common during periods of heavy precipitation which leads to increased levels of saturation in soils, or after cold conditions (**solifluction**) which leads to increased freeze-thaw activity and more rock falls (Figure 1.3).

Water is a vital ingredient in many slope failures. Water can weaken a slope by increasing shear stress and decreasing shear resistance. The weight of a potentially mobile mass is increased by:

- an increase in the volume of water
- heavy or prolonged rain
- rising water tables.

Figure 1.1 Increasing stress and decreasing resistance

Factors contributing to increased shear stress	
Factor	Examples
Removal of support through undercutting or slope steepening	Erosion by rivers and glaciers; wave action, faulting; previous rock falls or slides
Removal of underlying support	Undercutting by rivers and waves; subsurface solution
Loading of slope	Weight of water; vegetation; accumulation of sediments
Lateral pressure	Water in cracks; freezing in cracks; swelling; pressure release
Short-term stresses	Earthquakes; movement of trees in wind

Factors contributing to reduced shear strength	
Factor	Examples
Weathering effects	Disintegration of rocks; hydration of clay minerals; solution of minerals in rock or soil
Changes in soil and ground-water pressure	Saturation; softening of material
Changes of structure	Creation of fissures in clays; remoulding of sands and clays
Biological effects	Burrowing of animals; growth and decay of roots

Figure 1.2 Landslide at Oxford Castle mound, February 2007, probably triggered by intense rain and ground vibrations caused by redevelopment of the prison site

may be reduced if the water content becomes so high that the clay liquefies and loses its cohesive strength.

3. Vegetation binds the soil and as a result stabilises slopes. However, vegetation may allow soil moisture to build up and make landslides more likely.

Key term

Solifluction: in periglacial areas during the summer months the layer above the permafrost melts and the water-saturated soil becomes mobile, sliding easily down slope over the icy top layer of the permafrost below.

Figure 1.3 Seasonality and mass movements on the Isle of Thanet

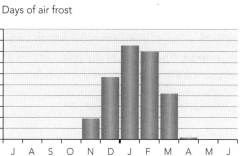

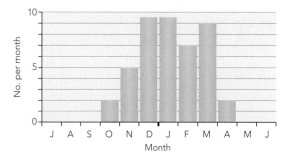

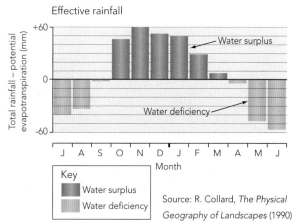

Source: R. Collard, *The Physical Geography of Landscapes* (1990)

Moreover, water reduces the cohesion of particles by saturation. Water pressure in saturated soils (pore water pressure) decreases the frictional strength of the slope material. This weakens the slope.

Under 'normal' conditions there are a number of factors which keep slopes in position. This is known as 'shear strength'. The down slope movement of slope material can be opposed by three main forces.

1. Friction will vary with the weight of the particle and slope angle. Friction can be overcome on very gentle slope angles if water is present. For example, solifluction can occur on slopes as gentle as 3 degrees.

2. Cohesive forces act to bind the particles on the slope. Clay may have high cohesion, but this

Activities

1. Define the terms shear strength and shear stress.
2. Study Figure 1.3 which shows seasonality of mass movements on the Isle of Thanet.
   a. What is meant by the term 'effective rainfall'?
   b. What is the relationship between the number of days of air frost and the number of rock falls per month?
   c. Suggest reasons for this relationship.

Types of mass movement

Mass movements include:

- very slow movement, such as soil creep
- fast movement, such as avalanches
- dry movement, such as rock falls
- very fluid movements such as mudflows.

A range of slope processes occur which vary in terms of magnitude, frequency and scales. Some are large and occur infrequently, notably rock falls, whereas others are smaller and more continuous, such as soil creep.

The types of processes can be classified in a number of different ways (Figure 1.4):

- type of movement: flows, slides, slumps
- rate of movement
- type of material
- water content.

One of the slowest forms of mass movement is soil creep. This is a slow, small-scale process which occurs mostly in winter in mid-latitude regions. It is one of the most important slope processes in environments where flows and slides are not common. Talus creep is the slow movement of fragments on a scree slope. Individual soil particles are pushed or heaved to the surface by wetting, heating or freezing of water. About 75 per cent of the soil creep movement is induced by moisture changes and associated volume change.

Falls occur on steep slopes, especially bare rock faces where joints are exposed. The initial cause of the fall may be weathering, such as freeze-thaw or disintegration, or erosion prising open lines of weakness. Once the rocks are detached they fall under the influence of gravity. Freeze-thaw on the Marsden Rock, a limestone arch near South Shields, triggered a series of rock falls in the winter of 1995–6 causing the collapse of the arch.

Figure 1.4 Classification of mass movement

Type of movement	Rate of movement	Type of material	Water content	Process
Flow	Very fast	Snow and ice	Very high	Snow avalanche; rock, rock debris; colluvium and soil will also make up debris flow
		Rock and rock debris	Low	Rock avalanche with rock debris
		Rock debris, colluvium and soil		Lahar
	Moderately fast	Soil	High	Mudflow
	Very slow	Ice	Very high	Rock glacier
Fall	Fast	Rock, rock debris, colluvium and soil	Low	Cohesive block fall or topple; individual block/grain fall/topple
Translation slide	Slow-fast	Rock	Low to medium	Plane, wedge slide
		Rock debris, colluvium and soil		Block slide, grain slide
Rotation slide	Slow	Rock	Low to medium	Circular slide
		Colluvium		Debris slump
		Soil		Soil slump
Creep/heave	Very slow	Rock	Low	Rock creep; cambering
		Colluvium and soil		Talus creep; solifluction; grain creep and rain splash

Figure 1.5 Landslide at Mam Tor, Peak District

Slides occur when an entire mass of material moves along a slip plane. These include:

- rockslides and landslides of any material, rock or regolith (weathered rock)
- rotational slides, which produce a series of massive steps or terraces.

Slides commonly occur where there is a combination of weak rocks, steep slopes and active undercutting (Figure 1.5). Slides are often caused by a change in the water content of a slope or by very cold conditions. As the mass moves along the slip plane it tends to retain its shape and structure until it impacts at the bottom of a slope. Slides range from small-scale slides that might be seen close to roads in hilly areas, to large-scale movements such as the Southern Leyte landslide in the Philippines in 2006 which killed 1126 people.

The slip plane can occur at a variety of places:

- at the junction of two layers
- at a fault line
- where there is a joint
- along a bedding plane
- at the point beneath the surface where the shear stress becomes greater than the shear strength.

Weak rocks such as clay have little shear strength and are particularly vulnerable to the development of slip planes. The slip plane is typically a concave curve and as the slide occurs the mass will be rotated backwards.

Slumps occur on weak rocks, notably clay, and have a rotational movement along a curved slip plane. Clay absorbs water, becomes saturated and thereby unstable. Frequently slumps occur when the base of a cliff has been undercut and weakened by erosion, as a result reducing its strength. The clay then flows along a slip plane. Human activity can also intensify the condition by causing increased pressure on the rocks, for example by water entering the soil from leaking pipes and drains and, on the coast, by the building of groynes starving the beach of sediment and making it easier for waves to erode the cliff. This was partly the case at the Holbeck Hall Hotel, Scarborough. By contrast, flows are more continuous, less jerky and are more likely to contort the mass into a new form. Fine grained materials, such as deeply weathered clays, become saturated with water, lose their cohesion and flow downhill as a very fluid mass (Figure 1.6). The flow is fastest at the surface and slows down at depth.

Activities

1 Describe at least two ways of classifying mass movements.
2 How useful are these methods of classification?

Figure 1.6 Mudflow near Plymouth, Montserrat

Case study of a flow/landslide | The Aberfan disaster, 1966

On 21st October 1966, a landslide involving a coal tip slag heap at Aberfan, South Wales, killed 147 people, 116 of whom were children at the Pantglas Junior School. Aberfan was overlooked by the tips of the Methyr Vale Colliery (Figure 1.7). The landslide involved over 100 000 cubic metres of colliery waste travelling at speeds of up to 30 km/hr. Earlier landslides had occurred in the vicinity. The National Coal Board believed that the speed of movement was slow enough to allow a warning to be given. Spoil heap number 7 was located above a spring. Water seaping through the sandstone emerged as a spring in the lower part of the tip. As the water passed through, it removed fine clay from the 'toe' of the tip thereby increasing its steepness.

> It was 9.15 and the school had just finished morning prayers when the million tons of coal waste, rocks and water crumpled 800 feet down the Aberfan mountain. The children were due to start their week's half-term holiday at noon. But many children never reached the school at all. Fog delayed a busload of 50 seniors and juniors from the village of Mount Pleasant nearby. Some decided to walk – and arrived late enough to miss the landfall. (22 October 1966)

> The National Coal Board (NCB) said last night that a build-up of rain water inside the tip at Aberfan had probably burst the base of the tip and caused 2 million tons of waste to slide down the valley. (22 October 1966)

> Already by Friday morning there had been much more rain during October in the area of South Wales which includes Aberfan than it gets on average during the whole month – and more than half of this had fallen in one torrential downpour two days before. (24 October 1966)

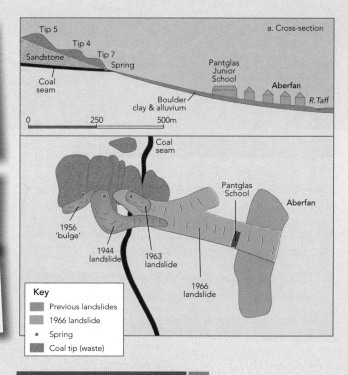

Figure 1.7 The Aberfan disaster

Case study of a flow/landslide | Holbeck Hall, 1993

On the 5th June 1993 a landslip along the upper sections of the boulder clay, at the 60 m cliffs at Start Bay, Scarborough, caused the destruction of the Holbeck Hall Hotel (Figure 1.8). The hotel had dominated the panorama since its construction in 1883. The coastline between Scarborough and Easington (Spurn Head) is among the most rapidly retreating areas in Britain and would cost millions of pounds to protect.

The Ministry of Agriculture, which supports coastal protection schemes, issued a statement blaming a succession of droughts making the area unstable. The boulder clay had become dry and cracked in previous years and then saturated by the rains in spring and early summer. The saturated clay became unstable, and slumped along a slip plane thereby causing an earthflow at the base of the slump (Figure 1.8). It is not thought erosion by the sea played a significant role. Little can be done except to let the slip stabilise itself when it reaches an angle of about 25 degrees.

Creep

Creep is the very gradual movement of regolith downslope. Evidence of creep includes tilted trees, fences, walls, and telegraph poles. A number of factors contribute to creep (Figure 1.9).

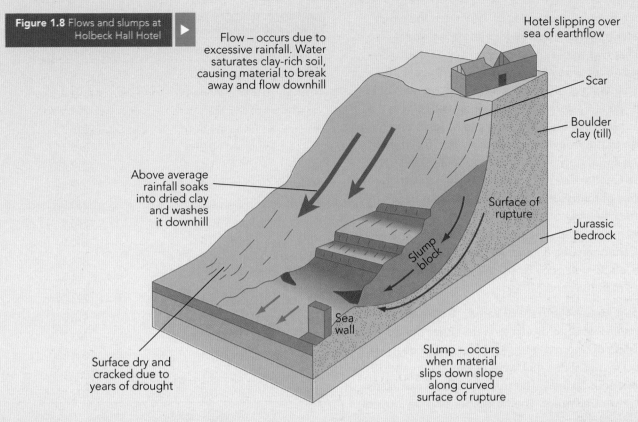

Figure 1.8 Flows and slumps at Holbeck Hall Hotel

Factor	Actions on soil
Frost heaving	Freezing and thawing, without necessarily saturating the regolith, causing lifting and subsidence of particles.
Wetting and drying	Causes expansion and contraction of clay minerals.
Heating and cooling without freezing	Causes volume changes in mineral particles.
Growth and decay of plants	Causes wedging, moving particles downslope; cavities formed when roots decay are filled from upslope.
Activities of animals	Worms, insects, and other burrowing animals displace particles, as do animals trampling the surface.
Dissolution	Mineral matter taken into solution creates voids in bedrock that tend to be filled from upslope.
Activity of snow	A seasonal snow cover tends to creep downslope and drag with it particles from the underlying ground surface.

Figure 1.9 Factors affecting soil creep

Research in Colorado USA suggests rates of 9.5 mm/year on slopes of 39 degrees, but just 1.5 mm on a 19-degree slope. Rates increase as soil moisture increases. However, in wet climates vegetation density increases and the roots bind the soil, thereby reducing movement. Rates in the UK for slopes as steep as 33 degrees are as low as 0.2 mm/year.

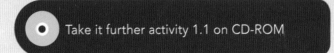

Take it further activity 1.1 on CD-ROM

Case study | The Venezuelan mudslides

The Venezuelan mudslides of 1999 were the worst disaster to hit the country for almost 200 years. The first two weeks of December saw an unusually high amount of rainfall in Venezuela (Figure 1.10). Precipitation was 40 to 50 per cent above normal in most of the Eastern Caribbean during 1999. On the 15 and 16 December an avalanche of rocks and mud began to pour down the slopes of the 2000 m-high Mount Avila, burying large parts of a 300 km stretch of the Central Coast (Figure 1.11). The rains triggered a series of mudslides, landslides and flash floods that claimed the lives of between 10 000 to 50 000 people in the narrow strip of land between the mountains and the Caribbean Sea (Figure 1.12). Over 150 000 people were left homeless by landslides and floods in the states of Vargas and Miranda.

Hardest hit was the state of Vargas. Countless mountainside slum dwellings were either buried in the mudslide or swept out to sea. Most of the dead were buried in mudslides that were between 8 m and 10 m deep (about 30 feet). The true number of casualties may never be known. The mudslides also destroyed roads, bridges, factories, telecommunications, buried crops in the fields and ruined Venezuela's tourist industry for the immediate future. The international airport of Caracas was temporarily closed and the coastal highway was destroyed or closed in many places. Flash floods halted operations at the seaport in La Guaira and hampered efforts to bring in emergency supplies. Container ships in the port were damaged and some hazardous materials leaked into the ground and into the sea. Economic damage was estimated at over US$ 3 billion.

Figure 1.10 Rainfall data for Venezuela, December 1999

	Daily rainfall (mm)	Cumulative rainfall (mm)
Dec 1	6.0	6.0
2	77.3	83.3
3	121.1	204.5
4	11.8	216.3
5	0.0	216.3
6	1.1	217.4
7	5.0	222.4
8	8.1	230.5
9	10.4	240.9
10	0.0	240.9
11	23.2	264.1
12	21.8	285.9
13	7.1	293.0
14	120.0	413.0
15	380.7	793.7
16	410.4	1204.1
17	2.9	1207.1
18	0.0	1207.1

Source: Venezuelan Ministry of the Environment and Natural Resources

The disaster was not just related to heavy rainfall. The government blamed corrupt politicians from previous governments and planners who had allowed shanty towns to grow up in steep valleys surrounding the coast and the capital, Caracas.

Figure 1.11 Venezuela showing the main areas affected by the mudslides of 1999

The immediate response was a search and rescue operation to find any survivors in the mudflows, landslides and buildings that had been damaged or destroyed. Few survivors were found. The other short-term response was to provide emergency relief – accommodation, water purification tablets, food and medicines to those in need. The relief operation was severely hindered by the poor state of the infrastructure.

Despite the tragedy, the government was determined that some good could come from the disaster. Up to 70 per cent of Venezuela's population live in the coastal area. A long-term plan had been to encourage people away from the over-crowded coast to the interior. The disaster may make this easier to achieve.

Government plans for rebuilding

The Venezuelan government announced plans to restore Venezuela's northern coastal region by rebuilding thousands of homes, expanding the country's main airport and constructing canals that can direct rivers away from communities.

Parts of Vargas were quickly rebuilt. First, the cities of Maiquetia, home to the country's main airport, and La Guaira, the largest port, were rebuilt, then the smaller towns, many of which had a tourist function.

The towns which were utterly devastated by the disaster and which were swept out to sea, will not be rebuilt. Instead, the land that they occupied has been turned into parks, bathing resorts or other outdoor facilities, for example Carmen de Uria, one of the villages hardest hit by the disaster.

2005 hazards

In 2005, floods and mudslides brought on by heavy rains in the northern and central coast of Venezuela caused 14 deaths. Some 18 000 people were affected, while 2840 houses were damaged and a further 363 destroyed. The same areas that were affected in the 1999 mudslides were also affected in 2005.

Activities

1. What were the causes of the Venezuelan mudslides?
2. Why were the impacts so great?

Discussion point

Natural hazards are an inevitable consequence of development, discuss.

Theory into practice

What evidence is there, in your home location, of mass movements?

Figure 1.12 Devastation caused by a mudslide in Caracas, Venezuela

1.2 What are the hazards associated with flooding?

Flood risk reflects a combination of physical and human factors, and these vary from place to place. Flooding has a variety of environmental and social impacts on the areas affected, which create a range of human responses to the hazard.

Case study | Flooding on the Yangtze River, China

The Yangtze River, or Chang Jiang, is the longest river in Asia and the third longest in the world. The river is about 6300 km long and flows from its source in Qinghai Province, at an altitude of over 5000 m in the Qinghai-Tibet plateau (Figure 1.13). It drains an area of over 1 800 000 km² and has an average discharge of 31 900 m³/s (Figure 1.14). It contains a very important element of China's economic and social geography (Figure 1.15).

Economic losses caused by flooding and meteorological disasters in China account for over 70 per cent of the total losses by all natural disasters. Each year the crop area damaged by meteorological disasters exceeds 50 million hectares, and the population affected is over 400 million. In addition, the total economic loss equals 3–5 per cent of GDP in China.

In 1998 flooding of the Yangtze (Figure 1.16) resulted in 3700 deaths, at least 15 million people losing their homes and economic losses of over US$ 26 billion. The floods inundated 25 million hectares of cropland. The flood prone area, accounting for only 8 per cent of the total national territory, is inhabited by nearly 50 per cent of the total population, where the gross industrial and agricultural product is estimated to be 66 per cent of the national total.

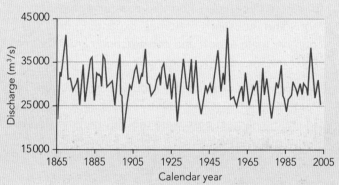

Source: Yang, S.L. et al., 'Trends in annual discharge from the Yangtze River to the sea (1865–2004)', *Hydrological Sciences Journal*, volume 50, October 2005

Figure 1.14 Annual discharge of the Yangtze at Datong (1865–2004)

Figure 1.13 Map of the Yangtze River basin – physical

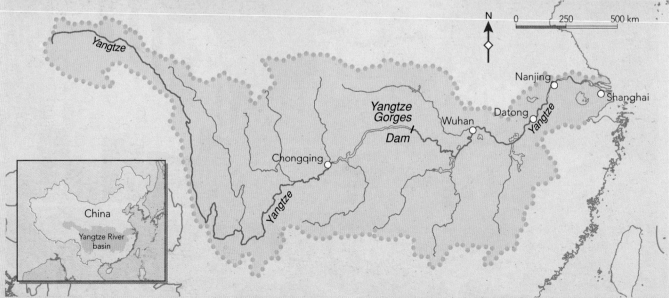

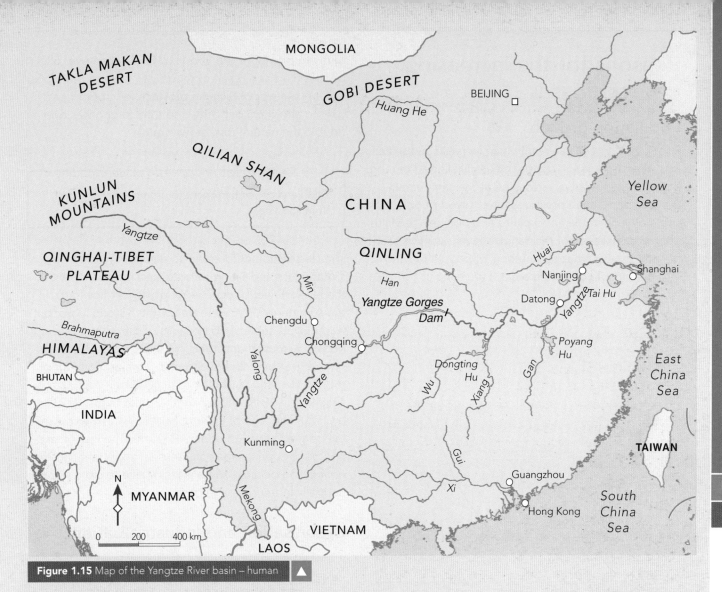

Figure 1.15 Map of the Yangtze River basin – human

There have been many major floods in China. In the early part of the twentieth century, monsoon rains caused over 100 000 deaths on both the Huang He and the Yangtze. At the end of the century and in the early twenty-first century, flooding caused 3700 deaths on the Yangtze. However, the biggest loss of life has been due to the destruction of levees by the military, resulting in over 1 million deaths in 1938.

The main natural causes of floods were heavy rainfall and rapid snowmelt. Rainfall along the Yangtze River from November 1997 to April 1998 was the highest ever recorded. In addition, the most serious snowmelt disaster of the century on the Qinghai-Tibet plateau occurred between November 1997 and February 1998.

Deforestation also contributed to the impact: up to 85 per cent of the watershed's original tree cover had been removed. Moreover, the flood plain was inadequately protected because of incomplete levees.

Figure 1.16 Flooding on the Yangtze River, 1998

Reasons for the increasing disaster losses

Population growth

Population growth can result in an increase in disaster loss. High population density means likelihood of more deaths in the event of a disaster; at the same time, growing population pressures have forced people to settle on vulnerable flood plains and hillsides. Satellite photographs taken of the Yangtze River over the past decade show increasing numbers of people moving into the most flood-prone areas (Figure 1.17).

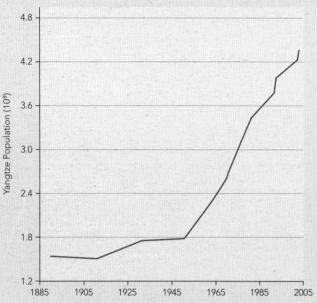

Source: Yang, S.L. et al., 'Trends in annual discharge from the Yangtze River to the sea (1865–2004)', *Hydrological Sciences Journal*, volume 50, October 2005

Figure 1.17 Population of the Yangtze River (1885–2005)

Environmental degradation

Environmental degradation increases the chance of flooding and landslides. Human activities in river basins like the Yangtze, such as deforestation, excessive land reclamation, large-scale road construction, mining in mountainous and hilly areas and wetlands destruction, degrades the soil and water, which results in floods or landslides. The soil erosion rate along the Yangtze River has risen by 40–50 per cent since the 1950s. Devastating landslides and floods have been exacerbated in recent years by deforestation and the construction of dams. There has been a large increase in the number of landslides along the shorelines of the reservoir created by the Three Gorges dam as the dam has created higher water levels. Dikes have hemmed in the Yangtze River and eliminated wetlands. As well as heavy rain, excessive logging of hillsides was partly to blame for the floods of 1998, when over 550 people died as a result of mudslides. During the 1950s the forest cover of Sichuan province fell from 19 per cent to about 6.5 per cent, and tree cutting was over double the natural tree growth rate.

Key term

Dike: artificial embankment to protect low-lying land from flooding.

The destruction of forests also results in natural water storage loss and silting of rivers and lakes, and raises the level of river beds. The Yangtze River flow rates in the summer of 1998 were below the historic highs, but water levels set records because of the silting of river beds – increased deposition of fine-grained material on the river bed. Dongting Lake, a major regulator of the Yangtze floods, shrunk from 6000 km² in 1700 to 4350 km² in 1949, and by another third to just 2820 km² by 1980. The area of Poyang Lake in Jiangxi Province fell from 5000 km² to under 4000 km². The loss of lake and pond surface area volume is an important cause of flooding because there is less space to store water naturally in the drainage basin.

Urbanisation and industrialisation

Rapid urbanisation results in bigger and more frequent floods on the one hand and more economic losses on the other. More importantly the damage to communications, water supply, electricity supply, oil and gas supply in one place may paralyse the whole economic system of a region. The Chinese rush toward economic modernisation has contributed to the severity of flood incidents. For example, in May 1992, the government of Hunan Province tore down one kilometre of dikes in order to open up new agricultural land. The result was that 80 000 hectares of farmland and 700 000 people lost their protection from flooding.

Land use and planning

Land reclamation for agriculture in flood-prone areas and accompanying human settlements increases flood vulnerability and losses of land by floods. Human activities, such as the filling in of lakes and rivers, have reduced the ability to regulate the floodwaters. China produces 70 per cent of its agricultural and industrial product within 10 per cent of its territory. Much of this densely populated area lies on the flood plain below the flood level of China's rivers, resulting in frequent and devastating floods. Owing to the reclamation of the flood plain and detention lakes (temporary holding lakes for flood waters), the detention area (temporary holding land for flood waters) for the lower Yangtze has been reduced to about a half of that in 1954.

Poor housing conditions

Poor housing is more likely to be affected by flooding than housing of the wealthy, which is sturdier and less likely to be in flood-prone regions. Thus flooding disproportionately affects poor people. Because of the shortage of suitable land for farmers, satellite photographs taken over the past decade show increasing numbers of land-poor farmers moving into the most flood-prone areas and building temporary accommodation. The encroachment by farmers onto natural flood-holding areas such as wetlands, lakes and even river beds is increasing the risk of disasters.

Inadequate prevention capacity and preparation

One reason for the Yangtze River flood of 1998 is that the dikes and reservoirs were only built to counter floods that might come once every 10 to 20 years. More silt and less water storage capacity in the Yangtze River basin has meant that even with less water, floods have became more serious. Increased silting deposits 150 million tonnes of mud on Hunan Province river and lake bottoms each year. The bottoms were one to two metres higher in 1998 than they were in 1949.

But although the 1998 water level was the highest ever, the rate of the flow of the river was not. For example, the peak flow rate of the Yangtze at Yichang at the height of the flood was 56 400 cubic metres per second. But according to records, the peak flow rate at Yichang has exceeded 60 000 cubic metres on at least 23 occasions.

Climate change

The seasonal floods that afflict China are caused in part by the Asian monsoon winds, which sweep rain clouds from the oceans toward China in the spring and summer. In 1998, however, the rains began earlier and were heavier than usual, a possible result of climate change, and the Chinese dikes were not prepared nor properly fortified to withstand the tremendous downpour.

> **Key term**
>
> **Monsoon winds:** a seasonal reversal of pressure and wind over land masses and neighbouring oceans. It also refers to the rains that accompany in-flowing winds (off the oceans). The summer monsoon is the wet season.

Management of the floods

In order to decrease the risk of flooding, Chinese authorities must reverse the destruction of forests, prevent encroachment onto the flood plains, and increase the water storing capacity (Figure 1.18) in dams, such as the Three Gorges Dam. A US$ 2 billion five-year reforestation programme is intended to reduce soil loss on the upper reaches of the Yangtze and Yellow River and so reduce silting at the Three Gorges Dam as well as flood risk. The Chinese government reforestation program includes ten shelter forest projects that will protect old growth forest (and the biodiversity of the forests) and develop sustainable forestry. In addition, the state timber company work force will be cut by 1 million workers and wood production will be reduced by 10 million cubic metres.

Figure 1.18 Dams are important for water storage, especially in areas where the natural vegetation has been removed

Most tools needed to reduce disaster vulnerability already exist, such as:

- risk assessment techniques
- better building codes and code enforcement
- land-use standards
- reparation plans for emergencies.

Mitigation of greenhouse gas emissions should also play a central role in response to human-related climate change, though this would not have an effect for several decades on the hazard risk of flooding.

Activities

1. What are the causes of flooding in the Yangtze drainage basin?
2. Why are the impacts of flooding so great there?
3. What can be done to manage the impacts of flooding on the Yangtze?

Storm surges in the southern North Sea

Flooding in coastal areas

Flooding of low-lying coastal areas can occur for a host of reasons, such as tsunamis, intense local precipitation, high river flows and storm surges. Some of these causes of flooding may interact with each other, as well as with other hazards such as human-induced subsidence. Coastal areas are characterised by growing concentrations of human population and socioeconomic activity, which means such floods can have severe impacts, including significant loss of life. Widespread efforts to mitigate coastal flood hazards are already apparent, and this need is likely to intensify throughout the twenty-first century.

Storm surges are generated by tropical and extra-tropical storms. The low air pressure and wind direction (inward-blowing winds) combine to produce large, temporary rises in sea level that have the capacity to cause extensive flooding of coastal lowlands. They are usually associated with strong winds and large onshore waves, which increase the damage potential. The largest surges are produced by hurricane landfalls, but extra-tropical storms can also produce large surges in appropriate settings. The southern North Sea region experiences significant surges due to extra-tropical storms, and the locations susceptible to flooding feature large populations and substantial investments (Figure 1.19).

Key terms

Extra-tropical storm: a temperate storm or depression formed outside of the tropics.

Hurricane landfall: the area of land where the hurricane first touches land.

Storm surge: rapid rise of sea level above normal high tide, in which water is piled up against the shore by strong onshore winds.

Storm surges

Surges are changes in sea level resulting from variations in atmospheric pressure and associated winds. They occur on top of normal tides, and when surges are added to high tides they can cause extremely high water levels and flooding (flooding is most severe when a surge coincides with a spring tide).

The magnitude of the surge is controlled partly by the intensity and track of the storm, and partly by the configuration of the coastline and seabed.

- Onshore winds serve to pile water against the coast and to generate surface currents and waves, which add to the maximum sea surface.
- A depression also reduces the atmospheric pressure, resulting in a rise in sea level (a fall of 1 mb in pressure results in a sea level rise of 1 cm).
- Coastlines fronted by a wide, shallow continental shelf experience larger surges than coastal areas with steeper slopes and greater water depths.
- Coastal configuration is also important. The southern North Sea, for example, is open to the north and nearly closed to the south, thus amplifying the potential for surges.

Key term

Continental shelf: the gently sloping margins of a continent, submerged beneath the sea. The continental shelf ends at the point where the seaward slope becomes very steep (the continental slope).

Figure 1.19 Floods in the Netherlands, 1953

Case study | Flooding in the Netherlands

Given appropriate conditions, surges due to extra-tropical storms can reach 2 to 3 metres in the southern North Sea, as happened in the storm surge of 31 January–1 February 1953. The surge killed up to 2000 people in the Netherlands and over 1600 km² were inundated (Figure 1.20). The areas most affected were concentrated between Rotterdam and the Scheldt and substantial areas remained flooded five months later. With improved flood defences major floods have been rare: the last time Hamburg in Germany was flooded was in 1962.

Flooding due to storm surges has a range of impacts, including property damage and destruction, human distress and health effects, and, in the worst cases, fatalities. Even under relatively mild surge regimes (< 1 m), significant property damage can occur. However, deep surges and fast-moving water can also lead to death by drowning.

Following the disaster of 1953, the Netherlands constructed the world's largest flood barrier in Zeeland, south-east Netherlands (Figure 1.21). Known as the Eastern Scheldt Storm-Surge barrier, it forms part of the Delta Project – a huge engineering operation that built a chain of flood defences to protect the Netherlands from a repetition of the 1953 catastrophe. The flood defences are built to a very high standard, able to withstand a 1 in 10 000 year storm-surge event. Equally important, there is a storm tide warning service that provides up to 36 hours' warning of a potential flood event. Collectively, these new measures have been extremely effective and there has been no flooding in the Netherlands since, even though the extreme water levels of the 1953 event have been repeated and even exceeded in some locations.

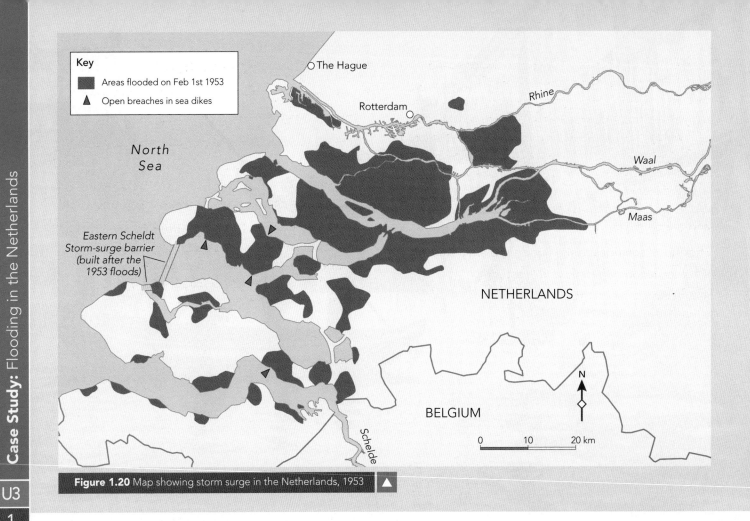

Figure 1.20 Map showing storm surge in the Netherlands, 1953

Responding to storm surges

Choosing change means accepting the hazard and changing land use, or even relocating exposed populations. Reducing losses includes trying to reduce the occurrence of the hazardous event or, more commonly, reducing the impacts of a hazardous event when it occurs. Both flood-protection and flood-warning systems are approaches to reduce losses. Accepting losses includes bearing the loss, possibly by exploiting reserves, or sharing the loss through mechanisms such as insurance. Hence the ability to recover from the disaster is of the utmost importance if losses are accepted.

Over time, technology is increasing the options that are available for hazard risk reduction, particularly those strategies that reduce losses. Examples of these approaches include warning systems, defence works and resistant infrastructure. This approach is most developed in urban areas around the North Sea, other parts of Europe such as southern Spain and Portugal, China and Japan, where flooding by surges claimed many lives up to the middle of the twentieth century.

Figure 1.21 Approaches to hazard reduction based on purposeful adjustment

Purposeful adjustment	Option
Choose change	• Change location of settlement • Change land use
Reduce losses	• Prevent effects of natural hazards • Modify natural hazard event
Accept losses	• Share loss of natural hazards • Bear loss of natural hazards

Climate change and sea-level rise represent an additional challenge around the world's coastal zones.

River flooding in the Netherlands

In addition to storm surges, the Netherlands is also subject to river floods, as in the case of the 1995 floods on the Rhine (Figure 1.22). During these floods, 27 people were killed and over 250 000 people were evacuated from their homes. Natural causes of the floods included heavy rain – Switzerland had over three times its average rainfall in January 1995; saturated soils – there was nowhere for the rain to soak away; and mild temperatures – snow melting in the Alps.

Human causes were also implicated. Much of the Rhine's flood plain had been built on; this meant that the impermeable surface increased the amount of rain reaching the river and increased the speed with which it did so. Where there was undeveloped ground, intensive farming had compacted the soil and increased overland runoff. Vegetation clearance also reduced interception. Channel straightening sped up the flow of water downstream, and dikes created faster and deeper flows.

Short-term solutions to the flood problem included:

- evacuation of people and livestock
- sandbags placed across doors
- moving downstairs furniture upstairs
- construction of temporary dikes
- clearance of underground car parks and subways.

In contrast, long-term measures included:

- the development of an early warning system
- dikes to increase the volume of water the river can hold – the Dutch flood protection scheme has spent over £1 billion on 600 km of dikes since 1995
- relief channels and basins to divert some of the water during the peak of the flood – however, this requires cooperation between a number of countries in the upper course to prevent flooding in the lower course of the river

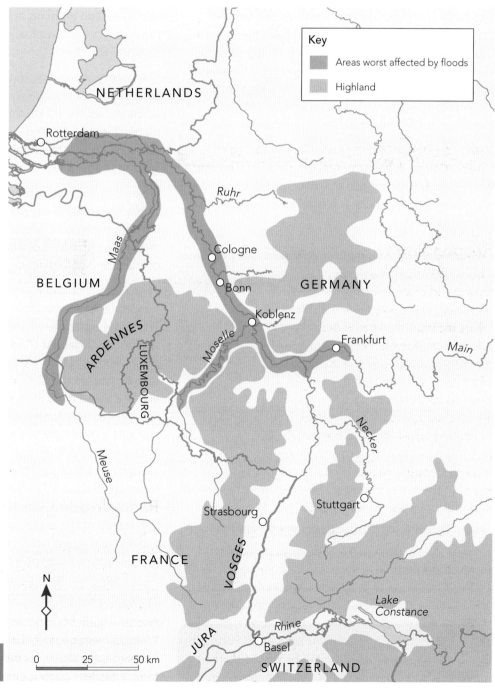

Figure 1.22 Rhine floods, 1995

- artificial flood plains – called forelands in the Netherlands – located within the dikes; these areas are allowed to flood and can be used for grazing and recreation when there is no risk of flooding
- the removal of sediment from the forelands – the sediment reduces the capacity of the forelands
- limiting residential and industrial development in flood plain areas.

Storm surges and coastal flooding

High death rates due to surges appear to be linked to land claim and substantial coastal modification, which has encouraged growth in vulnerable coastal populations without appropriate consideration of the potential for surges (for example, southern North Sea and Bangladesh). The death toll in surge events appears to have fallen substantially around the world as protection measures and forecasts/warnings are improved, including most recently in 1991 in Bangladesh when the number of deaths was reduced compared to the disaster in 1970.

However, there is no room for complacency, and the surge hazard will continue to evolve due to changing socioeconomic conditions, coastal land use and climatic risks.

Activities

1. What are the causes of storm surges?
2. Describe the impacts of a storm surge.
3. To what extent is it possible to manage the impacts of storm surges?

Discussion point

Human activities increase rather than reduce the impact of flooding. Discuss.

Theory into practice

What are the natural hazards of the drainage basin in which you live?

Take it further activity 1.2 on CD-ROM

1.3 What are the hazards associated with earthquake and volcanic activity?

Tectonic processes

Plate tectonics is a group of theories developed in the 1960s and 1970s which links sea-floor spreading, continental drift, earthquake activity, volcanic activity and mountain building. It helps to explain the past and present distribution of volcanoes, earthquakes and fold mountains.

The theory of plate tectonics states that the Earth's crust is not continuous skin but a series of rigid caps or plates, up to 100 km thick. There are seven major plates (Figure 1.23) and a series of minor plates. These move relative to one another. The plates ride on a semi-molten interior or mantle.

Plates lock into each other, rather like a jigsaw. However, where they meet there may be great instability because they are pushing in different directions. There are four main types of plate boundary: divergent, convergent, collision and conservative.

At divergent boundaries plates are moving apart and new material rises up from the mantle (Figure 1.24a). These are also called **constructive boundaries** or spreading ridges. At this type of plate boundary new ocean floor material is created, forming mid-ocean ridges, such as the Mid-Atlantic Ridge (Figure 1.23) and the East Pacific Rise. At these places fresh magma rises, cools, and solidifies. Further volumes of magma rise and periodically force their way into the mid-ocean ridge. As the ridge gives way, earthquakes and volcanoes occur. This process gradually pushes rocks on either side of the ridge apart and causes the sea floor to spread. Consequently, the continents on either side of the ocean gradually move apart. Volcanic activity near a mid-ocean ridge can create islands such as the Azores, Ascension Island and Tristan de Cunha. As the sea floor spreads, these islands can be carried away from the centre of the plate boundary.

At convergent boundaries two plates collide (Figure 1.24b). Where oceanic crust and continental crust converge, the denser oceanic crust plunges under the lighter, less dense continental crust. This creates

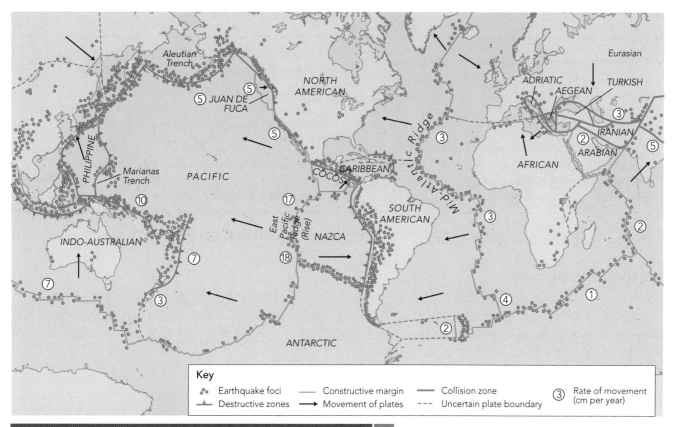

Figure 1.23 The world's main tectonic plates and types of plate boundary

a subduction zone. At a subduction zone there is a deep trench such as the Marianas Trench and the Aleutian Trench. As the oceanic plate descends it carries the crust back into the mantle. Much of the material melts at depths of between 100 km and 300 km and is completely destroyed by 700 km – hence the name **destructive boundary**. Parallel to the deep ocean trenches are lines of fold mountains or island arcs (Figure 1.24b). These are associated with large numbers of earthquakes and volcanic activity. As the plate melts, magma rises under very high pressure and escapes through fissures in the crumpled sediments of the continental crust. A good example is along the west coast of South America where the oceanic Nazca plate dives under the continental South American plate. The Andes, formed by the crumpling of sediments as the plates pushed together, is very active in earthquake and volcanic activity. Where the subduction takes place under the ocean (as opposed to under a continent), island arcs are formed, as in the case of Japan and the Philippines.

Collision boundaries occur when plates are too buoyant or thick to subduct. For example, where two continental plates converge, sediments that are trapped between them become folded and faulted (Figure 1.24c). This is thought to be the way in which the Alps and the Himalayas formed.

Conservative boundaries, or transform boundaries, occur when two plates move past each other with no formation or destruction of crust (Figure 1.24d). The most notable example is the San Andreas fault. These boundaries are associated with intense earthquake activity.

Figure 1.24 Processes at plate boundaries

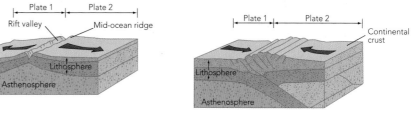

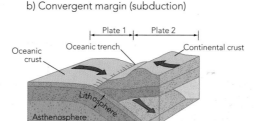

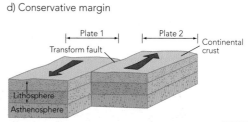

> **Key terms**
>
> **Collision boundary:** plate boundary in which two similar plates come together – neither is destroyed but both are folded to form fold mountains, e.g. the Eurasian and Indian plates which collide and form the Himalayas.
>
> **Conservative boundary:** plate boundary in which two similar plates move past each other – neither is destroyed but earthquake activity is common, e.g. the San Andreas fault where the North American plate and Pacific plate move in the same direction but at different speeds. Also called transform boundary.
>
> **Constructive boundary:** plate boundary at which new material is being created, e.g. Iceland. Also called divergent boundary or spreading ridge.
>
> **Destructive boundary:** plate boundary in which material is destroyed at a subduction zone, e.g. off the west coast of South America.

Earthquake activity

Case study | Earthquake in a MEDC – Northridge, Los Angeles, 1994

The earthquake at Northridge was magnitude 6.7, a large earthquake but not one of the largest (Figure 1.35, page 34). It lasted for just 15 seconds. However, it caused injuries to more than 9000 people and 51 people were killed. The epicentre was 32 km north-west of Los Angeles beneath the San Fernando Valley. The earthquake cost US$ 15 billion in damage (the most costly natural hazard up until that time). Nine bridges collapsed and portions of 11 major roads into Los Angeles had to close. Like many natural disasters in MEDCs, the loss of life was relatively small but the economic cost was very high.

> **Key terms**
>
> **Epicentre:** the point on the ground surface immediately above the focus.
>
> **Focus:** the point underground where the earthquake happens.

Case study | Earthquake in a LEDC – Bhuj, India, 2001

An earthquake measuring 7.9 on the Richter scale struck Bhuj and the surrounding area of Gujarat on 26 January 2001. The tremors lasted one minute. Estimates suggest that between 30000 and 100000 people died. An area the size of southern England was affected. The earthquake occurred on a national holiday, so many people were preparing for parades. Many of the buildings were poorly constructed – one quote from a TV reported stated that 'it is not earthquakes that kill people but buildings'. Hospitals, fire stations and civil administration buildings were all destroyed in the earthquake. There were very limited medical facilities and no power or communications in many parts of Gujarat. Up to 500000 people were made homeless. Aftershocks were felt for up to five days, and the rescue operation was unable to start in some areas for nearly three days. Poor sanitation and lack of drinking water led to the spread of disease such as cholera and typhoid. Hypothermia also set in, and there was much psychological trauma.

There is a wide range of hazards and impacts related to earthquakes (Figure 1.25).

Hazards	Impacts
Primary hazards: • ground shaking • landslides • faulting at the surface. Secondary hazards: • ground failure • liquefaction • rockfalls • mudflows and debris flows • tsunamis.	• Loss of life and injury • Total or partial destruction of building structure • Disruption of water supplies • Destruction of sewage disposal systems • Loss of public utilities such as electricity and gas • Floods from collapsed dams • Release of hazardous materials • Fires • Spread of chronic illness

> **Key terms**
>
> **Primary hazards:** the direct hazards associated with an event, for example ground shaking during an earthquake.
>
> **Secondary hazards:** the delayed hazards associated with an event, for example hypothermia that people may experience as a result of having to live outside after buildings have been destroyed.

Figure 1.25 Earthquake hazards and impacts associated with earthquakes

Tsunamis

Tsunamis are waves set off by submarine earthquakes, landslides into water, slumps and volcanic explosions under water. The word tsunami is a Japanese one meaning harbour wave, emphasising its impact on low-lying coasts rather than in the open ocean.

Most tsunamis consist of a series of waves generated by the rapid movement of the seabed. They differ from wind-generated waves in a number of ways:

◆ wavelength is very long – commonly 150–250 km but can be up to 1000 km

◆ velocities may reach 700–800 km per hour

◆ they have a low amplitude: 0.5–5.0 metres

◆ they have a long wave period: 15–60+ minutes

◆ the wave height is shallow in relation to the wavelength, so tsunamis are often undetectable in open oceans.

As a tsunami approaches the coastline, its speed and wavelength decrease rapidly, but the time between waves (wave period) remains the same, thus wave height increases.

The first waves to reach the shore are not usually the largest but may raise the sea level by 1–2 metres. The waves are followed by troughs which take the tide out very rapidly. The height of the waves and the depth of the troughs increases until immediately prior to the main wave, when the sea retreats far below the normal tide level. The main wave in a tsunami may reach 10–30 metres and may result from several waves in a sequence piling up together.

Case study | Asian tsunami, 2004

The cause of the tsunami was a giant earthquake and landslide caused by the sinking of the Indian plate under the Eurasian plate. Pressure had built up over many years and was released in the earthquake which reached 9.0 on the Richter scale. As the plates moved they caused the seabed to be pushed up and down, causing a column of water to be pushed away from the fault line as a series of giant waves.

Figure 1.26 shows the extent of the Asian tsunami. The main impact was on the western tip of the Indonesian island of Sumatra, the closest inhabited area to the epicentre of the earthquake. More than 70 per cent of the inhabitants of some coastal villages were reported to have died. Officials now estimate that about a third of the population of the provincial capital, Banda Aceh, died in the tsunami. By early March 2005, in Indonesia alone, over 111 000 people had died and more than 127 000 were still missing. The exact number of victims will never be known. A further 800 000 were reported as homeless.

Following the tsunami there were over 100 aid organisations operating in Indonesia alone providing emergency food, water and shelter to about 330 000 people. The Indonesian government estimated that reconstruction would cost US$ 4.5 billion (£2.4bn).

Apart from Indonesia, Sri Lanka suffered more from the tsunami than anywhere else (Figure 1.27). Southern and eastern coastlines were decimated. At least 31 000 people were known to have died, and thousands more were missing. The number of homeless people was put at between 800 000 and 1 million. Homes, crops and fishing boats were destroyed. At least 400 000 people lost their jobs. Sri Lanka's president, Chandrika Kumaratunga, launched a US$ 3.5 billion reconstruction drive. In Thailand nearly 6000 people were killed – including 2400 people from some 30 other countries.

India's Andaman and Nicobar Islands were also badly affected. Over 1850 of the islands' 400 000 people were killed and more than 5500 were missing – 4500 from Katchall Island alone. The economy was also badly affected. Salt water contaminated many sources of fresh water and destroyed large areas of arable land. Most of the islands' jetties were destroyed. Some aid agencies blamed the Indian government for refusing international aid in the first instance. The military built extra landing fields on the islands to help with relief. About 12 000 people were moved to relief camps on larger islands.

Elsewhere the impacts were more limited. Although Malaysia lies close to the epicentre, much of its

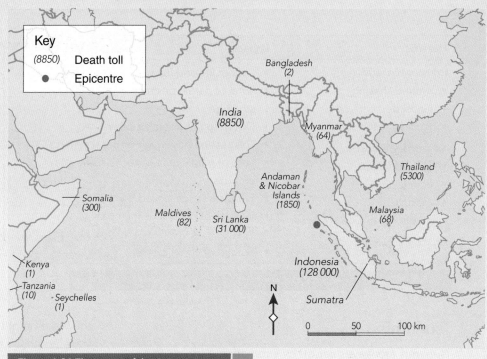

Figure 1.26 The extent of the Asian tsunami

Figure 1.27 Impact of the 2004 tsunami

coastline was shielded by Sumatra. Nevertheless, 68 people died. In Myanmar, the worst affected area was the Irrawaddy Delta, inhabited by subsistence farmers and fishing families. Over 60 people died, although the World Food Programme (WFP) says this may be an underestimate. Hundreds of migrant workers from Myanmar living in Thailand are also thought to have died. In Bangladesh, only two people were killed.

The impacts in Africa were varied. Somalia was the worst affected country, with damage concentrated in the region of Puntland, on the tip of the Horn of Africa. The water destroyed 1180 homes, smashed 2400 boats and left freshwater wells and reservoirs unusable. Between 150 and 200 Somalis died, thousands were made homeless and many fishermen were unaccounted for. In Kenya, there was only one death. The news of the tsunami reached Kenya and warnings were issued telling people to get away from the shoreline and head for high ground. Similarly, only one person was killed in the Seychelles.

Owing to the loss of life among tourists, it became a global disaster; people from nearly 30 different countries were killed.

Environmental impacts

There was a range of ecological impacts caused by the Asian tsunami. The force of the tsunami drove approximately 100 000 tonnes of water onto every 1.5 m of coastline and therefore caused considerable erosion in some parts and devastating deposition in others. The intrusion of salt water polluted fresh water sources and farmland, sometimes as far as several kilometres inland. Coral reefs were damaged and large areas of mangrove were uprooted. Of particular concern were the endangered sea turtles of Sri Lanka and some of the Indian islands. On the east coast of Sri Lanka almost all of the turtle hatcheries were destroyed when the sandy beaches were washed away.

Activities

1. What was the cause of the Asian tsunami?
2. Why was it considered to be a global catastrophe?
3. What were the short-term and long-term social and environmental impacts of the tsunami?

Case study | Kashmir earthquake, October 2005

The 7.7 magnitude earthquake was one of worst of the twentieth and twenty-first centuries (Figure 1.35 on page 34). There were over 22 aftershocks in the 24 hours after the main earthquake, one of which measured 6.2 on the Richter scale. The quake and its aftershocks were felt from central Afghanistan to western Bangladesh. Buildings were wrecked in an area spanning at least 250 miles, from Jalalabad in Afghanistan to Srinagar in Indian Kashmir.

The cause of the earthquake was the Indo-Australian plate moving against the Eurasian plate and the Iranian plate. The official death toll from the earthquake soared to more than 73 000. In Muzaffarabad district alone an estimated 1442 schools collapsed, 622 were considered too dangerous to use, and some 12 000 students and teachers died. More than 3 million people were left homeless and 69 000 seriously injured. Of the homeless, nearly 1 million had to sleep in the open.

In built-up areas, water and sanitation systems were broken, giving a high risk of disease outbreak. The devastating human impact of the earthquake, however, was blamed on lack of resources. International relief was chaotic and underfunded; and hundreds of thousands of survivors were at risk as the bitter Himalayan winter approached.

The UN had received just 12 per cent of the US$ 312 million pledged to its emergency appeal, in contrast with 80 per cent of pledges at the same stage after the Asian tsunami. There was a shortage of tents suitable to withstand the Kashmiri winter, as well as blankets, sleeping bags, warm clothes, medicine and food.

Some charities expressed concern that the public were suffering compassion fatigue after the Asian tsunami only nine months before. Also, the Kashmir earthquake did not affect people globally like the Asian tsunami did. Whatever the reason, the earthquake did not provoke the response from the rest of the world that it desperately needed.

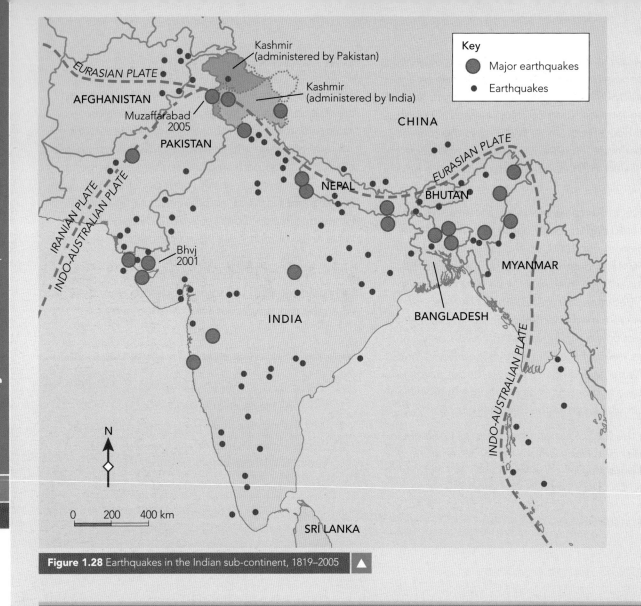

Figure 1.28 Earthquakes in the Indian sub-continent, 1819–2005

Activities

1. What was the cause of the Kashmir earthquake?
2. What were the primary and secondary effects of the Kashmir earthquake?
3. Suggest reasons why the amount of aid given to Kashmir, following the earthquake, was relatively low.

Volcanic hazards

One of the best known volcanic eruptions is that of Mount St Helens in 1980. Mount St Helens is on the North American plate, close to where the Juan de Fuca plate subducts. The volcano resumed activity after 123 years of being dormant. Scientists had predicted the eruption of Mount St Helens and so the area was closed to tourists and sightseers. On 18 May the volcano erupted killing 62 people, most of whom had gained lawful access to the restricted area. An earthquake of 5.1 on the Richter scale caused a massive landslide and avalanche. Pyroclastic flows and mudflows (lahars) caused huge amounts of material to be carried away from the volcano. An area of over 500 sq km was scorched, and the volcano created deposits nearly 50 m thick. Since then the volcano has been rebuilding a new dome, rich in acid lava. There has been renewed volcanic activity but not to the same extent and there have been no further deaths.

> **Key terms**
>
> **Pyroclastic flows:** super-heated flows of ash, cinders and pumice capable of travelling at speeds of up to 160 km and reaching temperatures in excess of 1000°C.
>
> **Lahars:** mudflows caused by the combination of water with volcanic ash and dust.

In contrast, a volcanic eruption in January 2002 near Goma, in the Democratic Republic of the Congo, lead to about 45 deaths and made about 500 000 homeless. Goma is located within the Rift Valley that extends from the Red Sea towards Lake Victoria. The airport was cut off although farmland was relatively unaffected. The human and physical geography of the region made the management of the disaster difficult. Some refugees from the disaster initially went to Kigali, in Rwanda, and then returned to Goma. Many refugees felt it was safer to stay in Congo than risk going to Rwanda. It was also more difficult to evacuate people because Goma is on the shores of Lake Kivu, a natural barrier to quick evacuation.

Case study | Volcanic hazards in Montserrat

One example of an ongoing volcanic eruption is the Soufrière Hills volcano on the Caribbean island of Montserrat. It has been erupting on and off for over a decade. It continues to have an impact on the island although there has been a change in the way the hazard is perceived and how the islanders have learned to cope.

The main hazards associated with volcanoes are the risk of death, loss of property, loss of economic activity and loss of infrastructure. There are a number of primary (immediate) and secondary (longer-term) hazards associated with the continuing eruptions on the island (Figure 1.29). These can also be classified as direct and indirect hazards. Volcanic activity has made over 60 per cent of southern and central parts of Montserrat uninhabitable.

Figure 1.29 Primary and secondary hazards associated with the continuing eruptions of Soufrière Hills, Montserrat

Primary hazards	Secondary hazards
• Rock bombs	• Mudflows (lahars)
• Lava flows	• Contaminated water
• Gases, especially sulphuric acid	• Fires
	• Landslides
• Heat	• Earthquakes
• Ash and dust	• Tsunami
• Pyroclastic flows	• Famine
	• Disease
	• Crop failure
	• Climate change

Figure 1.30 Redevelopment in the Davy Hill region of Montserrat

Case Study: Volcanic hazards in Montserrat

Plymouth was evacuated three times in 1995 and 1996. The volcano was responsible for 19 deaths – all of them farmers – caught out by an eruption during their return to the Exclusion Zone. Volcanic dust is another hazard, as it is a potential cause of silicosis and can aggravate asthma.

Redevelopment has been centred around the Davy Hill region (Figure 1.32). Much of the northern third of the island has seen considerable development – housing, schools, a hospital and other forms of infrastructure – as the islanders attempt to learn to live with the volcano in the south.

Continued volcanic activity in Montserrat threatens the country's economic performance, although there are signs that the island is beginning to turn the corner.

Montserrat is one of the islands that make up the Leeward Island chain in the Eastern Caribbean. There are at least 15 potentially active volcanoes spread along the length of the chain, of which the Soufrière Hills volcano is the one currently most active. The current period of eruption of the Soufrière Hills volcano began in 1995. Earthquakes started in 1992. In November 1995 lava was seen for the fist time. In December 1995, evacuation of southern Montserrat took place. In June 1997, due to a pyroclastic flow (Figure 1.33), 19 fatalities occurred within the Exclusion Zone. This was the biggest loss of life due to the volcano. A major dome collapse destroyed W H Bramble airport terminal building (Figure 1.34).

Long-term economic and social development

While the southern part of the island has been devastated and remains an Exclusion Zone, the northern part of the island is experiencing rapid development. The capital city, Plymouth, was abandoned following the 1997 eruption. Temporary government buildings have been built at Brades Estate, in the Carr's Bay/Little Bay area in north-west Montserrat. There has been a huge increase in the provision of housing in St John's, and there have been new schools, crèches and hospitals built. Much of the transport infrastructure has been improved, and there has been investment in the port facilities. FIFA have built a football pitch – this may rate as one of the most attractive pitches in the world. The

Figure 1.31 Montserrat – key facts

Population	9538 (2007)
Population growth rate	1.05% (8.43% in 2002)
Birth rate	17.51 per thousand
Death rate	7.02 per thousand
Infant mortality rate	7.03 per 1000 live births
Life expectancy	79 years (81.3 years for females, 76.8 years for males)

Figure 1.32 'Risk map' for Montserrat, September 1997

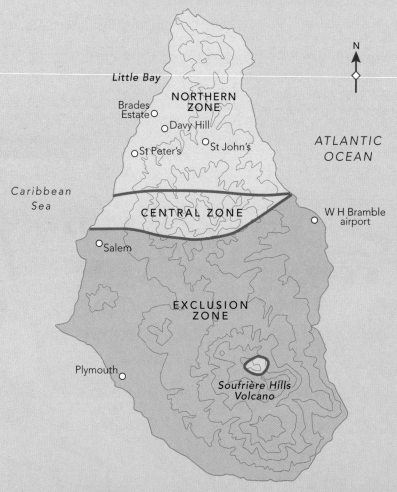

EXCLUSION ZONE No admittance except for scientific monitoring and national security matters
CENTRAL ZONE Residential area only, all residents on state of alert
All residents to have rapid means of exit 24 hours per day
Hard hat area, all residents to have hard hats and dust masks
NORTHERN ZONE Area with significantly lower risk, suitable for residential and commercial occupation

Source: Adapted from Potter et al., *The Contemporary Caribbean* (2004)

Montserrat government offices have been relocated to Brades Estate and rebuilt. New service industries have developed in Salem, and there has been an increase in ecotourism and adventure tourism.

Prospects for the economy depend largely on public sector construction activity. The UK launched a three-year US$ 122 million aid programme (known as the Country Policy Plan) to help reconstruct the economy. The EU agreed a £6.5 million grant to help relocate the capital from Plymouth to Brades Estate. The southern third of the island is expected to remain uninhabited for at least another decade. Agriculture accounts for 5.4 per cent of the GDP. The lack of suitable land means that it is unlikely that this figure will increase much in future years. By contrast, over 80 per cent of GDP is generated by services and this is likely to increase as the aid programmes continue.

Volcanic management includes monitoring and prediction. The Geographical Positioning System (GPS) is used to monitor changes in the surface of the volcano (volcanoes typically bulge and swell before an eruption). The development of 'risk maps' can be used to good effect (Figure 1.32).

Figure 1.33 Pyroclastic flows, Soufrière, Montserrat

Figure 1.34 W H Bramble airport, Montserrat, buried under a mudflow

Activities

1 What are the main hazards related to volcanic activity in Montserrat?
2 Why has most of the redevelopment of Montserrat been in the north of the island?
3 Suggest why Montserrat has seen a growth in population since 2000.

Activities

1 What are the primary and secondary hazards associated with earthquakes?
2 How and why do the impacts of natural hazards vary with level of development of a country?

Discussion point

To what extent are the secondary hazards associated with earthquakes and volcanoes worse than the primary hazards?

Take it further activity 1.3 on CD-ROM

1.4 Why do the impacts on human activity of such hazards vary over time and location?

Impact of earthquakes

The extent of earthquake damage is influenced by the following factors.

- *The strength of earthquake and number of aftershocks*: the stronger the earthquake the more damage it can do. An earthquake of 6.0 on the Richter scale is 100 times more powerful than one of 4.0; the more aftershocks there are the greater the damage that is done. Figure 1.35 shows the number of deaths and strength of earthquakes on the Richter scale (which is logarithmic).

- *Population density*: an earthquake that hits an area of high population density, such as in the Tokyo region of Japan, could inflict far more damage than one which hits an area of low population and building density, such as the Mid-Atlantic Ridge at Thingvellir, Iceland.

- *The type of buildings*: MEDCs generally have better quality buildings, more emergency services and the funds to cope with disasters. People in MEDCs are more likely to have insurance cover than those in LEDCs. An earthquake in California, for example, is likely to have different impacts and different responses to one in a LEDC, such as India (Figure 1.37).

- *The time of day*: an earthquake during a busy time – such as rush hour – may cause more deaths than at a quiet time. Industrial and commercial areas have fewer people in them on Sundays; homes have more people in them at night.

- *The distance from the epicentre of the earthquake*: the closer a place is to the centre of the earthquake, the greater the damage that is done.

- *The type of rocks and sediment*: loose materials may act like liquid when shaken; solid rock is much safer and buildings ideally should be built on flat areas of solid rock.

- *Secondary hazards* such as mudslides and tsunami, fires, contaminated water, disease, hunger and hypothermia.

Country	Year	Death toll (est.)	Richter scale
Indonesia	2004	248 000	9.1
China	1976	242 419	7.5
China	1927	200 000	8.3
China	1920	180 000	8.6
Messina, Italy	1908	160 000	7.5
Tokyo, Japan	1923	142 800	7.9
Turkmenistan	1948	110 000	7.3
Kashmir, Pakistan	2005	86 000	7.6
China	1932	70 000	7.6
Chengdu, China	2008	69 000	7.9
Peru	1970	66 800	7.7
Iran	1990	50 000	7.7
Quetta, Pakistan	1935	45 000	7.7
Turkey	1932	32 000	7.8
Avezzano, Italy	1915	30 000	6.9
Bam, Iran	2003	30 000	6.6
Chile	1939	28 000	7.8
Iran	1978	25 000	7.7
Armenia	1988	25 000	6.8
Guatemala	1976	23 000	7.5

Figure 1.35 The world's worst earthquakes by death toll in the twentieth and twenty-first centuries

Figure 1.36 Catastrophes in 2005

Event	Insured losses (US$ billion)	Location
Hurricane Katrina	45	USA/Bahamas
Hurricane Rita	10	USA/Cuba
Hurricane Wilma	10	USA/Caribbean
Winter storm Erwin	2	Europe
Hurricane Dennis	1	USA/Caribbean

Location	Event	Loss of life
Pakistan/India	Earthquake	73 000
Pakistan	Cold weather	2 000
Central America	Hurricane	1 600
USA/Bahamas	Hurricane	1 300
Indonesia	Earthquake	1 300

Earthquakes	Asia	23
	Africa	2
	Latin America (including the Caribbean)	3
	North America	1
Cyclones	Asia	53
	Africa	2
	Latin America (including the Caribbean)	45
	North America	2
Floods	Asia	14
	Africa	10
	Latin America (including the Caribbean)	8
	North America	1

Figure 1.37 Number of people killed in natural disasters, 1990–1999, per million population

Activities

1. Comment on the variation in the number of deaths due to earthquakes by continent.
2. Comment on the number of deaths due to earthquakes relative to the other hazards shown in Figures 1.36 and 1.37.

Factors affecting the impact of a volcano

Runny basaltic lava eruptions, such as at Mauna Loa, Hawaii, result in more continuous, less violent eruptions and generally less loss of life. In contrast, explosive eruptions, such as Mount St Helens and Pinatubo (Philippines) result in greater destruction of the environment, local economy and potential loss of life.

The impact of volcanic eruptions is also greater in areas of high population density. The potential impact of an eruption of Vesuvius (close to the urban area of Naples) or Popacatapetyl (close to Mexico City) is much greater than that of Mount St Helens or Hekla (Iceland), which are both in rural areas. The potential impact of the Soufrière volcano on Montserrat is now much less given that the population has been evacuated from the southern part of the island.

The size of the eruption is also important. The eruption of Krakatoa in 1883 was heard thousands of miles away whereas the small-scale eruptions and lava flows of Etna in 2000 (Figure 1.38) did not result in any loss of life. The potential impact of the 'super-volcano' located under Yellowstone National Park in the USA is unknown but likely to be considerable.

Frequent eruptions, such as the ones that occur in Hawaii or at the Soufrière mud springs on the island of St Lucia (Figure 1.39), reduce pressure within the magma chamber, and so an explosive eruption is unlikely. In contrast, where there is a big build-up of pressure, such as at Mount St Helens or in the Soufrière volcano on Montserrat, the likelihood of an explosive eruption increases.

Monitoring and prediction are vital to reduce the loss of life in a volcanic eruption. The Nyiragongo volcano in the Democratic Republic of Congo was not predicted whereas the eruptions of Mount St Helens and Pinatubo were. The ongoing monitoring of the Soufrière volcano means that future loss of life is likely to be low.

Some people are more at risk than others. One of the main secondary hazards of volcanoes is lahars or mudflows (Figure 1.40). On Montserrat, lahars were concentrated in the main valleys, often to significant depths.

Figure 1.38 Lava flows on Etna, 2000

The impacts will also depend on the attempts to manage the hazards. The impact of Eldfell volcano on the Icelandic island of Heimaey was reduced due to the impact of spraying the advancing lava flows with sea water. On Etna, diversion channels have been dug to divert lava flows away from settlements. Monitoring and prediction (for example, smell of sulphur, changes in water chemistry, physical swelling of the volcano and clusters of earthquakes) may suggest an imminent eruption.

The following case studies illustrate many of the factors discussed above. The first, the predicted volcanic eruption of Pinatubo in the Philippines, a NIC, compares well with the second, the unexpected earthquake at Kobe, Japan in 1995 (refer to the CD-ROM). Kobe, a large urban area in a MEDC was considered to be safe from earthquakes – not anymore!

Figure 1.39 Soufrière mud springs, St Lucia

Figure 1.40 Lahars on the island of Montserrat

Case study | Mount Pinatubo, 1991

Ring of Fire

Three-quarters of the Earth's 550 historically active volcanoes lie along the Pacific Ring of Fire. This includes most of the world's recent volcano eruptions, including Pinatubo. Indeed, the Philippines, an arc of islands found at the edge of an ocean, is beset by a variety of environmental hazards including cyclones, landslides, tsunami, earthquakes and volcanoes. However, without volcanic activity the Philippines would not exist: they comprise the remains of previous eruptions.

On 9 June 1991, Mount Pinatubo, a 1460 m volcano which had remained dormant for almost 600 years, began to erupt. The most serious eruptions scattered rock and ash over a 100-km radius and triggered a series of earth tremors of up to 5.4 on the Richter scale.

The eruption, just 80 km north of the capital Manila on the Philippines' main island, Luzon, combined with a tropical storm (typhoon) on 15–16 June, generating devastating mudslides and making up to 200 000 people homeless. The volcano also triggered numerous earthquakes. Nearly 350 people were killed, mostly by the collapse of buildings under the strain of rain-soaked ash and mud. The build-up of ash over several days caused many buildings to collapse. Indeed, only six confirmed deaths were attributed to the first few days – largely due to advance warning and evacuation procedures.

The mud storms and mudslides covered 50 000 ha of cropland, destroying all crops and over 200 000 homes. Supplies of electricity were cut for over three weeks, water became contaminated, and roads and telecommunications links were destroyed. An epidemic of respiratory and gastric diseases broke out in the temporary housing. The government estimated that 600 000 people lost their jobs.

Key terms

Pacific Ring of Fire: the Pacific Rim area in which several subduction zones causes a line of volcanic activity all around the Pacific Ocean. Some 75 per cent of the world's active volcanoes are located in this area.

Typhoon: an intense tropical storm in the West Pacific Ocean.

Evacuation

Over 200 000 people were evacuated as a result of the early warning system. By 10 June, 14 500 US Service Personnel and their families were evacuated from Clark Air Base to Subic Naval Base, 40 km south-west of Pinatubo. By 22 June, 20 000 dependants of US personnel had been evacuated to the USA.

Activities

1. Why is the Philippines prone to volcanic activity?
2. Explain why so few people were killed by the initial blast of the eruption but died in the following weeks.

 Please refer to the CD-ROM for a case study on the Kobe earthquake, 1995.

 Take it further activity 1.4 on CD-ROM

 Take it further activity 1.5 on CD-ROM

1.5 How can hazards be managed to reduce their impacts?

Hazard management and risk assessment

Ways of managing the consequences of a hazard include:

- modifying the hazard event, through building design, building location and emergency procedures
- improved forecasting and warning – for example, it is difficult to predict earthquakes, although in some cases there may be a 'recurrence interval' (an average time in years between large events)
- sharing the cost of loss, through insurance or disaster relief.

The impact of hazards varies over time. Volcanic and earthquake hazards are believed to build up over time. For example, in March 1980 Mount St Helens resumed volcanic activity after a dormant period which lasted some 123 years. The explosive activity of Montserrat's volcano occurred after a dormant period of nearly 400 years. Different types of volcano produce different impacts over time. Basaltic lava, such as that in the Hawaiian volcanoes, produce a more continuous, less explosive volcanic activity which is less destructive than the explosive, acidic eruptions, such as Mt Pinatubo. Earthquake activity also varies over time.

In the Tokai area of Japan and its vicinity there is a continuous earthquake observation network. In this area large-scale earthquakes have occurred repeatedly, as a result of the subduction of the Philippine and Pacific Plates beneath the Eurasian plate. In the area of the Suruga Bay more than 130 years have elapsed since the last earthquake. The possibility of a large-scale earthquake occurring in the near future is thought to be high.

Mass movements also vary over time. For example, rock falls on the Isle of Thanet are more likely in winter when there is increased frequency of freeze-thaw cycles and water efficiency (when total rainfall exceeds evapotranspiration). Rotational slumps and landslides are more likely after periods of heavy rain and increased marine erosion during storms.

Risk assessment involves consideration of:

1. the likely size and range of natural processes involved
2. the extent of the impacts
3. ways in which the impacts can be reduced.

Most risk assessment involves a statistical likelihood of the size and impact of the event. In addition, it is possible to map the distribution of risk and impact (Figure 1.32, page 32). It is also possible to identify

Pre-disaster planning	Post-disaster responses
• *Prevention*: action to reduce the severity of impact • *Mitigation*: action to reduce property damage and minimise economic impacts • *Preparedness*: action to increase speed and efficiency response	• *Response*: effectiveness depends on education, training and experienced of emergency response teams • *Recovery*: action to assist communities to return to pre-disaster conditions • *Redevelopment*: action to manage economic losses; there should be a long-term link between natural hazard and national economic activities

Figure 1.41 The natural hazard management cycle

Figure 1.42 Possible physical and social adjustments to hazards

Physical adjustments include:	Social adjustments include:
• building and construction techniques to withstand a hazard of given size and strength • identifying (mapping) and avoiding sites where hazards are likely to occur • predicting where and when a hazard might occur • preventing or altering the characteristics of a hazard.	• land-use zoning and restrictions for hazardous locations • establishing minimum building standards for hazardous locations • public awareness through education • issuing early warnings of imminent hazards • evacuation plans; preparations for emergency food and shelter • emergency preparedness programmes to protect life and property • spreading economic loss more fairly through insurance, taxation and grants • reforming a community so that it is less vulnerable to natural hazards and more aware of the conditions/factors that increase the risk of natural hazards.

Figure 1.43 Small-scale construction techniques used in flood control

- *Levee*: Raised concrete or earth embankment on the side of a river to prevent flooding.
- *Retaining walls*: Stone and concrete structures to protect against erosion on steep slopes.
- *Gabion*: Wire basket filled with rocks or stones to protect river banks and slopes against erosion (Figure 1.44).
- *Paved drains and culverts*: Concrete structures designed to move water quickly from roadsides and areas susceptible to flooding and erosion (Figure 1.45).
- *Check dams*: Small dams of variable height built across the width of small gullies. They are usually formed of rock and wire but wood and old tyres can also be effective.

Figure 1.44 Gabions used along a river bank in Barbados

vulnerability – by social group as well as by type of building and land-use.

Hazard preparedness is now included in hazard management as well as post-disaster relief (Figure 1.41).

Most of the techniques adopted for flood control are labour-intensive (i.e. requiring a lot of work by people), low-technology solutions (Figure 1.43).

Managing earthquakes

The impacts of earthquakes can be reduced in a number of ways including:

- land-use zoning
- building design
- stabilisation of steep slopes
- redevelopment of vulnerable sites
- improvements in warning and prediction.

Although there are building codes, they are difficult to enforce. This may be due to the increased cost of housing that is 'earthquake-proof'. In general MEDCs and NICs have better housing codes and better enforcement of building regulations, whereas in LEDCs the quality of housing is poorer and regulations may not be enforced especially in informal settlements.

Factors affecting the impact of an earthquake

A single-storey building has a quick response to earthquake forces. A high-rise building responds slowly, and shock waves are increased as they move up the building. If the buildings are too close together, vibrations may be amplified between buildings and increase damage. The weakest part of a building is where different connecting parts meet. Bridges, such as elevated motorways, are therefore vulnerable in earthquakes because they have many connecting parts.

Figure 1.45 Paved storm drain in St John, Antigua

Figure 1.46 Pipeline on rollers, Nejsjavellir, Iceland

Some built-up areas are very much at risk from earthquake damage – areas built on weak rocks, faulted (broken) rocks and on soft soils. How a site reacts to shaking depends on a number of factors including:

- magnitude of the shock wave
- depth of soil
- moisture content of the soil
- the nature of the rock – is it hard or soft?

Steep slopes and landforms formed by deposition are especially vulnerable to shaking caused by earthquakes. Liquefaction is a process in which sediments containing large amounts of water lose their strength and begin to behave like a fluid. This causes foundations to crumble and pipes to split. Deltas and alluvial fans (deposits formed when a river leaves an upland area and spreads out into a lowland area) are especially vulnerable. Many oil pipelines and water pipelines in tectonically active areas are built on rollers so that they can move with an earthquake rather than fracture (Figure 1.46).

Preparing for the next big one – lessons learned from the Asian tsunami, 2004

The memories of the cataclysmic events caused by the explosion of Krakatoa in 1883 have faded with time. After the event, people were eager to rebuild their lives and return to some kind of 'normality'. However, once the memories fade, the lessons learned from the disaster also fade.

The vivid images of the 2004 tsunami shown around the world may mean that the memory of this event will remain long enough to drive the development of a tsunami warning system and an education policy. Together, these should help to reduce the death toll in the event of another mega-tsunami in the region.

There is at present a real determination on the part of geologists, geophysicists and people in the region to develop an early warning system. Moreover, there is commitment from governments to put it in place. The interest is not just limited to countries surrounding the Indian Ocean, as the disaster affected people from many more countries. Along with the Indian Ocean countries that lost people, persons from a further 30 or so states were killed while on holiday, business or visiting friends and relatives.

Scientists believe that there is real scope for a multidisciplinary approach to hazard management. For example, by combining knowledge of volcanic eruption prediction with land-use management, planning regulations, construction regulations, effective education systems and warning systems, the risk from tsunamis in the region would be greatly reduced. Such a process relies on being ready for the event rather than reacting to the event. Nevertheless, getting the message across to people is one thing; getting them to act is another.

The World Bank estimated that losses caused by natural disasters in the 1990s could have been cut by US$ 280 billion (£150 billion) through advance spending on hazard preparation and management of just US$ 40 billion. For example, had there been a tsunami education plan in Banda Aceh, it is likely that the death toll from the tsunami would have been cut by tens of thousands. Similarly, had building inspectors in Bam, Iran, been trained and paid, the death toll from the earthquake in 2003 could have been much less.

Activities

1. In what ways is it possible to manage natural hazards?
2. What are the benefits of hazard preparation compared with disaster response?
3. Describe and explain how activities change during a disaster management cycle.

Discussion point

'Only small-scale events can be successfully managed'. How far do you agree with this statement?

Take it further activity 1.6 on CD-ROM

Case study | Managing a volcanic eruption – Eldfell, Heimaey, Iceland

In January 1973, an eruption occurred along a 2-km fissure in Iceland. Nearly all 5300 residents of the island of Heimaey were evacuated. Strong winds blew ash and cinder, burying homes in the main town and fishing port, Vestmannaeyjar. Massive lava flows threatened the town and the fishing port. This eruption is noted for being the first successful attempt to manage a large eruption. For nearly six months, sea water was sprayed onto the advancing lava flows in an effort to cool them down, stop them flowing and divert them away from the town and the port. The diverted lava flows added new land to the island and beneficially provided more protection for the port. It was the largest effort ever exerted to control volcanic activity. More than 30 km (19 miles) of pipe and 43 pumps were used to deliver sea water at rate up to 1 cubic meters per second. By the end of the eruption, 6 million cubic meters of water had been pumped onto the flows.

Activities

1. At what type of plate boundary is Iceland located? (Figure 1.23, page 25.)
2. What are the factors that allowed people to successfully manage the Eldfell eruption?

Knowledge check

1. What are the hazards associated with mass movement and slope failure?
2. What are the hazards associated with flooding?
3. What are the hazards associated with earthquake and volcanic activity?
4. Why do the impacts on human activity of such hazards vary over time and location?
5. How can hazards be managed to reduce their impacts?

Exam Café
Relax, refresh, result!

Relax and prepare

Student tips

What I wish I had known at the start of the year...

Hannah
"I used to be very confused over the different types of volcanic eruption. I could never remember which one produces the liquid runny lava that can cover a wider area and which one produces sticky lava. Actually, it is as simple as ABCD: B is for basic (free flowing, less explosive) which is common at Constructive margins; A is for acidic lava (viscous, more explosive) which is common at Destructive margins."

Anju
"I used to worry that I could not remember all of the facts and case studies and muddled up those for earthquakes with those for volcanoes. I now understand that at A2 level it is more important to keep a clear head and be willing to argue or evaluate a viewpoint than write down all of the facts I know."

Karl
"Remember that flooding has a wide range of origins – it isn't just rivers flooding following high rainfall or snowmelt! Don't forget coastal flooding (your work at AS may help you) and flooding as a result of tectonic action such as landslides blocking a river or flooding due to a volcanic eruption."

Sujit
"I always muddled up hazards with their impacts. When a question asked about reducing hazards I always started writing about how to reduce their impacts. I now appreciate the subtle difference. A hazard is the threat, rather than the impact, to humans and their buildings, etc. Without humans there would be no hazards, merely natural events."

Common mistakes – Sabina
"I never really appreciated the importance of my conclusion. As far as I was concerned it was the end part of an essay and if I ran out of time, so what! I now know that is very short-sighted as a conclusion allows me to draw together my thoughts and return to the question set to evaluate it effectively. I have 'saved' a number of poorly focused answers by a conclusion that has brought those points into line with the question I am answering. I now know that if I am going to miss anything out it won't be the conclusion, however pushed I am for time."

Refresh your memory

1.1 The conditions and processes that lead to mass movements

Cause	Reduction in shear strength and increase in shear stress if equilibrium disturbed
Physical	Climate – wet, lot of weathering, extremes of temperature Relief – steepness of slope Drainage – wet areas are lubricated Rock type – geology, structure, beds, porosity and tilt of rocks Vegetation – type and percentage of cover Animals – burrowing animals, walking on slopes Erosion of foot of slope, e.g. waves
Human	Over-steepening slopes, e.g. cuttings, quarries, adding waste Removing vegetation especially trees Irrigation – adding moisture to ground adds weight and lubricates Adding weight to slopes, e.g. buildings, walking
Processes	Creep – slow creep results in terracettes Avalanche – rapid; can be wet or dry Flow – rapid, highly fluid, saturated Slide – sliding material retains shape and cohesion. Slips on slide plane Slump – usually rotates along a slip plain

1.2 Mass movement has a range of environmental, economic and social impacts

Environmental	Relief – reduces slope angles, fills in valleys/hollows Drainage – may dam or divert rivers Vegetation – trees lean or fall Soil – collects at base of slope (catena effect) Rock strata – may bend the ends of beds (cambering)
Economic	Transport – road and rail distorted or broken causing disruption Cost of preventing or repairing damage Buildings/walls – collapse, lean, or have soil collect up slope slide Loss of farmland – soil slippage
Social	Disasters, e.g. Aberfan 1966, 144 killed
Human reaction	Short term – emergency rescue, shoring up buildings, clearing roads Long term – re-grade slopes, drainage, steel piles, steel nets, plant vegetation, protect slope foot from erosion

1.3 Flood risk reflects a combination of physical and human factors

Physical	Climate, e.g. snowmelt, heavy rain, low evaporation Relief, e.g. very flat, low-lying Drainage, e.g. density, regime, hydrograph, drainage pattern, channel type Vegetation, e.g. grass versus trees Rock type, e.g. permeability, porosity, water table
Human	Construction of impermeable surfaces, e.g. towns, roads Removal of vegetation – deforestation, farming Drainage – ditches, waste disposal, drains Changing rivers – dams, diversions, embankments

1.4 Flooding has a range of impacts which create a range of human responses

Short term	Deaths, destruction – buildings/possessions, pollution (sewage, oil, etc.)
Long term	Transport links broken, cost, loss of jobs, crops ruined, rehousing costs. May take over a year to dry out buildings
Responses	Short term: emergency aid, rescue, pumping out and drying out property Long term: • Individual, e.g. move, versus local authority versus national authorities' reactions • Planning, e.g. avoid floodplains, green sectors, e.g. parks • Planned retreat – leave certain areas to flood, e.g. Somerset Levels • Reduce surface flow, e.g. afforestation, contour ploughing • Channel modification, e.g. overflows, storage areas, dams, widen • Embankments – raise them, reinforce them, flood gates • Early warnings – better weather and river-level forecast

Exam Café

Refresh your memory

1.5 Earthquakes and volcanic eruptions are caused by plate tectonics

Converging	Destructive margin = deep earthquakes, explosive eruptions
Diverging	Constructive margin = shallow quakes, lava flows and shield volcanoes
Slide past	Conservative margin = shallow quakes, hot springs, few eruptions
Hot spot	Hawaiian – basic lava – shield volcanoes. Few quakes

1.6 Earthquakes have a range of impacts and create a range of human responses

Impacts	Contrast of disaster versus long-term impacts. Scale of impact crucial Primary: destruction, casualties, landslides, fires Secondary: disease, loss of infrastructure, housing, jobs, food, water Tertiary: cost of recovery, loss of crops, damage to mines/industries, trade
Responses	Short term: emergency aid (rescue, tents, medicines, food, water) – often military and charity Medium term: rehousing, rebuild hospitals, etc., repair infrastructure Long term: move population, improve warnings, emergency planning

1.7 Why do the impacts on human activity of such hazards vary over time and location?

Time	Level of development/technology Recurrence interval and frequency of event Time of day, season, year Build-up of events – warning interval
Location	Distance from plate margin, river, slope, etc. Population density, distribution, level of perception Highland versus lowland Urban versus rural Level of development – building type, ability to warn/evacuate Remoteness Type of hazard or mix of hazards Geology and rock structure

1.8 How can hazards be managed to reduce their impacts?

Prediction	Based on past events, monitoring of gas/pressure, etc., tiltmeters
Risk assessment	Calculate size and extent of risk – inform population, building design, location of vital buildings/facilities, e.g. power stations
Planning	Individual, e.g. store water, local authority, e.g. emergency centres, state or central, e.g. mobilisation of rescue services. Building controls
Prevention	Lubricate fault lines, dam or divert lava flows, flood volcanic vents
Warnings	Use of media, planned evacuations
Recovery	Emergency aid, insurance, state aid for reconstruction

Top tips...

> The specification refers to 'contrasting examples' and at 'least two contrasting types of earth hazards' when considering their impacts. Contrasting examples is an invitation to compare differing locations, typically MEDC versus LEDC or highland versus lowland or inland versus coastal, but it is most effective to keep to the same type and scale of hazard. Earthquakes are ideal as you can look at the impacts of quakes of the same magnitude (on the Richter scale) on very different types of area, for example California versus Kashmir. When looking at two types of hazard it is best to take ones that are distinctly different in terms of impact over time and space. Hence it is more effective to compare a sudden, localised flash flood or avalanche with the more measured longer term lava or ash eruptions.

Exam Café

Get the result!

Example question

Examine the view that in MEDCs the economic impacts of earthquakes are more important than social impacts. [30 marks]

Student answer

Earthquakes produce a range of primary and secondary impacts. In October 1989, the Loma Prieta earthquake (6.9) 95 km south of San Francisco resulted in $600 billion of damage but also 67 people died. Clearly this had a huge economic impact but deaths, injuries and the resulting trauma to millions had a large social impact. It is important to compare the short- and long-term economic and social impacts to assess their relative importance.

Examiner says

Clear focus on impacts with a good use of an example to illustrate the scale of impacts on a MEDC. This demonstrates a high level of knowledge and understanding. Good to see social not just deaths.

Examiner says

The last sentence is key, as it sets up where the rest of the answer is going and clearly shows that an evaluation will occur that refers to impact over time. This would create a good initial impression.

Examiner's tips

▷ When considering the impacts of earth hazards, such as an earthquake, remember that the impacts may be directly the result of the quake (primary) such as building collapse, landslides and fires, or may be secondary, such as disease and tsunamis which have resulted from the initial quake. There are also longer-term impacts (tertiary) which may have wide repercussions such as the loss of employment when factories are destroyed or the long-term cost to the economy of rebuilding.

▷ Remember that all the examinations at A2 are synoptic. This means that there is an expectation that you will draw on your knowledge and understanding from across the subject and from wider background reading or the media. Synoptic is all about synthesising your experience of the subject to bring out connections, linkages and a variety of viewpoints. Answers with a clear synoptic approach should score well.

» Exam Café

Chapter 2
Ecosystems and environments under threat

The Siberian tiger is an endangered species: there are only 450 left in the wild. The tiger is threatened by poaching and the logging of the Korean Pine. The pine is vital to the tiger because it provides pine nuts for wild boar, elk and deer which are the tiger's main prey. The tree is now protected by law and a planting programme is underway to restore links in the food chain and help preserve the tiger's numbers. This example shows that human activities pose threats to physical environments in a number of ways. An appreciation of how ecosystems function is vital to the conservation and the sustainable management of physical environments.

Questions for investigation

- What are the main components of ecosystems and environments and how do they change over time?
- What factors give the chosen ecosystem or environment its unique characteristics?
- In what ways are physical environments under threat from human activity?
- Why does the impact of human activity on the physical environment vary over time and location?
- How can physical environments be managed to ensure sustainability?

Consider this

Ecosystems are dynamic and subject to short- and longer-term natural changes. What will this oak woodland look like in six months, and 50 years' time? Is there any evidence in the photo that this woodland is managed?

2.1 What are the main components of ecosystems and environments and how do they change over time?

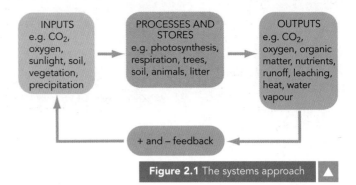

Figure 2.1 The systems approach

The systems approach

Ponds, woods and salt-marshes are all examples of **ecosystems**. An ecosystem is a group of organisms which interact with each other and the surrounding environment so that matter is exchanged between the **biotic** and **abiotic** parts of the system. Ecosystems vary in size from a single oak tree to an entire rainforest. They can be viewed as systems (Figure 2.1) which have inputs such as sunlight, weathered rock and rainfall and outputs including waste products, carbon dioxide and oxygen. Processes operating within the system include photosynthesis and respiration and the stores include trees. A state of dynamic equilibrium exists when the inputs and outputs are in balance, but disturbance can produce a **positive** or a **negative feedback**. For example, when food is plentiful, the snowshoe hare population increases; this initiates a rise in numbers of its main predator, the lynx, thereby creating a positive feedback. Ecosystems are regarded as **open systems** because energy and matter are exchanged with the surrounding environment. In contrast, in a **closed system** energy alone is exchanged.

Stores and flows in an ecosystem

The two major functions within an ecosystem are the movement of energy through the system and the recycling of nutrients within it.

Energy flows

Figure 2.2 shows how energy flows through an ecosystem. Sunlight is absorbed by chlorophyll in the leaves of green plants, blue-green algae and some types of bacteria. These organisms also take in water and carbon dioxide and together with the sunlight produce carbonates and oxygen by **photosynthesis**. The oxygen escapes into the air, or into water if the ecosystem is a pond, while 20 per cent of the carbonate is used up for **respiration** which releases heat energy. The majority of the carbonate is converted into starch and cellulose, producing in green plants new leaves and branches and thicker stems. The rate at which an ecosystem produces new growth or **biomass**, together with the carbonate used for respiration, is called its **gross primary productivity**. If respiration is excluded, then the term **net primary productivity** is applied.

Key terms

Abiotic: inorganic matter – soil, air, water.

Biomass: living matter (plants, algae, bacteria) calculated as total dry weight in a given unit area at one point in time, usually measured in kg/m^2.

Biotic: organic living matter – plants, animals, bacteria.

Closed system: system in which energy, but not matter, is exchanged across boundaries.

Gross primary productivity: the amount of new biomass created in a given area in one year including that used for respiration, measured in $kg/m^2/year$.

Net primary productivity: excludes that used up through respiration.

Open system: system in which energy and matter are exchanged across boundaries.

Photosynthesis: the process by which solar energy is absorbed by green plants, blue-green algae and some bacteria in the presence of water and carbon dioxide to create carbonates and oxygen.

Positive and negative feedback: change in one part of the ecosystem which brings about change in another part.

Respiration: the release of energy from carbohydrates and fats which gives out carbon dioxide.

Organisms which create energy by fixing sunlight are called **autotrophs** or **primary producers**. Primary producers are eaten by **herbivores** or **primary consumers**. Herbivores are then eaten by **carnivores**, which are secondary consumers, which in turn are eaten by tertiary consumers. Herbivores and carnivores are both heterotrophic; that is, they are incapable of producing their own food. **Saprophytes** are organisms which feed on dead or decaying matter. Many of these are bacteria and fungi which release digestive enzymes onto decaying matter which dissolve it and then they absorb the soluble products. Slightly larger fragments of detritus are eaten by **detritivores** such as earthworms and woodlice. Bacteria, fungi, earthworms and woodlice are all examples of **decomposers** in that they break up organic matter into simpler substances which releases nutrients.

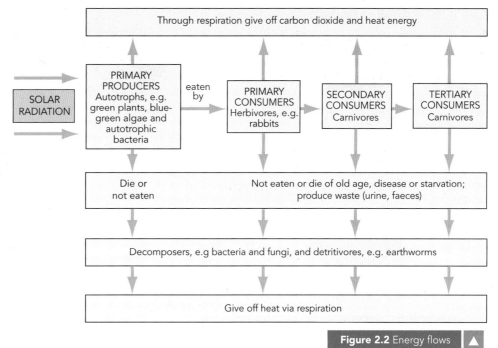

Figure 2.2 Energy flows

Key terms

Autotrophs: organisms which create energy from inorganic matter. Most are phototrophic, i.e. they create energy from photosynthesis and include green plants, blue-green algae and some bacteria. A few bacteria are chemotrophic, i.e. they obtain their energy without sunlight from inorganic matter such as hydrogen sulphide.

Carnivore: meat eater.

Decomposer: organisms which break down dead animal or plant matter to release nutrients, principally fungi and bacteria but also earthworms and woodlice.

Detritivore: an animal which eats detritus, i.e. small fragments of dead and decaying matter.

Herbivore: plant eater.

Primary consumer: an organsim which feeds on a primary producer.

Primary producer: first link in a grazing food chain (see autotroph).

Saprophyte: organism which feeds on dead animals and plants, particularly bacteria and fungi.

The flow of energy from one set of organisms to the next produces a **food chain** (Figure 2.3). Ecosystems contain both grazing and detrital food chains. Organisms which all obtain their food in the same way, such as herbivores, form a **trophic level**. Most ecosystems rarely have more than four or five tropic levels and energy is lost at each stage because:

- not all of the food is digested
- some plants and animals are not eaten by herbivores and carnivores
- activities such as chewing, mating and catching prey consume energy
- plant and animal respiration releases heat energy.

The number of individuals in each trophic level can be represented diagrammatically as a pyramid of numbers. This type of pyramid, however, ignores the size of individuals so that, for example, a single oak tree carries the same weight as an aphid. In contrast, other types of trophic pyramid which are based on biomass or energy take size into account (Figure 2.4).

Grazing food chains are very simple models of ecosystems which ignore the fact that many animals feed on what is available at the time, i.e. they are opportunists. Some animals are also **omnivores**, such as the badger, or are herbivorous when young but become carnivorous when adult, for example tadpoles and frogs. **Food webs** attempt to show more complex feeding patterns (Figure 2.5), but they too have drawbacks in that energy losses and the relative importance of individuals are not taken into account.

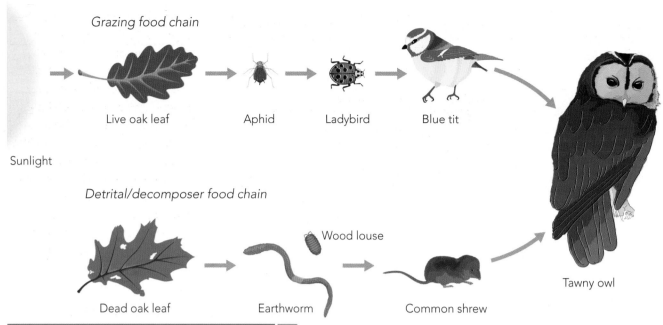

Grazing food chain

Live oak leaf → Aphid → Ladybird → Blue tit → Tawny owl

Sunlight

Detrital/decomposer food chain

Dead oak leaf → Earthworm → Wood louse → Common shrew → Tawny owl

Figure 2.3 Woodland grazing and detrital food chains

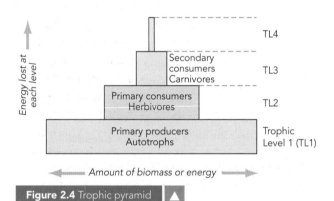

Figure 2.4 Trophic pyramid

Key terms

Food chain: a chain of organisms through which energy flows.

Food web: a group of interacting food chains within a community of organisms.

Omnivore: plant and meat eater.

Trophic level: group of organisms which have the same method of feeding, e.g. herbivores.

The impact of removing an organism from an ecosystem depends on the species involved and the complexity of the food web. Providing herbivores can switch to alternative food sources, the removal of one plant will have little effect, but the loss of a top carnivore has a greater impact on the ecosystem. Eventually, however, the equilibrium will be restored because once all the vegetation has been eaten, the herbivores, and in time the carnivores, will die of starvation, which gives the plants time to recover.

Activities

1. Look at the food chain below and suggest what would happen in the short and long term if the lion was removed?

   grass → wildebeest → lion

2. Look at the food web in Figure 2.5. What would happen in the short and long term if the ladybirds were removed?

3. Although there are fewer animals in the higher levels of a trophic pyramid they tend to be larger. Suggest why this is the case.

4. Give reasons why total biomass declines towards the top of the trophic pyramid.

5. a The data below shows selected species collected from an oak tree. With the help of the woodland food web shown in Figure 2.5, classify the species below as producers, primary, secondary or tertiary consumers and then construct a pyramid of numbers.

Species	Number of species
Sparrowhawk	1
Treecreeper	4
Common shrew	4
Spider	5
Bank vole	6
Hairstreak caterpillar	45
Dunbar moth caterpillar	30
Oak tree	1

   b How might a pyramid of biomass differ in shape from that of a pyramid of numbers?

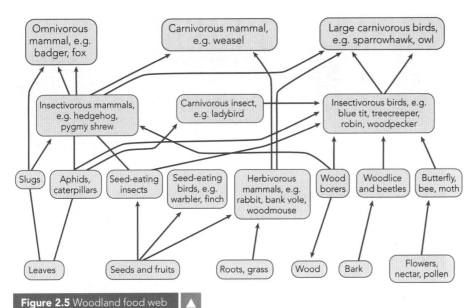

Figure 2.5 Woodland food web

trees, dead twigs and branches together with dead animals such as rabbits and mice and their waste products, comprise the litter store. Earthworms, beetles, woodlice, bacteria and **mycorrhizal** fungi all help to break down or decompose litter to produce nutrient-rich **humus**. Rock weathering also releases nutrients such as calcium, magnesium, phosphorus, sodium, potassium, iron and aluminum, which become attached to small **clay-humus complexes** within the soil.

Nutrient cycles

In addition to the carbonate created by photosynthesis, plants also need essential nutrients such as phosphorus, nitrogen and iron as well as small qualities of trace elements including boron. The nutrients are recycled through the ecosystem between the biomass, litter and soil stores (Figure 2.6). Nutrient losses, such as surface runoff and **leaching**, are made good by inputs from rock weathering and precipitation. The size of each store and nutrient flow reflects the volume of nutrients held or transferred.

Understanding how nutrient recycling works can be explained by investigating the flows in a deciduous woodland (Figure 2.7). Autumn leaves falling from

The plant roots take up the nutrients in solution from the clay-humus complexes by a process called **cation exchange**. Nutrients taken up through the roots enable the plant to grow which increases its biomass. Tree trunks, roots, stems and twigs make up a large proportion of the biomass which explains why this is the largest of the three nutrient stores, although in other ecosystems this is not always the case. Eventually some of the biomass dies and the nutrients are again recycled. Some nutrients are lost via surface runoff, although interception by the tree canopy prevents some of this in woodlands. Nutrients are also lost through leaching. Nutrients released by rock weathering and brought in by precipitation compensate for losses.

Humans can alter nutrient cycles in various ways. Harvesting crops removes nutrients from the system, but losses can be made good by adding farmyard manure or applying nitrate, phosphate and potash mineral fertilisers. Burning vegetation releases ash which is rich in phosphorous and potassium. Grazing livestock produces waste products which decompose

Figure 2.6 Nutrient cycle

(Inputs) Precipitation, sunlight → Litter, decomposing leaves
(Ouput) Surface runoff removes nutrients
Biota, e.g. trees and animals
Soil
(Ouput) Leaching removes nutrients
(Input) Nutrients released from weathered rock

Key
● stores
→ flows

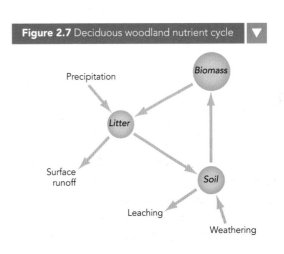

Figure 2.7 Deciduous woodland nutrient cycle

to produce humus, but when the animals are removed for slaughter the biomass store sharply declines. Logging also reduces the size of the biomass store and in time that of the litter and soil store as well. Loss of the forest canopy reduces interception, and instead precipitation reaches the ground to become runoff which carries away nutrients. Some water also infiltrates the soil and nutrients are lost through leaching. Root wedging, a physical weathering process which normally helps to break down rocks to release nutrients, ceases, while the removal of trees also reduces transpiration and makes the soils wetter.

Figure 2.8 Deciduous woodland in spring

> **Key terms**
>
> **Cation exchange:** the process by which nutrients such as calcium, which are attached to the clay-humus particles, are exchanged for hydrogen ions on the plant root and nutrients are taken up into the plant.
>
> **Clay-humus complexes:** tiny fragments of humus and clay which have negative surfaces onto which positively-charged nutrients become attached.
>
> **Humus:** well-decomposed, dead plant and animal matter with no original structures visible.
>
> **Leaching:** the movement of nutrients, soil mineral particles and humus, held in solution and suspension down through the soil.
>
> **Mycorrhiza:** an association between plant roots and fungi which is beneficial to both partners.

Change in ecosystems and environments

Ecosystems and environments are subject to short- and long-term changes. Seasonal fluctuations in temperature, precipitation and light levels bring about short-term change, for example in tropical grasslands, acacia trees shed their leaves to conserve moisture in the dry season. In temperate climates, deciduous trees lose their leaves in winter in response to declining light levels and the increasing difficulty of extracting water from cold soils. By shedding leaves they reduce transpiration to offset the effect of **physiological drought**. In spring, before the tree foliage is fully developed, light reaches the woodland floor stimulating primroses and then bluebells to flower and seed (Figure 2.8). After the 'spring window' closes, the canopy becomes denser and only a few plants can flower.

Ecosystems and environments in the longer term are rarely static and produce primary and secondary successions.

Primary successions

A sequence of plant communities which occupies ground which has not been vegetated before is called a **primary succession** (Figure 2.9). That which forms on bare rock is known as a lithosere, while a succession which develops on sand dunes is called a psammosere.

> **Key terms**
>
> **Physiological drought:** condition created when soils are too cold for roots to extract water, hence the plant suffers from drought.
>
> **Primary succession:** plant succession that develops on land which has not been vegetated before.

On newly exposed ground uncovered by a retreating glacier, the first species to colonise the surface are often blue-green algae, lichen, moss and avens or *Dryas* (Figure 2.10). Species diversity is low but there are lots of individuals. The early colonisers, or **pioneers**, can withstand the harsh conditions imposed by exposure and low nutrient levels. On death they

Figure 2.9 Plants colonising bare ground

and soil, and all the ecological niches are filled. At this point no further change takes place and the **plant succession** is said to have reached its **climatic climax**. If woodland is the last stage in the succession, there is sometimes a slight fall in species diversity towards the end because the canopy shades out the ground layers.

- **Key terms**

  Climatic climax: plant community which is in equilibrium with its regional climate and soils, and in which all the ecological niches are filled and no further change takes place.

  Pioneers: the first plants to colonise an area. Plants are small and have low nutrient requirements. Many are annuals and legumes (can fix atmospheric nitrogen).

  Plant succession: series of plant communities which progressively occupies a site.

contribute organic matter and when alive their roots help to bind together a developing soil.

By providing shelter, stability and nutrients, the pioneers improve the environment and encourage new species to invade and grow. By this process, known as facilitation, a new community becomes established which eventually displaces the pioneers. Each new community in turn improves the physical conditions for its successor. Plants become taller and more structurally complex and species diversity increases. Eventually a situation is reached such that the ecosystem is in equilibrium with its regional climate

Secondary successions

Secondary successions occur on formerly vegetated surfaces. A naturally occurring event such as a forest fire, strong wind, or an old tree collapsing may be responsible for the clearance. Removal of the tree canopy allows more sunlight to reach the woodland floor, stimulating light-loving flowering plants such as willow-herb and butterflies, which are attracted by the

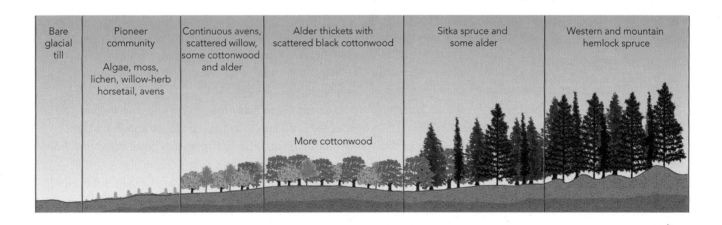

Figure 2.10 Primary succession on glacial till

plants. Gradually, taller and more diverse communities colonise the clearing. Birch saplings, which like sunlight, become established and underneath these oak seedlings begin to germinate. Oak requires shade to begin to grow, but eventually the trees overtop and shade out the birch. In some areas, beech eventually replaces oak to become the climatic woodland.

Humans can also initiate secondary successions by clearing and later abandoning land. Where a succession is more permanently prevented from developing, a **plagioclimax community** is created. One example seen in southern Britain is lowland heathland which is maintained by continuous grazing and periodic burning (Figure 2.11). Left to itself, it would be invaded by bracken, gorse and birch and eventually be replaced by oak woodland.

Figure 2.11 Lowland heathland

Key terms

Eutrophication: the nutrient enrichment of water bodies causing algae to proliferate.

Plagioclimax community: plant community permanently arrested by human activity.

Secondary succession: plant succession which develops on land that has been vegetated before.

Other human activities causing change

Human activity	Effects
Drainage of wetlands and coastal salt-marsh	This lowers water-tables and alters habitats.
Trampling	Light trampling increases species diversity, but heavy pressure damages plants. Some species are more resistant to trampling than others, which is why tolerant plants such as plantain are often found along footpaths.
Fertiliser application by farmers	Nitrate and phosphate fertilisers increase crop yields, but excess fertilisers not taken up by the plant often find their way into ditches and lakes causing nutrient enrichment or **eutrophication**. Algae thrive in the nutrient-rich water and shade out waterweed. On death, the algae sink to the floor of the lake where they are decomposed by bacteria which consume oxygen and this de-oxygenates the water.
Species removal	In the 1920s the wolf was deliberately removed from Yellowstone Park in the USA because it was considered a threat to elk and buffalo herds. Elk numbers subsequently rose and consumed young aspen, willow and cottonwood trees. Wolves have recently been reintroduced into Yellowstone to control the elk numbers and restore the vegetation.
Introduction of alien plants and animals (accidental and deliberate)	This brings about change, particularly on islands where many species are endemic and have no fear of predators. On Hawaii, for example, the introduction of the mongoose and the rat has posed a major threat to flightless, ground-nesting native birds. Fireweed, a plant introduced from the Canary Islands, has also displaced native species on volcanic soils.
The burning of fossil fuels	This creates acid rain which adversely affects lakes and coniferous forests. Acid rain which falls on already acid soils developed under coniferous woodland, leaches nutrients such as potassium and magnesium from the soil. Acidity also increases the solubility of normally insoluble minerals such as iron and aluminium, which make the soil toxic. Pine needles turn yellow and die and trees lose their vigour and become vulnerable to insects. Acidity also alters plant compositions often causing a decline in species diversity. Heavy metals, such as mercury, kill sensitive species such as trout in lakes.

Figure 2.12 The effects of human activity

Activities

Look at the plant succession in Figure 2.10.

1. Why are legumes such as alder important in the succession?
2. Why might species diversity decline under the climax forest?
3. Suggest naturally occurring events which might prevent the succession reaching its climax.
4. How might the soil pH change as the succession changes?

Discussion point

Should lowland heathlands be managed?

Theory into practice

Look out for local examples of colonisation and succession, such as plants developing on paving stones and walls, or on derelict land, or on farmland which has been taken out of production (set-aside).

2.2 What factors give the chosen ecosystem or environment its unique characteristics?

A **salt-marsh** is a vegetated mudflat found on the coast. It develops between the mean low water of neap tides (MLWN) and the mean high water of spring tides (MHWS). Marshes form in sheltered estuaries and on the landward margins of spits and bars (Figure 2.13), where the water is shallow and the coastal gradient is low. Creeks cross the marsh surface which fill and drain with each incoming (flood) and outgoing (ebb) tide (Figure 2.14). The marsh is broadly divided into a lower area which is covered and exposed twice daily by tides, and an upper zone which is submerged by spring tides which occur only a few times each month.

Figure 2.13 Salt-marsh backed by sand dunes

Key term

Salt-marsh: coastal vegetated mudflat.

Stores and flows within a salt-marsh ecosystem

The salt-marsh ecosystem contains a variety of different organisms. Rooted plants such as glasswort grow on the marsh surface, while *Enteromorpha*, a green alga, lives on the mudflats. Bacteria live on and within the mud and decompose the algae and decaying plant matter. Crustaceans and molluscs live within or on the surface of creeks and mudflats which are periodically covered by the tides. Brent geese graze the marshes in winter, while wading birds, such as redshank, breed on the marshes in summer and feed on crustaceans, worms and molluscs on the mudflats and within creeks. Fish, such as the goby, live in the creeks and crabs scavenge for food on the mudflats.

Figure 2.14 A salt-marsh creek

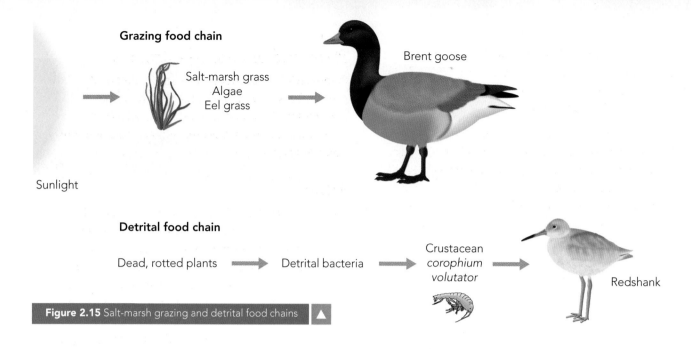

Figure 2.15 Salt-marsh grazing and detrital food chains

Energy within salt-marshes is transferred by grazing and detritus food chains (Figure 2.15). The detritus food chain is particularly important and contributes to the high productivity of the ecosystem. Phototrophic and **chemotrophic bacteria** decompose plant matter and algae. The products of their decomposition and waste are consumed by organisms including crustaceans, worms and molluscs, which in turn are eaten by other molluscs, crustaceans, birds and fish. A food web summarising feeding patterns is shown in Figure 2.16.

> **Key term**
>
> **Chemotrophic bacteria:** bacteria which obtain their energy from inorganic substances without using sunlight.

Nutrients washed in from rivers and the sea also contribute to the high productivity of the ecosystem.

These, together with precipitation, sunlight and decaying vegetation, contribute inputs into the nutrient cycle. Nutrients washed out by the tides, or leached down through unsaturated soil in the upper marsh, constitute losses or outputs. The biomass is composed of salt-marsh plants, invertebrates, fish and birds.

Physical factors influencing the marsh

Drainage

The twice daily ebb and flood of the tide, with variations in its height and penetration, has several important effects on the marsh ecosystem. Tidal movements exert a mechanical drag, uprooting plants, removing seeds and dislodging animals from their habitats. Submergence reduces photosynthesis

Figure 2.16 Detrital and grazing salt-marsh food web

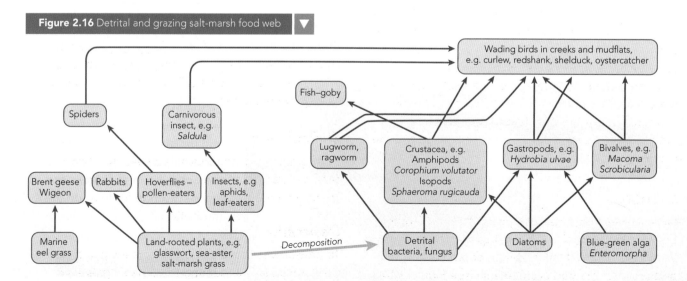

because the supply of carbon dioxide declines and mud stirred by the tides cuts down light levels. Tidal fluctuations also alter salinity levels causing **osmotic** problems for animals and plants.

As the tides recede, plants and animals have to tolerate the effects of desiccation. On hot summer days, plants uncovered by the tide have to endure high temperatures and scorching solar rays, while on winter nights, exposed surfaces can fall below freezing. Meanwhile, periods of high rainfall leach out salt from the upper marsh, while low rainfall and high temperatures increase soil salinity.

> **Key term**
>
> **Osmosis (osmotic):** the diffusion of water across a semi-permeable membrane as the result of differences in concentration of solution.

Relief

The low marsh gradient means that the ebbing tide is slow to drain and soils remain waterlogged for long periods. Water-logging produces anaerobic conditions and land plants find it difficult to respire through their roots. Minor depressions on the marsh surface trap pools of water forming salt-pans. Rainwater accumulating in these hollows may dilute the salinity, but high temperatures cause evaporation leading to hyper-saline conditions. Creeks which cross the marsh fill at high tide and sometimes overtop the banks depositing mud to create levees. Drainage and soil aeration is better on these raised areas than on the surrounding marsh.

Soils

The soils on the marsh surface are **gleys**, i.e. they are frequently waterlogged and contain varying amounts of clay and silt. Bacteria living on or within the mud consume large amounts of oxygen and release hydrogen sulphide, turning the sediment black. When the tide ebbs, the top layer of the mud becomes aerated, but even during a flood tide the soil water-table does not completely rise to the surface and the presence of air helps plants.

Climate

Salt-marshes fringe coasts above 30° N and S of the Equator. In the tropics they are replaced by mangroves. Variations in rainfall influence the degree of leaching on the better drained upper marsh (see Figure 2.16).

Plant and animal adaptations

Plants and animals have evolved a number of mechanisms to survive the harsh conditions imposed by fluctuations in salinity, periods of submergence and exposure, shifting mud and low nutrient levels.

Salinity

Plants adapted to saline conditions are called **halophytes**. Some have salt glands or shed their leaves when salinity becomes excessive, such as cord grass, seablite, sea-aster, sea-lavender and sea-purslane. Succulents, such as glasswort, absorb and store salt to counter the problem of living in saline soils. Normally plants take up water and nutrients via their roots, but if the soil has a higher concentration of salt than the plant, then water passes from root to mud. Halophytes must therefore be able to absorb and store some salt in their tissues in order for water and nutrients to pass from the saline soil.

Animals seal the entrances of their burrows when the tide comes to minimise fluctuations in salinity levels, while others regulate the composition of their internal body fluids by a process called **osmoregulation**, to counter the osmotic pressures caused by changes in salinity, for example amphipod crustaceans.

> **Key terms**
>
> **Gley:** soil where all the pore spaces are occupied by water, i.e. waterlogged.
>
> **Halophyte:** a species which tolerates saline conditions.
>
> **Osmoregulation:** process of changing bodily salt concentration to prevent osmotic stress.

Water-logging

To counter the effects of water-logging, glasswort, cord grass and sea-lavender have tissues with large air spaces on their stems called aerenchyma, which convey oxygen to their roots.

Shifting mud

Cord grass has deep roots to aid stability, while sea-lavender has small leaves to reduce drag when immersed by the tide.

Desiccation

To prevent desiccation at low tide, many plants have small, rolled leaves (common salt-marsh grass), fine hairs (sea-purslane), a thick, leaf cuticle and stomata sunk into deep grooves (cord grass) which reduce transpiration. Others, such as glasswort, have no leaves and instead photosynthesize on their stems. Cord grass has a C4 metabolism which allows carbon to be stored for use in photosynthesis when conditions are favourable. Some animals burrow to escape the effects of desiccation, while weevils and gall midges live within the tissues of halophytes.

Nitrogen deficiency

Marshes are deficient in nitrogen and to counter this cord grass, common salt-marsh grass and sea-lavender have nitrogen-fixing bacteria on their roots.

> **Discussion point**
>
> Why do different salt-marshes contain different plant communities?

Succession – natural change over time

Species vary in their tolerance to submergence and salinity, which produces a zonation across the marsh (Figure 2.17). Eel grass, a marine plant, grows in shallow water which is permanently covered by the tide, while the green alga *Enteromorpha* forms fine filaments across the surface of stable mudflats. The first land species to colonise the bare mudflat is glasswort (Figure 2.18) and sometimes cord grass. These pioneer plants can withstand twice daily submergence in salt-water, desiccation, shifting mud and have low nutrient requirements. Species diversity is low, but individuals are numerous because they have no competitors. Their roots help to stabilise the mud and trap sediment which increases the height of the marsh. When the plants die their organic matter provides vital food for decomposers.

The pioneers provide shelter, soil stability and nutrients for the establishment of a taller more structurally complex second plant community which commonly consists of sea-aster, annual seablite and common salt-marsh grass. The pioneers meanwhile are displaced and migrate to newly exposed mudflat closer to the seashore.

Successive plant communities replace each other as sedimentation and organic matter raise the level of the marsh, until it is inundated only at the very highest spring tides. The soils become less waterlogged and plant transpiration further helps to dry out the marsh. As the marsh becomes higher, precipitation leaches out soluble nutrients which are carried down through the soil profile.

Creeks which cross the mudflats become more entrenched as the marsh surface becomes higher. Sea-purslane grows on the more aerated soils of levées which border the creeks. Highly saline salt-pans are occupied by glasswort.

The low marsh communities thus described are replaced on the upper marshes by sea-lavender and then red fescue and sea-rush. Species diversity increases and the vegetation becomes taller. Few marshes eventually grade into climax woodland; many instead end in a seawall and are reclaimed.

Figure 2.17 Plant zonation across salt-marsh

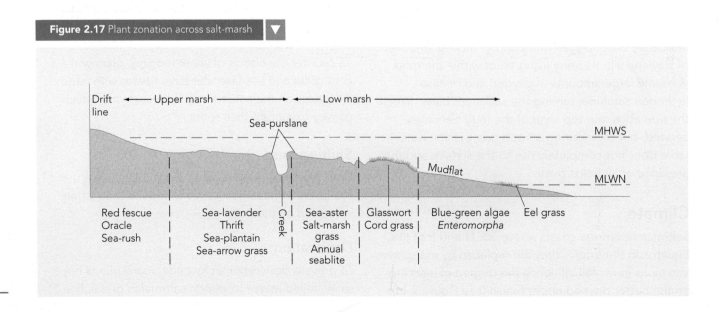

Figure 2.18 Sea-lavender and glasswort

Activities

1 Present the following data, which shows changes in the percentage cover of species across a salt-marsh, as a kite diagram or similar.

	Distance from shore (m)										
Species	1	3	5	7	9	11	13	15	17	19	21
Cord grass	75	80	88	75	57	24	6	–	–	–	–
Glasswort	13	21	30	51	53	56	55	50	15	–	–
Enteromorpha	2	6	6	6	4	3	2	2	–	–	–
Sea-aster	–	–	–	–	–	2	2	–	–	–	–
Salt-marsh grass	–	–	3	20	48	75	82	57	56	10	10
Sea-purslane	–	–	–	1	2	5	20	53	98	95	9
Sea-lavender	–	–	–	1	2	5	17	10	5	–	–
Sea-plantain	–	–	–	–	–	–	13	10	8	3	–
Sea-arrow	–	–	–	–	–	3	10	14	14	–	–
Common reed	–	–	–	–	–	–	–	–	–	10	18

2 Plot changes in salinity (as measured by a conductivity meter) against distance from shore on a scatter-graph and add a best-fit line. Then carry out a Spearman's Rank Correlation Coefficient Test to measure the strength of any observed relationship.

Distance (metres) inland from sea	Conductivity (micro-siemens)
1	4800
3	4300
5	4600
7	3060
9	2800
11	3000
13	480
15	460
17	250
19	345
21	330

Human influences on salt-marsh ecosystems

Salt-marsh is relatively rare in Britain and covers only about 40 000 ha. Human influences which have altered the ecosystem over time include agriculture, settlement and industry, pollution, invasive species and conservation.

Agriculture

Sheep and cattle have lightly grazed the upper marshes for centuries. Light grazing diversifies plant communities and improves habitats for invertebrates and breeding birds such as redshank. Heavy grazing, however, damages plants especially species such as sea-lavender and sea-purslane on which many invertebrates depend. When grazing ceases, coarser, more competitive grasses invade, reducing plant diversity.

Marshes around The Wash in East Anglia were drained as early as Roman times, but large-scale reclamation occurred between 1400 and 1700 to create farmland. As a consequence of drainage, wildfowl which grazed on the marshes and wading birds which fed within the creeks lost their habitats. Moreover, even where wetland was not reclaimed, drainage of the surrounding cultivated land lowered the marsh water-table.

Settlement and industry

In the past, on marshes such as those in north Kent, clay was dug to make bricks and salt was extracted from salt-pans. More recent threats to marshes have included:

- marina development (see Solent Case study opposite)
- port developments (see Solent Case study opposite)
- channel dredging and offshore extraction of sand and gravel (this increases tidal speeds and erodes mudflats which fringe salt-marsh; see Solent Case study opposite)
- airport development
- reservoirs built across rivers upstream of estuaries (this deprives the salt-marsh of silt and clay)
- coastal barrages, such as that proposed across the Severn estuary
- reclamation for industry and settlement. On Teesside, for example, 83 per cent of the inter-tidal area has been reclaimed for oil refineries, power stations, steelworks and petrochemical plants. Reclaimed salt-marshes are attractive to industry because they provide large flat areas on which to build, nearby water for processing and waste disposal, and with access to the sea to import and export materials.

Pollution

As towns and industry have grown, salt-marshes have increasingly become threatened by a variety of types of pollution. Discharge from industrial and domestic effluent, coal-powered power stations, oil spillages, leaking pipes, waste tipping and air pollution all adversely affect habitats. Marshes fringing the Severn estuary, for example, have been polluted by metals such as zinc and lead carried in the wind from nearby smelters and chemical works. Heavy oils spilled from tankers and pipelines smother plants, while lighter varieties seep into plant tissues impeding gaseous exchange, light absorption and transpiration. Anti-fouling paint, which prevents micro-organisms attaching themselves to the underside of boats, harms shellfish and other invertebrates. Fertiliser which drains from agricultural land, together with sewage, causes eutrophication.

More indirectly, rising sea-levels, attributed to global warming, are eroding marshes many of which cannot migrate inland because they are backed by seawalls which protect farmland. As a result of this 'coastal squeeze', marshes are becoming smaller, threatening wildlife habitats.

Invasive species

Spartina anglica, an invasive alien type of cord grass, first appeared on the south coast of Britain in 1870 and has since spread to many marshes in Britain, displacing low marsh plants. Attempts have been made to remove it by cutting or treating with herbicides, but now natural die-back appears to be occurring, particularly in areas such as The Solent where it first appeared.

Case study | Threats to The Solent

The salt-marshes which fringe The Solent (Figure 2.19) are internationally important for wildfowl such as Brent geese and waders including dunlin and redshank. Early activities on the marshes included duck shooting and salt production, but between 1600 and 1900, 90 per cent was reclaimed for agriculture. More recent threats have been from the following:

- Road construction – the M275 into Portsmouth and industrial development built on reclaimed salt marsh in Portsmouth Harbour.
- Urban expansion of Portsmouth and Southampton – creates sewage and risk of eutrophication.
- Port development, for example a container terminal at Southampton. Risk of oil spills from tankers coming into Fawley Oil Refinery. Dredging of the Solent to accommodate larger ships increases tidal speeds eroding mudflats.
- Recreation, for example the construction of leisure marinas at Lymington and Hythe. Recent decline in eel grass in Chichester Harbour is attributed to anti-fouling paints and eutrophication. The harbour is a major dinghy and sailing centre.
- Agricultural runoff – Langstone Harbour suffers from algal growth caused by agricultural runoff and sewage.
- Invasive species – the marshes are dominated by the invasive cord grass *Spartina anglica*, although since 1950 natural die-back has occurred leaving mudflats vulnerable to erosion.
- Sea-level rise is causing coastal squeeze, as seen around Langstone Harbour.
- Thermal power stations along The Solent raise water temperatures at their outfalls changing salt-marsh ecosystems.

Conservation measures include the establishment of a number of reserves (Figure 2.19), the banned use of anti-fouling paints, and a boat speed limit of 8 knots in Chichester Harbour.

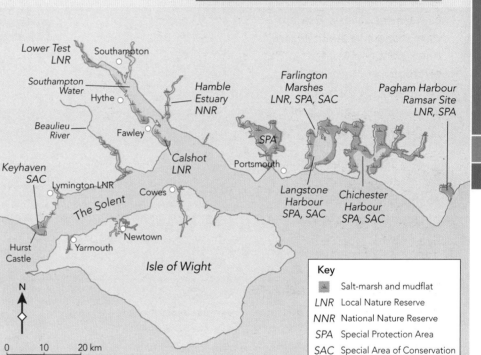

Figure 2.19 Salt-marshes around the Solent

Take it further activity 2.1 on CD-ROM

Conservation

Today, salt-marshes are recognised as performing a number of important functions. As highly productive ecosystems, they support diverse flora and fauna. In winter, large flocks of Brent geese and wigeon graze on marshes bordering the Solent, the Essex coast and Morecambe Bay, and many of these areas have been designated National and Local Nature Reserves (NNRs, LNRs) and Sites of Special Scientific Interest (SSSIs). Creeks and mudflats support commercially important cockle and oyster industries, and the marsh surface filters out pollutants draining from the land to the sea. Marshes attract visitors who come to watch the birds and sail in small boats. They also act as a natural form of coastal defence by dissipating wave energy and trapping sediments, and for this reason in some locations seawalls have been dismantled to allow the sea to spread in and re-establish marshland (as shown in the Essex Case study overleaf).

Case study | Essex marshes

Essex originally had 35000 ha of salt-marsh, but drainage for agriculture and other forms of development has meant that only 2000 ha remain. Sea-level rise as a result of global warming, together with natural subsidence in the North Sea Basin, is causing the salt-marsh and mudflat to retreat about 2 m annually. The marshes are internationally important for over-wintering wildfowl such as Brent geese and wading birds including grey plover, knots and dunlins. In recognition of their value, some have become nature reserves and SSSIs.

At Abbotts Hall Farm on the Blackwater estuary, and more recently on Wallasea Island on the River Crouch, seawalls have been dismantled and the farmland behind flooded to re-establish salt-marsh. In 2006, six breaches were made in the seawall on Wallasea Island and the sea flooded 115 ha of former wheat fields (Figure 2.20). Before flooding, the area was landscaped so that bull-dozed hollows became shallow lagoons, and artificially raised areas became islands for nesting and roosting birds. Pollutant-free sediment from Harwich was deposited on mudflats to encourage salt-marsh to form, which has already been colonised by glasswort and cord grass. The scheme, costing £7.5 million, also involves the construction of a seawall further inland along which a new public footpath will be constructed for visitors to observe wildlife on the marshes.

Activity

Design an interpretation board which explains the Wallasea Project to visitors.

(For useful websites, go to www.heinemann.co.uk/hotlinks, enter the express code 7627P and click on the relevant links.)

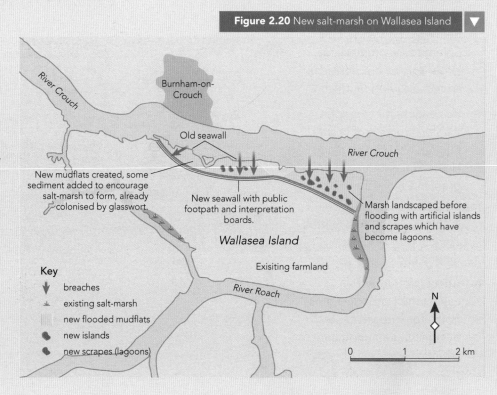

Figure 2.20 New salt-marsh on Wallasea Island

2.3 In what ways are physical environments under threat from human activity?

Human activity poses threats to physical environments in both planned and unintended ways. Salt-marshes, as already described, have been degraded by a variety of human activities and now we will explore the threats to woodlands.

Human impacts and threats to woodlands

Since 1945, the main threats to deciduous woodlands have been from agriculture, settlement, industry, transport, quarrying and recreation. To a lesser extent, threats have come from pollution, invasive species, disease and conifer plantations. (For threats prior to 1945, see the Case study on pages 67–9.)

Agriculture

Intensification of agriculture after 1945 destroyed a lot of **ancient** and **ancient semi-natural woodland**,

and remaining fragments often became too small to support viable populations of plants and animals. Field drainage, nutrient enrichment from nitrate fertiliser and agricultural sprays have all changed species' compositions along woodland edges.

Settlement/industry/infrastructure/quarrying

New roads and the widening of existing ones, airport and quarry expansion and the construction of industrial estates, housing, superstores, telecommunication towers and leisure centres have resulted in a direct loss of woodland habitats (Figure 2.21). Even where trees have been spared, adjacent new developments have altered local hydrology and sometimes caused air pollution which has affected nearby woodland.

> **Key terms**
>
> **Ancient semi-natural woodland**: woodland which contains original native trees and shrubs, but may have been managed by coppicing or felling.
>
> **Ancient woodland**: land which has been woodland since at least 1600.

Recreation

In popular recreational areas, such as the New Forest, horse-riding, off-road parking and footpath trampling damage soils and vegetation. High traffic volumes on roads which cross the forest cause noise and air pollution and disturb wildlife, especially ponies. Dogs chase wild animals and disturb ground-nesting birds, while picnicking increases the fire risk.

Pollution

Prior to the Clean Air Acts of the 1950s, smog reduced light levels in urban areas such as London, which encouraged deciduous trees to shed their leaves prematurely. Coniferous trees, which grow in acid soils, are particularly prone to the damaging effects of acid rain because a decline in pH makes the soil toxic, but deciduous trees can also be damaged by air pollution. Plant nutrients such as calcium, magnesium and potassium are leached from the soil by acidic water. Micro-organisms, which normally decompose organic matter, are also less active in acidic conditions. Tolerance of air pollution varies: species such as London Plane, which is often planted in cities, is resistant, while birch is more susceptible. Deciduous trees growing near roads and airports are damaged by emissions of nitrous oxide, un-burnt fine carbon particulates (smoke) and carbon monoxide. Heavy metals emitted from smelters turn leaves yellow and slows down micro-biological activity.

Invasive species

Sycamore, an invasive species first introduced from central Europe many centuries ago, has spread throughout woodlands displacing native trees.

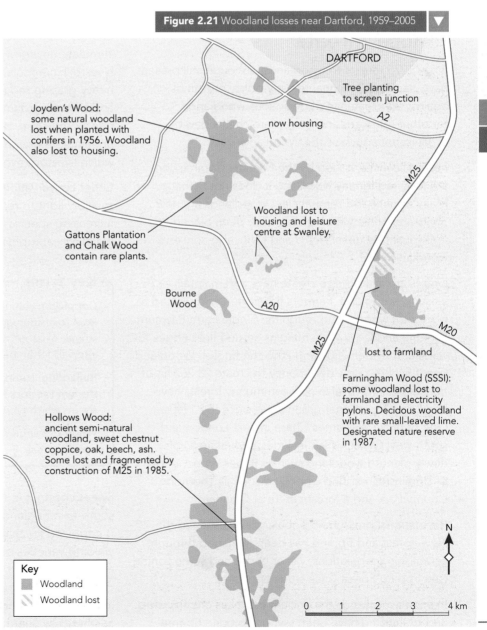

Figure 2.21 Woodland losses near Dartford, 1959–2005

Disease

Dutch elm disease, introduced accidentally from North America in imported logs in the 1970s, altered the structure and composition of woodland and continues to affect them. As the trees died, canopies were opened up and species, such as alder, filled in the gaps. More recently, Sudden Oak Death, a fungal disease first identified in California in 1995, has spread to English oak woodlands.

Conifer plantations

Up to the 1980s, conifers were often planted within deciduous woodlands. The trees acidified the soil, altered species' compositions and their dense, spreading branches prevented light reaching the woodland floor, eliminating shrub and herb layers.

Conservation measures to reduce threats

Since 1975 threats to woodlands, although still present, have declined. National forestry policy presumes against the clearance of deciduous woodlands for other purposes and threats are challenged by organisations such as the Woodland Trust.

Felling licences, issued by the Forestry Commission, are required for any woodland it does not manage. Many ancient and semi-ancient woodlands are now SSSIs and preservation orders have been placed on individual rare trees. NNRs and LNRs also conserve woodland.

Grants are available to create new, and maintain existing, woodland. Farmers have been encouraged to plant new woodland through the Single Farm Payment Scheme and leave wide margins around field edges for wildlife. Between 2001 and 2005 the English Woodland Grant Scheme provided money to create 28 262 ha of new woodland. Twelve new community forests have been developed near urban areas since the 1990s. One of these is Thames Chase in east London and south-west Essex, which includes several areas of newly-planted woodland such as Pages Wood, north of Upminster, existing woodland such as Thorndon County Park and a forest centre at Cranham.

Recreational pressure in popular forests such as the New Forest and Epping has been managed through establishing regulations, visitor codes and some zoning (see Epping Forest Case study opposite).

In some woodlands the ancient practices of **coppicing** and **pollarding** have been revived to open up tree canopies and stimulate ground cover. In the past, trees were coppiced, or cut at ground level, to stimulate shoot growth, which was regularly harvested for timber and to make charcoal. Coppicing declined in the 1800s when coal replaced charcoal as a fuel source and consequently trees have become old, top-heavy and diseased. Trees were pollarded, or cut at head height, to prevent animals grazing the leaf foliage, while the shoots which grew from the stump were also harvested for fuel and timber. Like coppicing, this practice declined in the 1800s when woodlands were no longer used for **wood grazing**. The recent reintroduction of coppicing and pollarding in woodlands such as Epping Forest (see Epping Forest Case study opposite) has allowed more light to reach the floor. This has benefited species such as bluebells and butterflies, and provided more ground cover for nightingales, small mammals and grazing deer.

Wood grazing, another ancient practice, has also been revived in woodlands such as Epping Forest to stimulate new growth and increase species diversity. Livestock numbers are, however, controlled because heavy grazing reduces soil infiltration capacity and trampling damages leaves which cannot then photosynthesise. Heavy grazing also leaves behind unpalatable grasses, as well as bracken and gorse which livestock avoid.

Other conservation techniques include creating glades to allow light to reach the woodland edge, thus increasing species diversity, and leaving dead wood in place for saprophytes.

Key terms

Coppicing: cutting trees, especially hazel, at ground level to encourage stems, or 'spring', to grow from the stump, or 'stool', which is then harvested at regular intervals for timber and charcoal.

Pollarding: cutting the branches at head height to prevent livestock eating the foliage and harvesting the shoots which grow from the trunk for fuel and timber.

Wood grazing: grazing cattle, sheep and pigs beneath trees.

Sweet chestnut is still commercially coppiced in south-east England for hop poles and fencing. Hazel is coppiced for thatching and making charcoal. High-yielding varieties of willow and poplar are coppiced every three to four years for bio-fuel. Plantations currently occupy only a few hectares; should this increase, large blocks of willow might be perceived as aesthetically unattractive.

Case study | Epping Forest: human impacts on the ecosystem

Epping Forest is located on slightly high ground between the River Roding and the River Lea in north-east London and covers an area of 2400 ha (Figure 2.22). About 70 per cent of the area is covered by deciduous woodland and many of the trees are over 400 years old. It is a Special Area of Conservation (SAC) because its beech woods are developed on acidic soils, with holly and yew in the shrub layer. The forest supports rare species such as the moss *Zygodon forsteri* and stag beetles. Dead wood is deliberately left in place to support saprophytes.

Epping Forest was established as a royal hunting ground in the Norman period. Local people or commoners were allowed to graze their livestock under the trees, which were pollarded. Trees were also coppiced for fencing and fuel (Figure 2.23). In the 1850s, nearby Hainault Forest was largely cleared for farmland and development. Some 100 000 trees were felled in just six weeks. Concern that landowners might also enclose common land in Epping Forest led to the 1878 Epping Forest Act. The Act gave ownership to the Corporation of London to protect the forest for recreation, but the commoners lost their grazing rights and pollarding ceased. The Wildlife and Countryside Act of 1981 further strengthened protection and now 66 per cent is designated as SSSI and SAC.

By the 1970s, many trees in the forest had become old, top-heavy and diseased, and the ground flora had been shaded out. Re-pollarding of hornbeam trees began in 1990 to open up the canopy. A programme of crown reduction was also started to prevent old beech and oak trees collapsing. Secondary woodland along woodland edges was coppiced to stimulate shrub growth for nesting birds, such as the nightingale. Grazing was reintroduced in 2002 when Longhorn cattle were released into the woodland. Some 720 ha of buffer land adjacent to the forest was purchased by the Corporation of London to further protect the woodland (see Figure 2.22). A Green Arc Partnership was set up in 2003 with the aim of linking open spaces on the borders of north and east London; the Woodland Trust acquired the first land for this purpose in 2006.

The forest remains a popular recreational area for cyclists, horse-riders and walkers. Footpath trampling

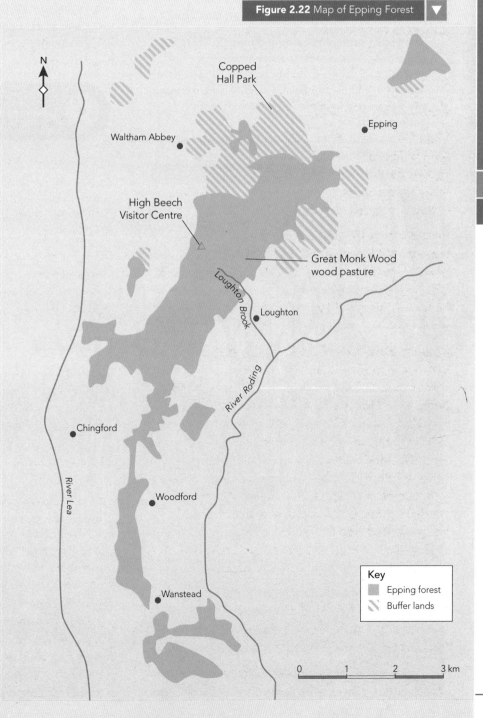

Figure 2.22 Map of Epping Forest

has reduced sensitive species such as heather and moss, and fly tipping has led to the closure of car parks at night. Byelaws prohibit cycling and horse-riding in sensitive areas such as Loughton Brook and on easy access paths, for example around High Beech Visitor Centre. Horse-riders must also obtain a licence from the forest authorities before they can use the woodland. Visitors should not drop litter as this may kill animals and cause fires. Dog owners are urged not to allow their dogs to disturb deer and ground-nesting birds. A visitor centre at High Beech provides information about the forest and offers educational courses.

Theory into practice

Small-scale, local physical environments such as commons, parks, ponds, rivers and hedgerows are affected by a variety of human activities. Think about habitats in your local area and identify human impacts and threats and any steps taken to conserve these environments.

Activities

1. Create a diagram summarising the threats to woodlands.
2. Using the Woodland Trust website, follow links to woodlands currently under threat and draw up a table of examples of woodlands lost, saved and currently threatened by different types of development. (To access the Woodlands Trust website, go to www.heinemann.co.uk/hotlinks, enter the express code 7627P and click on the relevant link.)
3. Design a visitor leaflet explaining the ecology of Epping Forest or woodland you are familiar with and how it is managed.
4. Compare recent 1:50000 Ordnance Survey maps with earlier editions to identify changes in woodland coverage.

Take it further activity 2.2 on CD-ROM

Figure 2.23 Coppiced woodland

2.4 Why does the impact of human activity on the physical environment vary over time and location?

Human activities have both positive and negative impacts on physical environments. The creation of a national park represents a positive impact, while footpath trampling is an example of a negative impact. The degree of impact depends on a number of factors which include the following:

- *Nature of the environment* – rainforests are very fragile because species diversity is high and the organisms are highly specialised. Also, much of the biomass is stored above the ground which means that when the trees are felled a large proportion of the nutrients is removed. Grasslands, in contrast, are more resilient because most of the biomass is stored underground.
- *Type of human activity* – for example, large-scale, open-cast mining causes more damage than footpath trampling.
- *Numbers involved* – large numbers of people cause more damage to an ecosystem than the occasional visitor.
- *Length of impact* – impacts may initially be small but increase over time.
- *Level of development* – as countries develop they have a greater, and more varied, level of impact on their physical environments, although greater prosperity and awareness of their actions may lead in time to conservation and sustainable development.

Human impacts on MEDCs over time

In more economically developed countries (MEDCs) (i.e. those towards the top of the development continuum) the impact of human activities on physical environments has been long and varied. Many landscapes have been totally altered, although conservation now protects sensitive sites, as the following examples show.

Case study | Human impacts over time

United Kingdom

Initially, small numbers of **Mesolithic** hunter-gatherers with primitive hunting tools caused little impact. **Neolithic** farmers began clearing the wildwood, which contained a variety of trees including oak, elm and small-leaved lime, for agriculture about 4000 BC using stone axes and fire. By the Iron Age in 500 BC, half of England had been cleared of wildwood, and only one-fifth remained by the Domesday Survey in 1086. The Normans established Royal deer hunting forests, while medieval peasants continued to carry out ancient practices such as coppicing, pollarding and grazing livestock in wood pasture. Oak trees were felled from managed woodland for shipbuilding, while oak bark was used for leather tanning. Meanwhile, coppiced woodland was used to make charcoal to smelt iron.

After 1750, as death rates declined, the population rose which placed greater demands on the countryside to produce food and common land was enclosed by hedges. The felling of timber to build naval ships and ocean-going trading ships peaked between 1750 and 1860. When coal replaced charcoal as a fuel source, coppicing declined. Woodlands were no longer perceived as useful and some were converted to coniferous plantations or cleared for agriculture.

> **Key terms**
>
> **Mesolithic:** the Middle Stone Age, which in Britain was between 10 000 and 4000 BC, and was associated with hunter-gatherer communities.
>
> **Neolithic:** the New Stone Age, which in Britain was between 4000 and 2000 BC, and was associated with the first farmers.

Meanwhile, wetlands continued to be reclaimed for agriculture while some coastal sand dunes, which were also perceived as being of little value, were converted to commercial conifer plantations.

Woodland covered only 5 per cent of Britain in 1900. After the First World War, the Forestry Commission was created to develop a strategic reserve of wood, and extensive areas of moorland in northern Britain and Scotland were forested with pine, spruce and larch plantations. These altered moor and peat bog habitats and caused a decline in species diversity.

The intensification of agriculture, brought about by improvements in machinery and use of nitrate fertilisers after the Second World War, further reduced natural habitats such as chalk grassland. The use of herbicides and pesticides and removal of hedgerows to create larger fields also caused species to decline.

The growth of the motorways and urban expansion after 1950 further destroyed natural habitats such as lowland heathland. Large-scale, open-cast limestone quarries and the construction of reservoirs also removed valuable ecosystems. The greater availability of cars opened up the countryside for recreation and leisure, and as a consequence popular areas such as the Lake District suffered from road congestion. The recent opening of the cable railway to the top of the Cairngorms typifies how improvements in technology, coupled with more leisure time, have placed pressure on environments.

As pressure on physical environments has increased, so has the need to conserve and protect sensitive areas. Early pioneers in the field of conservation were the Royal Society for the Protection of Birds (RSPB) set up in 1889 and the National Trust founded in 1895. Today, the National Trust protects a variety of environments, such as Wicken Fen, a wetland in Cambridgeshire. The 1949 Access to the Countryside Act led to the establishment of SSSIs and NNRs, and many have since also been designated under various UNESCO and EU schemes as Biosphere Reserves, Ramsar wetland sites, Special Protected Areas (SPAs) and SACs (see Figures 2.24 and 2.25). The 1981 Wildlife and Countryside Act created marine nature reserves and strengthened the legal protection of threatened habitats such as SSSIs. The 1949 Act also led to the creation of **National Parks** of which there are now twelve in England and Wales and two in Scotland. These provide the public with access to the countryside and protect landscapes.

Key term

National Park: an extensive area of high landscape value which is protected and managed for public enjoyment.

More recently farmers have been given incentives to conserve the countryside by replanting hedgerows and woodlands, returning reclaimed land to wetlands and farming organically. New salt-marshes have been established to protect the coast against rising sea-levels in Essex, while conifers have been removed from sand dunes to create more open and diverse habitats, for example at Ainsdale in Lancashire.

Figure 2.24 Preserving Lopham Fen, Suffolk

Europe

Like the UK, other western European countries have experienced recent threats to their physical environments from tourism and pollution. Over 100 million tourists visit the Mediterranean coast each year and high intensity developments have destroyed natural habitats. Pollution threatens species such as the Mediterranean monk seal and loggerhead turtle. Meanwhile, acid rain has damaged lakes and coniferous forests in Scandinavia, Germany, Poland and the Czech Republic.

The USA

In the USA, economic progress and technological advances have also had significant impact on physical environments. In the 1700s, 60 million buffalo roamed the Great Plains in the USA and were sustainably hunted by Native Americans. The arrival of the Union Pacific Railway and buffalo shoots in the late 1800s decimated the population so that by 1900 there were only a few hundred left. Since then, however, captive breeding and release schemes have caused their numbers to increase in protected areas such as Yellowstone National Park. More recently, in the north-east, acid rain carried on the wind from the coal-powered plants in the Ohio valley has killed fish in the Adirondack Lakes. In the south-west, large-scale, irrigated, agriculture projects have resulted in salinisation of soils in Arizona and changes to the Colorado delta ecosystem.

More positively, the USA established Yellowstone, the first national park in the world, in 1872 and many more have since been created. Moreover, some past mistakes, such as the straightening of the Kissimmee River in the 1950s in the Florida everglades, which resulted in a huge loss of biodiversity, have now been reversed in a plan costing US$ 8 billion.

Discussion point

Do you think the area under woodland in the UK will increase or decrease in the next 50 years. Why?

● Take it further activity 2.3 on CD-ROM

Figure 2.25 Examples of protected environments in the UK

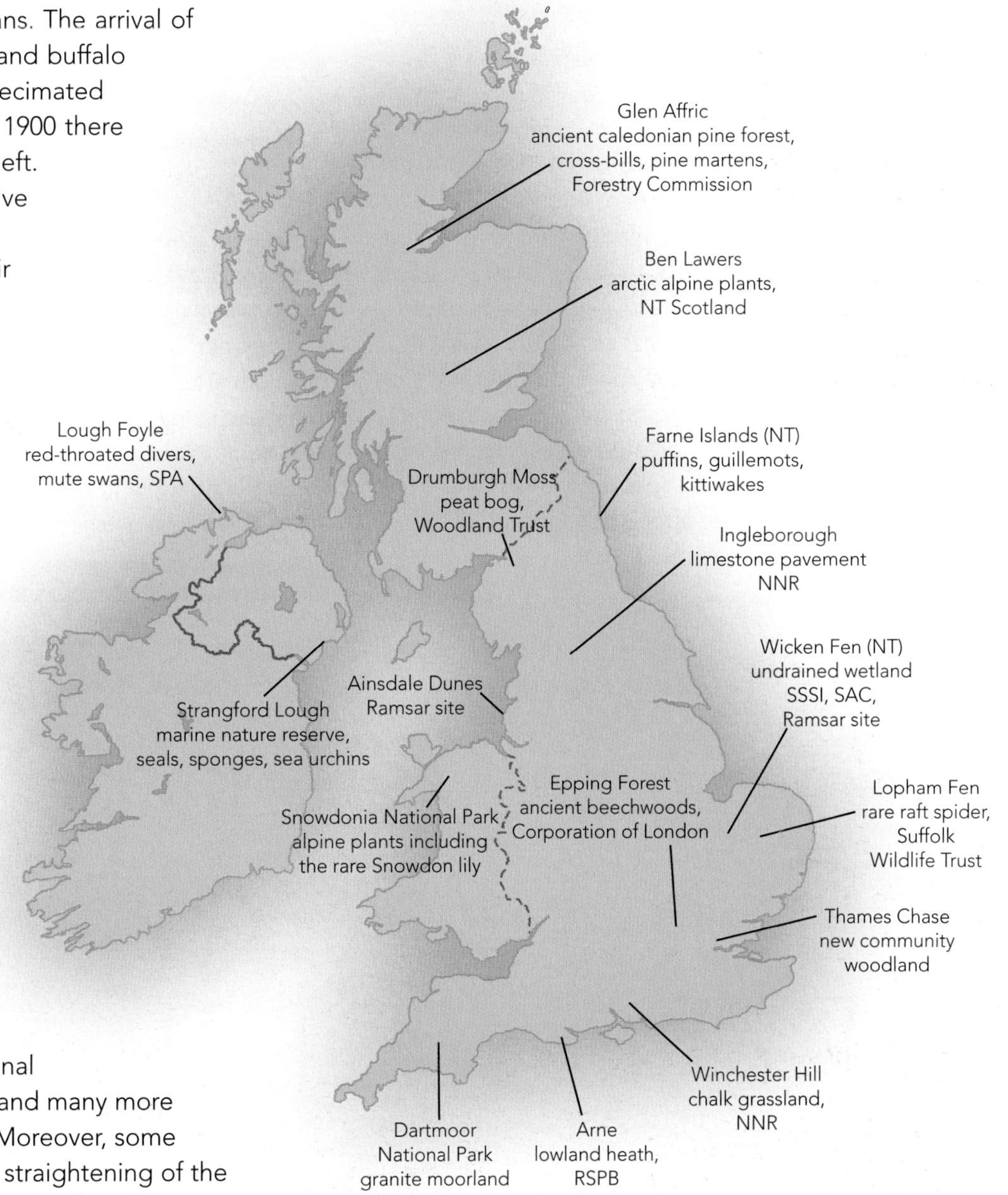

- Glen Affric ancient caledonian pine forest, cross-bills, pine martens, Forestry Commission
- Ben Lawers arctic alpine plants, NT Scotland
- Lough Foyle red-throated divers, mute swans, SPA
- Farne Islands (NT) puffins, guillemots, kittiwakes
- Drumburgh Moss peat bog, Woodland Trust
- Ingleborough limestone pavement NNR
- Strangford Lough marine nature reserve, seals, sponges, sea urchins
- Ainsdale Dunes Ramsar site
- Wicken Fen (NT) undrained wetland SSSI, SAC, Ramsar site
- Snowdonia National Park, alpine plants including the rare Snowdon lily
- Epping Forest ancient beechwoods, Corporation of London
- Lopham Fen rare raft spider, Suffolk Wildlife Trust
- Thames Chase new community woodland
- Dartmoor National Park granite moorland
- Arne lowland heath, RSPB
- Winchester Hill chalk grassland, NNR

Human impacts on LEDCs over time

Although significant impacts on the physical environments have occurred more recently in less economically developed countries (LEDCs), change is now occurring at a rapid pace. Hunter-gatherer communities in the rainforests of Brazil and nomadic tribesmen in east Africa have sustainably managed their environments for centuries. Colonial expansion in the nineteenth century resulted in clearance of some areas for plantation crops such as rubber and coffee, but since the 1960s rising populations and exploitation of resources by multinational companies have placed severe pressures on the environment, as the following examples show.

Case study | Human impacts on Indonesia

Until recently, human activities have had little impact on Indonesia's physical environments. The Portuguese established a Spice Trade in the 1500s which was taken over by the Dutch in the 1600s, who also set up sugar and coffee plantations in the 1800s. In the early 1900s, foreign-owned companies established rubber plantations and began drilling for oil.

Since the 1960s, multinational companies have exploited the rainforest for timber, particularly from Borneo and Irian Jaya (Figure 2.26). Rainforest still covered 82 per cent of Indonesia in the 1960s, but by 2005 it had fallen to 49 per cent (Figure 2.27). Exploitation has provided the Indonesian government with valuable foreign earnings to develop its economy. Concerns about the level of deforestation expressed by organisations such as the World Bank have, however, led to a ban on the export of raw logs. Despite this, illegal logging still occurs, even in national parks. Some deforested areas have been replanted with fast-growing species such as pine and eucalyptus, but these species cannot replace the diversity of the original ecosystem. Indonesia has recently suggested that developed countries could offset their carbon emissions by paying for the forests to be conserved rather than felled.

A second major threat is oil-palm plantations. Between 1980 and 2000 the area under oil-palm increased from 400 000 to 3 000 000 ha. Most has been planted on previously logged or 'degraded' forest, which, although not so biologically diverse as the primeval rainforest, nevertheless did support the Sumatran elephant, tiger and orang-utan (Figure 2.28).

Lesser, but still significant, threats to the Indonesian rainforest have been caused by mining, oil and gas exploitation and other plantation crops including cocoa and pepper. Since the 1970s, international companies have developed large-scale, open-cast mines such as Grasberg in Irian Jaya. Mercury and cyanide, which are used to concentrate the gold, have contaminated rivers, killing fish and polluting the sea-floor. Although some open-cast pits have been in-filled after use and

Figure 2.26 Map of Indonesia

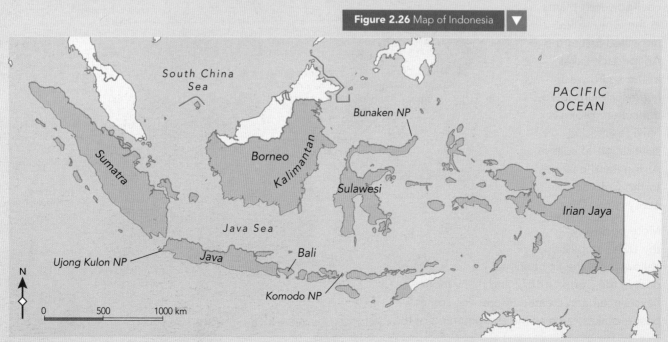

Figure 2.27 Rainforest destruction in Indonesia

replanted with fast-growing trees, others still scar the landscape. A recently proposed gold mine threatens the Bunaken National Marine Park off the island of Sulawesi, which contains 400 species of coral and thousands of fish including rare varieties of pygmy seahorses.

Rainforest has also been cleared for settlements, roads, agriculture and for fuel wood to support Indonesia's expanding population, which in 2006 totalled 225 million and was growing by 1.1 per cent per annum. The practice of **shifting cultivation**, which in the past had little impact on the environment, has been extended so that crops are grown for longer periods, which exhausts the soil.

Indonesia's coral reefs are also threatened by human activities, including cyanide poisoning which is used to stun and capture fish and dynamite fishing (although forbidden). Sediment released from logging and urban development has also clouded the water, preventing algae, which have a symbiotic relationship

Figure 2.28 The Orang-utan faces extinction if the destruction of its natural habitat continues

> **Key term**
>
> **Shifting cultivation:** a type of agriculture practised in tropical areas where soil fertility is maintained by alternating periods of cropping with periods of fallow.

with coral, carrying out photosynthesis. Coral is also removed for building, liming and ornamental purposes, while sewage and agricultural runoff is causing eutrophication.

Concerns about the environment have led the government and conservation groups such as World Wildlife Fund for Nature (WWF) to examine ways in which it might be protected. The WWF have set up several projects in Indonesia to conserve threatened species such as the Sumatran tiger, elephant, rhino and Borneo orang-utan. There are 50 national parks in Indonesia which conserve 20 per cent of the rainforests, including Ujong Kulon which is a World Heritage Site (WHS). Komodo National Park, which contains the world's largest lizard, is also a WHS and a UNESCO biosphere reserve. The park is zoned into intensive areas which include villages and tourist enclaves, wilderness zones (where tourism is limited to trails and camps), and sanctuary zones where access is permitted only for research.

In other LEDCs, such as Sudan, pastoralists migrating with their herds have until recently caused little impact on tropical grasslands, but more sedentary farming practices in the twentieth century have resulted in overgrazing near settlements. Acacia trees have been cut for fuel to meet the needs of an expanding population and grassland has been cultivated to grow cash crops such as sugar cane. Deforestation, overgrazing and **monoculture** have resulted in soil erosion and **desertification**.

Key terms

Desertification: the conversion of marginal land to a desert as the result of climatic change and mismanagement of resources.

Monoculture: growing one species in a given area.

National reserves and parks have been set up in many LEDCs to protect physical environments. In Kenya, for example, the Masai National Reserve protects big game from poaching, and entry fees have contributed to conservation, research and captive breeding programmes. Tourism in the parks has, however, also disturbed the wildlife. In Ghana, Kakum National Park was opened in 1994 to protect the rainforest from timber exploitation and to encourage ecotourism. Sustainable ecotourism is also well-developed in countries such as Costa Rica and Peru. In 2007, Gola National Park was established to protect rainforest in Sierra Leone, in a scheme funded by the European Commission to conserve forests and cut carbon emissions.

In conclusion, it can be seen that economic, social and technological developments have placed increasing pressure on physical environments although some are now protected or sustainably managed. Economic pressures will, however, continue to drive exploitation, as seen in the recent expansion of the project to extract oil from tar sands in northern Canada where there are concerns about deforestation and water contamination.

Activities

Using the information in this section, write an essay to answer the following question:

'To what extent do positive and negative impacts on the environment increase or decrease with economic, social and technological development?'

 Take it further activity 2.4 on CD-ROM

2.5 How can physical environments be managed to ensure sustainability?

When human activities impact on physical environments, they may need to be managed in order to be sustainable. Some of the ways this can be achieved are by conservation, having visitor codes and planning controls, and restricting use.

Restricting use

Visitor overcrowding in fragile physical environments can be controlled by imposing limits on the number of visitors accessing sites at any one time, charging entrance fees and temporary closures. Entrance fees also generate revenues for restoration and conservation. Temporary closure gives ecosystems a chance to recover, but may result in overcrowding at other times.

Another way of restricting use is **zoning**, i.e. limiting different human activities to specific spaces or times. One approach is to preserve sensitive areas for research and conservation while developing other sites to withstand high visitor pressure by constructing paved roads, car parks, toilets, information centres and picnic areas. In this way developed areas, or **honeypots**, alleviate pressure on more fragile areas. Zoning is now widely adopted in many national parks including the Great Barrier Reef off Australia and Arches National Park in the USA (see case studies on pages 73–4 and 76–7). The opposite approach is to spread visitors across the environment to lessen pressure on any one area.

> **Key terms**
>
> **Honeypots:** sites designed to accommodate large numbers of visitors so that the impact on the surrounding environment and local people is minimised.
>
> **Zoning:** confining different human activities to specific spaces or times.

Case study | Sustainable management of the Great Barrier Reef

The Great Barrier Reef off the Queensland coast in Australia is the largest coral reef system in the world (Figure 2.29). Stretching for 2000 km, it is larger in area than the British Isles. It contains 2900 separate coral reefs, seagrass beds, rocky reefs and sand-flats. The ecosystem supports 1500 different types of fish, 350 types of coral, 400 sponges, 40 000 molluscs and 6 species of turtle (Figure 2.30). It became a marine park in 1975 and was designated a World Heritage Site in 1981.

Acting for the Australian government, the Great Barrier Reef Marine Park Authority (GBRMPA) has the task of balancing the needs of tourists and fishermen with those of conservation and research. A key management strategy has been to zone the marine park into areas where different types of activity can take place. In 2004, the zones were revised and the size of highly protected areas, where for example no fishing or collecting is permitted, was increased from 4.6 to 33.3 per cent.

The park has seven zones which are: a General Use Zone, a Habitat Protection Zone, a Conservation Zone, a Buffer Zone, a Scientific Research Zone, a Marine Park Zone and a Preservation Zone. Most activities are allowed in

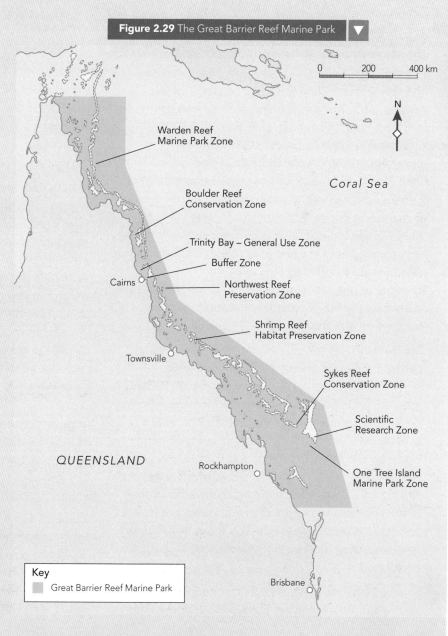

Figure 2.29 The Great Barrier Reef Marine Park

Figure 2.30 A coral reef supports a large number of organisms

the General Use Zone including boating, fishing, shipping, tourism, fish farming and the harvesting of sea cucumber. In contrast, only limited and heavily regulated research is permitted in the Preservation Zone. No fishing or collecting is allowed in the Marine Park Zone and other activities require a permit. In general, areas closer to the coast tend to coincide with General Use Zones, while the outer reefs fall within Marine Park Zones (Figure 2.29). Park authority vessels and planes patrol the reef area to enforce the zonation and monetary penalties are imposed on those who breach the regulations. Prospective developments are subject to environmental assessments before permission is granted.

The reef attracts 1.9 million tourists and 4.9 million recreational users annually. Tourism generates AU$ 5.1 billion each year and creates 54 000 full-time jobs. An environmental management charge is levied on visitors participating in tourist activities, which in 2005 generated AU$ 7 million for management, education and research. In popular recreational and tourist areas, such as Blue Pearl Bay, which is in the Marine Park Zone, overcrowding is prevented by restricting boat length to 35 m and visitor group size to 40. Eight public moorings have been installed and reef protection markers delineate where anchoring is prohibited.

In areas where fishing is permitted, agreements such as the Coral Reef Finfish Fishery Programme limit the number of commercial vessels in operation and the size of the catch. The landing of threatened species, such as the Queensland Grouper, is also forbidden.

Although the GBRMPA has jurisdiction over the marine park, pollution from the adjacent coastline is more difficult to control. Land cleared for sugar cane production, as well as the removal of mangrove for tourist developments, has resulted in sediment entering the sea which has adversely affected the coral. Agricultural runoff has also caused eutrophication which has damaged seagrass communities. Targets to limit pollution from the 26 river catchments near the reef have, however, been recommended. The discharge of sewage and other effluents is also regulated and the practice of recycling effluent is increasing.

Another threat to the reef is the crown-of-thorns starfish which is killing the coral. The increase in the starfish population may be the result of overfishing of starfish predators. Eutrophication could also be responsible for increasing microalgae which may benefit starfish larvae. The biggest threat the reef currently faces is, however, coral bleaching. This occurs when zooanthellae (algae), which live within the polyp and carry out photosynthesis, are expelled, causing the coral to die. The cause of coral beaching is uncertain; it might be the result of rising sea temperatures associated with global warming.

Visitor codes

Establishing visitor codes of conduct, such as prohibiting the removal of plants, also helps to conserve environments. Codes are, however, difficult to enforce and monitor.

Planning controls

Proposed new developments in areas of high landscape value such as national parks are subject to strict planning controls. Buildings must be designed and built in materials which blend in with their surroundings. Developers are also often required to carry out environmental impact assessments (EIAs). These should highlight the direct and indirect environmental impacts of any proposed development and offer strategies to minimise damage. A common method of assessing impacts is to measure physical, environmental and perceptual carrying capacity.

Physical carrying capacity

This is the maximum number of visitors an environment can hold at any one time. In environments where most people arrive by car, this is usually determined by counting the number of parking spaces (Figure 2.31). Throughput or daily capacity is calculated by multiplying the average length of stay by the length of time the site is open. Capacity can be increased by enlarging the size of existing car parks or adding new ones. Alternatively, car parks may be temporarily or permanently closed to reduce pressure.

Environmental carrying capacity

This is the maximum number of visitors an environment can support before vegetation, soil and rock is badly degraded and there is significant water, air and noise pollution. The amount of damage which occurs depends on how many people visit the site, when they come and what activities they pursue, as well as the nature of the relief, soil, geology and vegetation. Motorcyclists, for example, inflict more damage than walkers. Steep slopes are more susceptible to erosion than lower gradients and some plants are more vulnerable to trampling than others. Badly eroded footpaths, excessive noise, fumes, litter and wildlife disturbance are all indications that environmental carrying capacity has been reached.

Perceptual carrying capacity

This is the number of people or activities an environment can support before visitors feel their

Figure 2.31 Yosemite National Park – car park spaces indicate carrying capacity

enjoyment is spoiled by the presence of others. In wilderness areas, where people seek solitude, tolerance of others is lower than, for example, around visitor centres.

The concept of carrying capacity has been applied to the evaluation of resources and visitor experiences within national parks in the USA. Arches National Park (see case study below) was the first to adopt a Visitor Experience and Resource Protection (VERP) programme. This involved measuring variables including visitor numbers at honeypot sites, parking availability, water quality, footpath erosion, wildlife disturbance and visitor experience.

Case study | Arches National Park

Arches, in south-east Utah, became a national park in 1969. It contains over 2000 natural stone arches (Figure 2.32) as well as ancient rock art. The desert ecosystem supports collared lizards, jack rabbits, rattlesnakes, mule deer, bighorn sheep and the rare biscuitroot plant.

The park receives about 800 000 visitors a year, most of whom arrive between May and September. Many tour by car along the scenic drives stopping off along the way to walk along short trails to visit the arches. Others hike in the backcountry, camp, and explore the area using horses and mountain bikes. The main problems are overcrowding at popular sites, off-road parking, litter, trampling the delicate **cryptobiotic soil crusts** (Figure 2.33), and graffiti which is carved on the sandstone rock formations.

Key term

Cryptobiotic soil crust: hard, grey-brown soil crust composed of micro-fungi, algae, cyanobacteria and moss. Filaments of cyanobacteria help to bind soil particles together and prevent erosion and they also fix nitrogen. The crust also helps to conserve moisture.

Figure 2.32 Landscape arch in Arches National Park

excluded. Cycling is permitted only on designated roads. Vehicles are only allowed to park in allocated spaces, which reduces pressure on popular sites such as Delicate Arch.

Park regulations are enforced by patrols and visitors are fined for breaking the rules. Plants, animals and rock art are protected by state and federal law. All backcountry overnight hikers must have a permit and the gathering of firewood is prohibited. Drivers are not allowed to stop on roads and visitors must not feed wild animals. A road speed limit is enforced and visitors must stay on trails to protect the delicate cryptobiotic crust. Displays at the visitor centre help to raise public awareness of the importance of conserving the landscape and desert ecosystem.

Figure 2.33 Cryptobiotic soil crust

Income from entrance fees, which in 2006 totalled approximately US$ 460 000, together with fees from backcountry hiking permits, campground fees and ranger-led tours, supports management and restoration projects. These include: the removal of tamarisk, a bushy plant which smothers other species; widening and maintenance of existing trails and re-routing of others; enforcement of park rules; monitoring of water quality and vegetation; and wildlife surveys.

The park has been delimited into nine zones (Figure 2.34). These range from a 'developed zone' around the visitor centre and campgrounds, to a 'sensitive resource protection zone' where the public are

Figure 2.34 Arches National Park showing zones

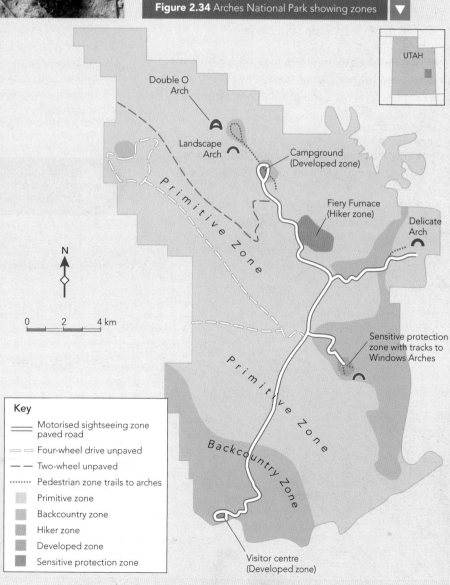

Conservation

Totally protecting environments through **conservation** has drawbacks as well as benefits.

- The creation of national reserves and parks, particularly in LEDCs, has sometimes led to the displacement of local people. The Maasai, for example, were removed from their homelands when the Maasai Mara National Reserve was created in Kenya.

- Over-protection of animals, for example elephants in Kenya, has meant their numbers have increased causing damage to crops. Total suppression of natural fires, for example in the Great Smoky Mountains National Park in the USA, has altered forest compositions and been unfavourable to plants which are dependent on fire for their regeneration.

- Conserved areas, such as rainforest, even when protected may not survive unless they can be shown to have an economic benefit.

A preferable option, therefore, may be **sustainable development** rather than conservation. Ways of achieving this in areas such as rainforests include:

- ecotourism

- helping local people to improve their farming methods so that soils are not degraded

- replanting cleared areas with high-yielding, fast-growing species to protect the native forest from exploitation

- setting up schemes with industrialised countries to protect the forest to offset their carbon emissions

- encouraging corporations and individuals in wealthy countries to adopt and conserve small areas of rainforest.

Key terms

Conservation: managing environments so that they are protected from change.

Sustainable development: developing an environment using methods and techniques which ensure that biodiversity is preserved.

Activities

1. Using websites such those listed at www.heinemann.co.uk/hotlinks (enter the express code 7627P and click on the relevant link for this activity), find images of Arches National Park and design a visitor brochure highlighting the park attractions and the ways in which the public protect the area.

2. Drawing on the information contained in this section, write an essay to answer the following: 'With reference to case studies, describe and explain how physical environments may be managed to ensure sustainability'.

Take it further activity 2.5 on CD-ROM

Knowledge check

1. What are open and closed systems?
2. What is an ecosystem?
3. How are energy and nutrients transferred through an ecosystem?
4. What human and physical factors bring about change to an ecosystem?
5. What are primary and secondary successions and plagioclimax communities?
6. What are the main stores and flows in a selected local ecosystem/environment?
7. How do physical factors such as soil, drainage, relief, microclimate and climate influence the selected ecosystem/environment?
8. How does the chosen local ecosystem/environment change over time?
9. What are the human impacts on a chosen local ecosystem/environment?
10. What are the main threats to a selected local ecosystem/environment? How can conservation reduce these threats?
11. In what ways can human activity impact positively and negatively on physical environments?
12. Why are some physical environments more at risk than others?
13. Why might the impacts increase/decrease as a country develops?
14. In what ways can environments be managed sustainably?

Exam Café
Relax, refresh, result!

Relax and prepare

Student tips

What I wish I had known at the start of the year...

Imogen

"I always try to watch the news or read at least the Sunday paper as a lot of the geography at A2 comes from keeping up to date on world events. Recent programmes on the threat of global warming on the wildlife of Antarctica have been excellent for both this module and the climatic hazards one. Up-to-date examples always impress examiners."

Bruno

"Try to choose very local examples if you can. You will know more detail if it is local to where you live and if you do make a minor mistake the examiner is less likely to spot it. For this unit I use our local wood which is partly a nature reserve. Not only do I know it well but we also did some fieldwork on the effect of different types of trees on the extent and type of undergrowth."

Michael

"The word 'evaluate' or the phrase 'to what extent' are very common at A2. Such questions are trying to get you to discuss the 'fors' and 'againsts' of a viewpoint or statement. Remember to support your view with examples but also you should comment that this viewpoint would vary with circumstances. Your opinion might be very different if you lived in a city or the countryside, were rich or poor, were young or old, or lived somewhere else. Always try to think of some exceptions to your argument or point of view."

Common mistakes – James

"I sometimes forget that this is a synoptic paper so I can bring in material from the units I did at AS. This topic is especially good for re-using or applying those studies at AS of local ecosystems such as salt-marshes and dune transects and larger biome-type studies of tundra and hot desert environments. It certainly broadens my choice of examples and allows me to develop some effective evaluations, as these differ in emphasis from the examples I studied at A2."

"It is easy to forget the systems part of 'ecosystems'. I try to focus my answers so I make clear the inputs, outputs and stores. This provides a structure which helps develop my discussions and evaluations more effectively."

Refresh your memory

2.1 Ecosystems and environments are open systems	
Stores	Biomass (plants, animals, etc.), litter and humus, soil
Flows	Leaf fall, death, decomposition, nutrients drawn up
Inputs	Water, sunlight, minerals from weathering, litter/minerals washed/ blown in, deposition from elsewhere
Outputs	Leaching from litter and soil, erosion, evapo-transpiration

2.2 Various factors interact to create distinctive environments	
Physical	Micro or local climate – precipitation, temperature, humidity, wind, sunshine hours
	Relief – slope angle/direction, altitude, roughness
	Rock type – porosity, minerals, structure, pH
	Drainage – free draining versus waterlogged
	Soil – depth, structure, texture, minerals, pH
	Biotic – plants, animals, etc.
Human	Agriculture – chemicals, adding species, drainage, irrigation, clearance
	Settlement – micro-climate, drainage, change in habitats
	Industry – buildings, pollution, extraction of raw materials
	Pollution – change in water/air quality, noise, visual, landfill
	Conservation – nature reserves, parks, afforestation

2.3 Human activity poses a range of threats to and creates impacts on physical environments	
Planned	Agriculture – removal of woodlands for crops, overgrazing, agri-chemicals, drainage of wetlands, waterlogging
	Mining and quarrying – removal of hillsides, subsidence, dust
	Construction – creates impervious surfaces, noise, dust
	Settlement – removes natural vegetation, changes drainage
	Forestry – strips land, reduces precipitation, increases runoff
	Transport – fumes pollute, creates barriers to wildlife, noise
Unplanned	Fire – removes moisture, destroys habitats, increases carbon emissions
	Disease – may reduce or wipe out species, e.g. myxomatosis
	Disruption of food chain or nutrient flows – change in species
	Floods – increased erosion, waterlogging, increased deposition
	Pollution – acid rain, global warming, change in species
Impacts	Temporary – migration of species, fall in numbers, rise in numbers, change in food chain
	Long term – destruction, extinction of species, introduction of new species, adaptation, change in nutrient supply

Refresh your memory

2.4 Why does the impact vary over time and location?	
Time	Level of development
	Short versus long term
	Changes in technology
	Changes in population
Location	Development differs
	Technology differs
	Population number and type differ
	Incomes and wealth differ
	Cultures differ
	Political attitudes differ

2.5 Types of environmental management to ensure sustainability	
Sustainability	Is it possible given the rising population pressure? Most schemes try to manage to increase sustainability
Government	National parks
	Sites of Special Scientific Interest (SSSI) – over 4000 in England covering 8 per cent of area
	Areas of Outstanding Natural Beauty (AONB)
	National nature reserves – over 200 in England covering nearly 1 per cent of area
	Special areas of Conservation and special Protection areas (EU designation)
Local Authorities	Local nature reserves, parks, etc.
	Planning – land-use zoning, green belts, etc.
Private	Charities, e.g. National Trust, RSPB
	Individual landowners, farmers, etc.
Military	Manage environment as bi-product of military ranges
Corporate	Large companies, e.g. salt-marshes near power stations or refineries

Top tips...

The opportunity to study 'at least one local ecosystem' is an invitation to do some fieldwork or at least an investigation into the pressures on and repercussions for that ecosystem or environment. Examiners will not distinguish between 'ecosystem' and 'environment' but the former helps focus students onto the systems and structure more. There is nothing to stop you drawing on some of the AS examples, such as sand dunes or riverside environments. 'Local' refers to size and scale rather than location. This prevents candidates using the tropical rainforest yet again, although this could be used to illustrate the impact of human activity over time.

Exam Café

Get the result!

Example question

Study Figure 4, a photograph of a game park in South Africa, and Figure 5, a table showing visitor numbers. Suggest the possible issue(s) indicated and suggest appropriate strategies that could be used to manage its impact. [10 marks]

Figure 4

Figure 5

Year	Visitor numbers (000)	Income generated (million Rands)
2007	178	167
2000	156	101
1998	98	75
1978	121	79
1970	56	33

Student answer

Figure 5 shows that visitor numbers have increased over three times since 1970 and income generated by five times, but it has not been a steady increase as numbers fell in 1998. Figure 4 shows the way safaris disturb wildlife, so the main issue is the increasing pressure on the environment but also the area is increasingly becoming dependent on income from visitors. This could be reduced by limiting access to the park via toll gateways, which in turn could raise entrance fees. This would ration numbers of visitors and keep local income high. This in turn could be invested in further conservation measures.

Examiner says
Good reference to the data but also an appreciation of anomalies.

Examiner says
Main issue identified but also a subsidiary or connected issue.

Examiner says
Sound strategy offered with some practical considerations well linked to the two issues and extended to suggest other gains.

Examiner's tips

Examinations are based on the need to assess certain Attainment Objectives (AOs). The overall balance is set out in the specification. At A2 it is different from AS.

▷ 30 per cent of the marks come from AO1 — knowledge and understanding of the content, concepts and processes, so less is based on what you have been taught. Most of these marks can be accessed by effective use of examples and geographical concepts.

▷ 40 per cent of the marks come from AO2 — analyse, interpret and evaluate. This is tested by how you analyse and interpret information (data) and viewpoints and how you apply understanding in unfamiliar contexts, i.e. not straight from your notes. Evaluation is the most demanding of these aspects as it asks you to make a judgement or justify an opinion based on the information.

▷ The remaining 30 percent is AO3, which covers the use of a variety of methods, skills and techniques to investigate issues, reach conclusions and communicate findings. Hence this is especially important in the skills paper where it represents 50 per cent of the available marks.

This shows that A2 has a different emphasis from AS, where knowledge and understanding were more important than its application and evaluation. This partly explains why some candidates perform and achieve very differently at the two levels. Grade A at AS is no guarantee of an A grade at A2.

Chapter 3
Climatic hazards

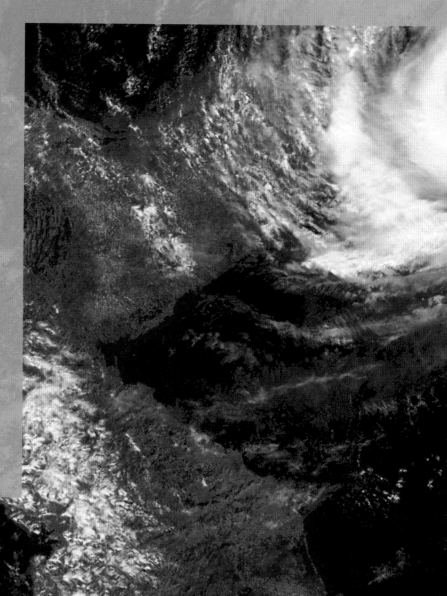

Weather and climate is one aspect of physical geography that affects us daily, for example it dictates what clothes we wear. But there are extreme weather conditions, such as too much or too little rainfall, too much heat or not enough, that have major impacts on society. Climate hazards are varied and would appear to be increasing in frequency and intensity, or at least the reporting of climatic hazards is increasing.

Questions for investigation

- What conditions lead to tropical storms and tornadoes and in what ways do they represent a hazard to people?
- How do atmospheric systems cause heavy snowfall, intense cold spells, heatwaves and drought and in what ways do they represent a hazard to people?
- Why do the impacts of climatic hazards vary over time and location?
- What can humans do to reduce the impact of climatic hazards?
- In what ways do human activities create climatic hazards?

Consider this

According to the United Nations, climate change is a greater threat to humanity than international terrorism. Suggest ways in which this may be the case.

3.1 What conditions lead to tropical storms and tornadoes and in what ways do they represent a hazard to people?

Tropical storms

Hurricanes, typhoons and cyclones are among the most violent storms that affect the world. Figure 3.1 shows that they affect a very large area. Their different names reflect their location: hurricanes in the Atlantic; typhoons in the western Pacific; cyclones, used fairly generically but also more specifically, in southern Asia.

Hurricanes are **compound hazards** which include heavy rainfall, strong winds, high waves and can cause other hazards such as flooding and mudslides. They are intense hazards which affect a large area but are difficult to predict accurately. The onset of any individual hurricane is rapid. They may travel slowly at first but their path is erratic (Figure 3.2). Hence it is not always possible to give more than 12 hours' notice. This is not enough for proper evacuation and precautionary measures.

Hurricanes develop as intense low pressure systems over tropical oceans. Winds spiral rapidly around a calm central area known as the **eye** (Figure 3.3). The diameter of the whole hurricane may be as much as 800 km although the very strong winds which cause most of the damage are found in a narrower belt up to 300 km wide.

Hurricanes normally develop in the westward flowing air just north of the Equator (known as an **easterly wave**). They begin life as small-scale tropical disturbances or tropical **depressions**, which are localised areas of low pressure that cause warm air to rise. Tropical disturbances create thunderstorms which persist for at least 24 hours. These may develop into tropical storms, which have greater wind speeds of up to 117 kilometres

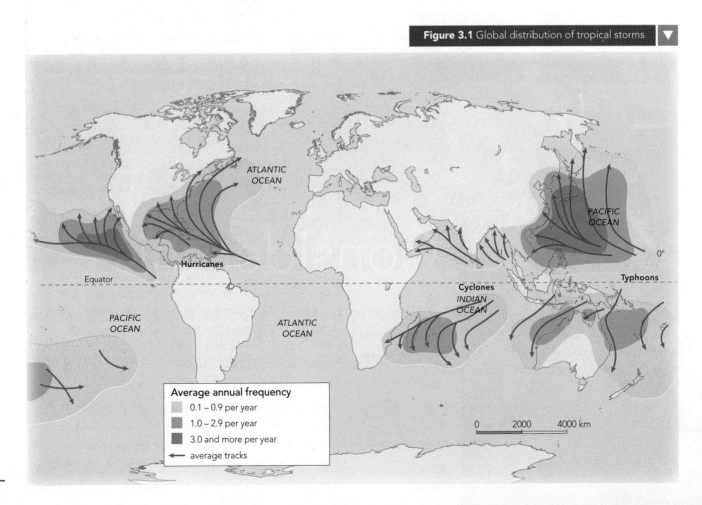

Figure 3.1 Global distribution of tropical storms

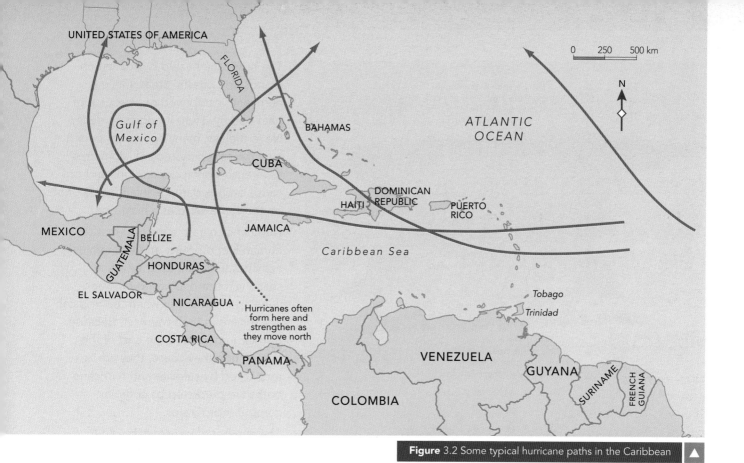

Figure 3.2 Some typical hurricane paths in the Caribbean

per hour (73 mph). However, only about 10 per cent of tropical disturbances ever become hurricanes – storms with wind speeds above 118 km/h (above 74 mph).

For hurricanes to form, a number of conditions are needed (Figure 3.3).

- Sea temperatures must be over 27°C (energetic evaporation from warm water carries latent heat energy into the atmosphere. When condensation occurs, this energy is released, helping to generate the very strong winds).

- The low pressure area has to be far enough away from the Equator so that the Coriolis effect creates rotation in the rising air mass – if it is too close to the Equator there is insufficient rotation and a hurricane would not develop.

Hurricanes are also characterised by high-intensity, high-volume rainfall – up to 500 mm in 24 hours, invariably causing flooding on land.

Once the rising air has become established (a convection cell), the system is self-perpetuating as long as conditions remain favourable. The rising air releases large quantities of heat during condensation. This reinforces the instability (rising of air) within hurricanes. At the eye of the hurricane air descends from the top of the system. As it does so it is warmed and therefore able to hold more moisture. Condensation is reduced and the eye remains cloudless.

Key terms

Compound hazard: an event which produces many hazardous impacts such as a tropical storm produces heavy rain, strong winds and high waves.

Coriolis effect: deflecting force produced by the rotation of the Earth, deflecting objects to the right of their path in the northern hemisphere and to the left of their path in the southern hemisphere.

Depression: low pressure system caused when warm air masses rise over colder air masses.

Easterly wave: an area of low pressure which travels from the east within the Trade Wind zone.

Eye: the central area of a tropical storm, characterised by calm conditions or light variable winds.

Hurricane winds can cause 15-metre waves in the open ocean. Peak heights reaching land can be as high as 6 metres. In a mature hurricane, pressure may fall to as low as 880–970 millibars. This, and the strong contrast in pressure between the eye and outer part of the hurricane, leads to strong gale-force winds. A mature hurricane is typically 200–500 km in diameter, with clouds up to 12 km in height (Figure 3.4).

Hurricanes that affect the Caribbean and the USA mostly form in the east Atlantic and travel westwards with the easterly trade winds. Their exact path, however, is dependent upon localised conditions and is difficult to predict.

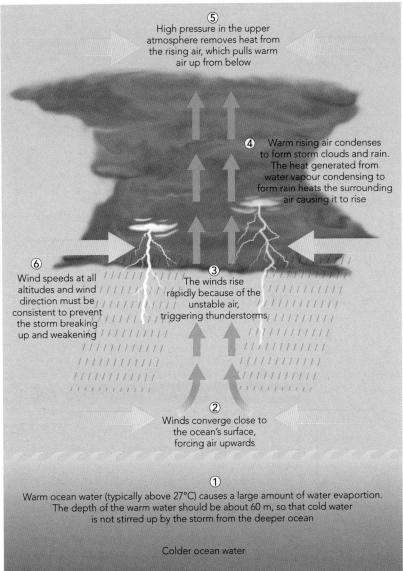

Figure 3.3 Formation of a hurricane

① Warm ocean water (typically above 27°C) causes a large amount of water evaportion. The depth of the warm water should be about 60 m, so that cold water is not stirred up by the storm from the deeper ocean

② Winds converge close to the ocean's surface, forcing air upwards

③ The winds rise rapidly because of the unstable air, triggering thunderstorms

④ Warm rising air condenses to form storm clouds and rain. The heat generated from water vapour condensing to form rain heats the surrounding air causing it to rise

⑤ High pressure in the upper atmosphere removes heat from the rising air, which pulls warm air up from below

⑥ Wind speeds at all altitudes and wind direction must be consistent to prevent the storm breaking up and weakening

The hurricane hazard is greatest on islands and coastal areas. Once a hurricane is deprived of its source of heat and moisture it begins to decay. Nevertheless, many hurricanes persist over land for hundreds of kilometres as intense frontal depressions and can cause serious damage inland, albeit not on the same scale as a hurricane.

Hurricanes create a major threat to human life, property and economic activities (Figure 3.5). They are a seasonal hazard, peaking between June and November when the tropical oceans are at their warmest. Because of their impact, and the cost of the destruction they cause, they are monitored by satellite and hurricane paths are predicted by complex computer programs.

The number of people in the USA living in 'at risk areas' is increasing: the proportion of people living within 50 km of the coast is set to rise from 50 per cent in 1990 to 75 per cent by 2010. In 1992, a US survey suggested that between 80 per cent and 90 per cent of the population had never experienced a Category 3 hurricane. This created a false sense of security and the population tended to underestimate the risk of hurricane damage.

(Source: Musk (1988) *Weather Systems*)

Figure 3.4 Winds in a hurricane

Activities

1. Describe the distribution of hurricanes as shown in Figure 3.1.
2. Comment on the possible impacts of different category hurricanes as shown in Figure 3.5 for a LEDC and a MEDC.
3. Suggest different reasons why the impacts of hurricanes increase with increasing energy.

Hurricane strength – the Saffir-Simpson Hurricane Scale

Type	Category	Damage	Pressure (mb)	Windspeed (km/h)	Stormsurge (m)
Depression	—	—	—	>56	—
Tropical storm	—	—	—	63–117	—
Hurricane	1	Minimal	>980	119–153	1.2–1.5
Hurricane	2	Moderate	965–979	154–177	1.8–2.4
Hurricane	3	Extensive	945–964	178–209	2.7–3.6
Hurricane	4	Extreme	920–944	210–249	3.9–5.5
Hurricane	5	Catastrophic	< 920	>249	>5.5

Hazards and impacts of different strength hurricanes

Category One hurricanes	Winds reach 119–153 km/h and a storm surge is generally 1.2–1.5 m (4–5 ft) above normal. No real damage to building structures. Damage primarily to unanchored mobile homes. Also, there may be some coastal road flooding and minor pier damage.
Category Two hurricanes	Winds reach 154–177 km/h. Storm surges are generally 1.8–2.4 m (6–8 ft) above normal. Some damage to roofing material, doors and windows. Considerable damage to vegetation, mobile homes and piers. Coastal and low-lying escape routes flood 2–4 hours before arrival of the hurricane eye. Small craft in unprotected anchorages break moorings.
Category Three hurricanes	Winds reach 178–209 km/h. Storm surges are generally 2.7–3.6 m (9–12 ft) above normal. Some structural damage occurs to small residences and utility buildings. Mobile homes are destroyed. Flooding near the coast destroys smaller structures with larger structures damaged by floating debris. Land below 1.5 m above mean sea level may be flooded inland up to 13 km. Evacuation of low-lying residences within several blocks of the shoreline may be required.
Category Four hurricanes	These are characterised by winds of between 210 and 249 km/h. Storm surges are generally 3.9–5.5m (13–18 ft) above normal. Some complete roof structure failures occur on small residences. Complete destruction of mobile homes occurs. There is extensive damage to doors and windows. Land below 3 m above sea level may be flooded and require massive evacuation of residential areas as far inland as 10 km.
Category Five hurricanes	Winds are greater than 249 km/h. Storm surges are generally greater than 5.5 m (18 ft above normal). Complete roof failure happens on many residences and industrial buildings. There is also some complete building failures with small utility buildings blown over or away, and complete destruction of mobile homes. Severe and extensive window and door damage is likely. Low-lying escape routes are cut-off by rising water 3–5 hours before arrival of the centre of the hurricane. Major damage to lower floors of all structures located less than 4.5 m above sea level and within 457 metres of the shoreline. Massive evacuation of residential areas on low ground within 8–16 km (5–10 miles) of the shoreline may be required.

Figure 3.5 Hazards associated with differing hurricane strengths

Monitoring hurricanes

As yet there are no known atmospheric conditions that automatically lead to hurricane formation. Forecasters can only describe its path once a hurricane has formed. The National Hurricane Center in Miami, Florida searches out potential hurricanes in their early stages and tracks them through their life cycle until they decay and die.

Satellites detect hurricanes in their early stages of development and can help to provide early warning of imminent hurricanes. The ability to look down from space at weather systems has been a great benefit to weather forecasters ever since the first weather satellite was launched in 1960. Previously there were many areas of the globe, particularly the great oceans, where weather systems could develop without being identified by observers at the surface. Satellites have changed all that. Now they provide vital information about the location and movement of weather systems, as well as information about the vertical structure and composition of the atmosphere. This is especially important for hurricanes which develop and mature over tropical oceans.

Types of satellite

There are two types of satellite providing weather data – geostationary and polar orbiting. The **geostationary** type orbits at a height of 35780 km above the earth, and takes 24 hours to complete each orbit. This means that a satellite in orbit over the Equator will appear to be stationary and 'hang' over the same spot on the Earth's surface all the time.

Polar-orbiting satellites pass over the Earth from pole to pole. The orbit is much lower than that of the geostationary satellites, the images provide detailed information about the cloud structure.

Visible and infrared images

Satellites carry meters which measure **visible light** and provides visible images. What is being viewed is sunlight which has been reflected from the Earth or clouds. In general, the brighter the cloud appears the thicker it is. The disadvantage of this approach is that visible images are only available during daylight.

Another type of meter is an **infrared** meter. These measure the temperature of the cloud or the Earth's surface. The images are usually prepared in such a way that cold surfaces appear white, and warm ones darker. Since temperatures in the lower part of the atmosphere normally decrease upwards, high cloud (with low temperatures) appears white with low cloud or the Earth's surface appearing darker. Infrared images are available even when there is no daylight.

A combination of visible and infrared images is very useful. Two examples will illustrate the point.

- If both images are bright in a particular area there is likely to be thick, high cloud.
- If the visible image shows areas of bright cloud, but the infrared image in the same area is dark, then there is low cloud (or perhaps fog).

Reinforced aircraft fitted with instruments fly through and over hurricanes, and weather radar can locate storms within 322 km (200 miles) of the radar station.

Key terms

Geostationary: a satellite that remains over the same spot on the Earth's surface.

Infrared: heat energy radiated from a heat source, typically of wavelength 9–100 μm.

Polar-orbiting satellites: satellites that pass over the Earth from pole to pole.

Visible light: light visible to the naked eye, typically of wavelength 3–9 μm.

Activities

1. What are the differences between a geostationary satellite and a polar orbiting satellite?
2. How do satellites help monitor and track the progress of a hurricane?

Discussion point

With increased technology it is easier to predict successfully the path of a hurricane. Discuss.

Case study | The impact of Hurricane Ivan on Grenada, September 2004

Hurricane Ivan struck Grenada in the afternoon of Tuesday 7 September 2004 (Figure 3.6). The eye of Ivan passed about 10 km south of the airport at Point Salines. Ivan, a Category 4 hurricane, was one of the most powerful hurricanes to hit the Caribbean region for a decade. It ravaged Grenada with winds of 220 kilometres per hour.

The relative lack of rain coupled with the hurricane's arrival during daylight hours served to reduce the loss of life and limit damage to road and drainage infrastructure. Had it passed during night time hours, the death toll would undoubtedly have been much higher.

Hurricane Ivan affected the entire island. The vast majority of the damage was experienced in the southern portion of the island. Visible damages included the partial or total loss of buildings, broken and uprooted trees, broken utility poles and damaged vehicles.

As Ivan was a relatively dry hurricane, damage from flooding and mudslide was not extensive. Streams did flood, particularly near bridges and culverts where debris built up. Roadways were blocked by debris and fallen trees, but generally remained intact with little evidence of landslides or washouts. The storm surge associated with Ivan was relatively small and sea defences appeared to have resisted the wave action forces without damage.

Some 37 people died and approximately 90 per cent of the houses were damaged or destroyed and 50 per cent of the population was made homeless.

The agricultural sector was decimated. Of particular concern was the destruction of **cash crops**, e.g. nutmeg. (Nutmegs account for 80 per cent of agricultural exports, and it takes at least seven years for nutmeg trees, when replanted, to grow and bear fruit.) Tourism and agriculture are the 'twin pillars' of the Grenadian economy – over 60 per cent of employment in the tourism industry was lost.

> **Key term**
>
> **Cash crop:** crop that is grown for sale rather than for subsistence use within a household.

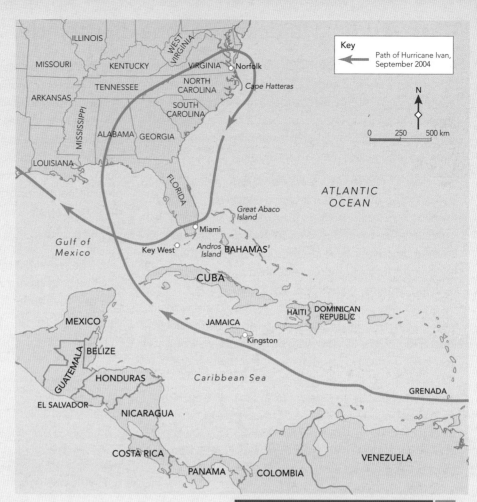

Figure 3.6 The track of Hurricane Ivan

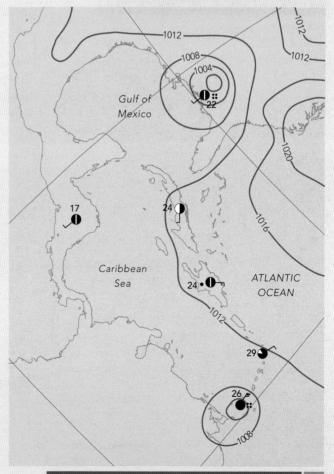

Figure 3.7 Synoptic weather map of Hurricane Ivan

Activities

1. Suggest reasons why the impacts of Hurricane Ivan were limited.
2. What were the long-term impacts of Hurricane Ivan?

Case Study: The impact of Hurricane Ivan on Grenada

Case study | Hurricane Jeanne, September 2004

Also in September 2004, Hurricane Jeanne hit the Dominican Republic and Haiti (Figure 3.8). Hurricane Jeanne's slow journey across the Caribbean, amassing huge quantities of water, contributed to the torrential rainfall over Haiti bringing devastating mudslides and flooding, and killing over 3000 people.

Two days of rain sent torrents down mountains in Haiti's Artibonite and Nord-Ouest provinces, causing rivers to burst their banks. Heavy rainfall and rising sea levels submerged poorly defended urban areas, causing massive loss of life through drowning. The floods tore through the coastal town of Gonaives and outlying districts, covering crops and roads. Much of Gonaives was under waist-deep water and aid workers were finding it difficult to evacuate all those in need. At least 550 people died when a three-metre wall of water and mud destroyed large areas of the town. More than 100 other people died in the region. Gonaives, a city of 250 000 people, was declared a disaster area with an appeal for international aid. The US government pledged an immediate US$ 60 000 (£33 500) in aid. More than 100 000 people in Gonaives needed shelter, food or medical aid.

The scale of the disaster was blamed on deforestation, which has left communities vulnerable to flash floods. The mountainous country which was once heavily forested now has less than 2 per cent tree cover. This has led to severe soil erosion which allows water to rush off the steep slopes. Most trees have been cut down to make charcoal for cooking. A recent UN environmental report described Haiti as 'one of the most degraded countries in the world'. Haiti is the poorest country in the Western hemisphere with some 80 per cent of the population living below the poverty line. The torrential rainfalls associated with Hurricane Jeanne were devastating, but Haiti being such a poor country meant that it was less prepared than other islands to cope with the hazard, thereby making the impact of the hazard so much worse.

Activities

1. Suggest why Hurricane Jeanne had such a large impact in Haiti.
2. What were the hazards associated with Hurricane Jeanne?

Figure 3.8 The track of Hurricane Jeanne

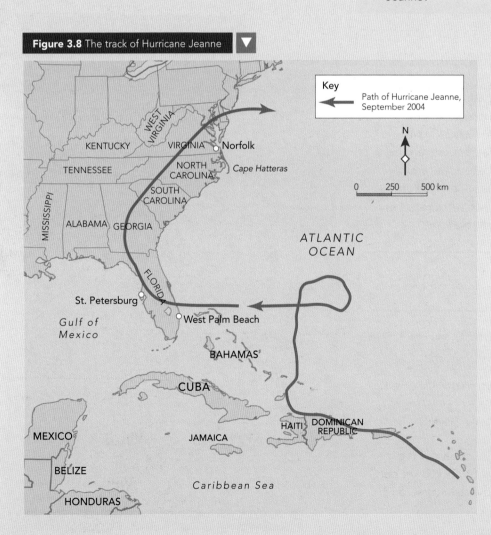

Tornadoes

Tornadoes are among the world's most violent, extreme weather conditions. They are a complex phenomenon and are difficult to study. They are particularly common in parts of the USA and in the UK. The UK holds the record for the greatest density of tornadoes per km², although they tend to be small-scale. Damage is related to the strength of the tornado and the materials that buildings are constructed of. Management of tornadoes is very difficult.

Tornadoes are small and short-lived but highly destructive storms. Because of their severe nature and small size, measurement and observation of them is difficult. A tornado consists of an elongated funnel (**vortex**) of cloud which descends from the base of a well-developed cumulonimbus cloud, eventually making contact with the ground. In order for a vortex to be classified as a tornado, it must be in contact with the ground and the cloud base. Tornadoes contain rotating violent winds, perhaps exceeding 100 metres per second. Pressure gradients in a tornado can reach an estimated 25 mb per 100 m. This compares to the most extreme pressure gradients of about 20 mb per 100 km around a larger-scale cyclone!

> **Key term**
>
> **Vortex:** whirling and rising mass of air and cloud in a tornado.

Moisture, instability, lift, and wind shear are the four key ingredients in tornado formation. Most tornadoes, but not all, rotate anticlockwise in the northern hemisphere and clockwise south of the Equator. Tornadoes form from anvil clouds during very severe electrical storms, known as supercells. Supercells are significantly longer, rotating thunderstorms, typically lasting several hours and over 10 km in diameter.

During a supercell, there is an updraft of moist air rising through the electrical storm. If the wind intensity varies with height, the updraft begins to rotate in the mid-levels of the atmosphere. This rotating updraft is known as a mesocyclone.

In order for a tornado to develop, a strong current of cool air must move downward. This downward current occurs due to the density of colder air in comparison to warm air, which causes cold air to sink as warm air rises. As the speed of the downdraft increases to speeds of up to 160 km per hour, a tornado forms between the downdraft and the updraft (Figure 3.9).

Many tornadoes have a short life – they can last from several seconds to more than an hour. Once contact with the ground is made, the track of a tornado at ground level may frequently extend for only a few kilometres, though there are examples of sustained tracks extending over hundreds of kilometres. The diameter of the funnel is rarely more than 200 m; track length and width are therefore limited.

Tornadoes, being associated with extreme atmospheric instability, show both seasonal and locational preference in their incidence. 'Favoured' areas are temperate, continental interiors in spring and early summer, when **insolation** is strong and the air may be unstable, although many parts of the world can be affected by tornado outbreaks at some time or another. The Great Plains of the USA, including Oklahoma, Texas and Kansas, are particularly prone at times when cool, dry air from the Rockies overlies warm, moist 'Gulf' air. In some areas of the USA tornadoes tend to follow paths from a specific direction, such as north-west in Minnesota or south-east in coastal south Texas. This is because of an increased frequency of certain tornado-producing weather patterns, for example hurricanes in south Texas or north-west-flow weather systems in the upper Midwest of the USA.

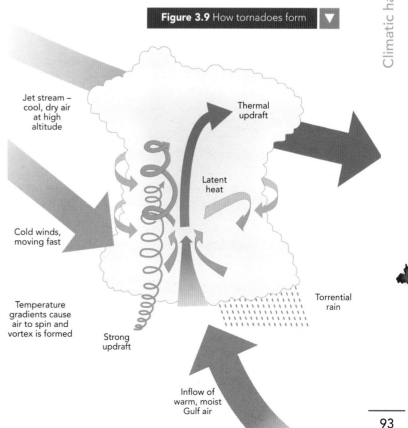

Figure 3.9 How tornadoes form

> **Key term**
>
> **Insolation:** contraction of 'incoming solar radiation', meaning exposure to the sun's rays.

Tornado damage

About 1000 tornadoes hit the USA yearly. On average, tornadoes kill about 60 people per year – most from flying or falling debris. A tornado's impact as a hazard is extreme. There are three damaging factors at work. First, the winds are often so strong that objects in the tornado's path are simply removed or very severely damaged. Second, strong rotational movement tends to twist objects from their fixing, and strong uplift can carry some debris upwards into the cloud. Third, the very low atmospheric pressure near the vortex centre is a major source of damage. When a tornado approaches a building, external pressure is rapidly reduced, and unless there is a nearly simultaneous and equivalent decrease in internal pressure, the walls and roof may explode outwards in the process of equalising the pressure differences.

Although winds from the strongest tornadoes far exceed those from the strongest hurricanes, hurricanes typically cause much more damage individually and over a season, and over a far larger area. Economically, tornadoes cause about a tenth as much damage per year, on average, as hurricanes. Hurricanes tend to cause much more overall destruction than tornadoes because of their much larger size, longer duration and the variety of ways they cause damage (storm surge, flooding, winds). Tornadoes, in contrast, tend to be much smaller, last for minutes rather than hours and days, and the strength of wind is the main cause of damage.

Tornado damage scale

Dr T. Theodore Fujita developed a damage scale for winds, including tornadoes, which is supposed to relate the degree of damage to the intensity of the wind (Figure 3.10). This F-scale should be used with caution as it does not take into account differences in building structure and materials.

A new Enhanced F-Scale has been used since 2006. The Enhanced F-scale is a much more precise way to rank tornado damage than the original, because it classifies damage F0–F5 calibrated by engineers across over 20 different types of buildings. A team of meteorologists and engineers has worked on this for several years. The idea is that a 'one size fits all' approach does not work in rating tornado damage, and a tornado scale needs to take into account the typical strengths and weaknesses of different types of construction. This is because the same wind does different things to different kinds of buildings. In the Enhanced F-scale, there will be different, customised standards for assigning any given F-rating to a well-built, well-anchored wood-frame house compared to a garage, school, skyscraper, unanchored house, barn, factory, utility pole or other type of structure. In a real-life tornado track, these ratings can be mapped together more smoothly to make a damage analysis.

Activities

1. Under what conditions are tornadoes likely to form?
2. Why are the impacts of hurricanes normally greater than the impacts of tornadoes?
3. Why are tornadoes difficult to predict?

Figure 3.10 Fujita Tornado Damage Scale

Category	Damage	Description
Category F0	Light damage (<117 km/h)	Some damage to chimneys; branches broken off trees; shallow-rooted trees pushed over; sign boards damaged.
Category F1	Moderate damage (117–180 km/h)	Peels surface off roofs; mobile homes pushed off foundations or overturned; moving cars blown off road.
Category F2	Considerable damage (181–253 km/h)	Roofs torn off frame houses; mobile homes demolished; boxcars overturned; large trees snapped or uprooted; light-object missiles generated; cars lifted off ground.
Category F3	Severe damage (254–332 km/h)	Roofs and some walls torn off well-constructed houses; trains overturned; most trees in forest uprooted; heavy cars lifted off ground and thrown.
Category F4	Devastating damage (333–418 km/h)	Well-constructed houses levelled; structures with weak foundations blown off some distance; cars thrown and large missiles generated.
Category F5	Incredible damage (419–512 km/h)	Strong frame houses lifted off foundations and swept away; car-sized missiles fly through the air in excess of 100 metres (109 yards); trees debarked; incredible phenomena will occur.

Case study | The Indiana/Kentucky tornado, November 2005

Indiana's deadliest tornado for more than 30 years flattened a mobile home park outside Evansville and struck other communities early in November 2005, killing at least 16 people and injuring 200 others. The 'twister' was the deadliest to strike the state since the 'super outbreak' of 21 tornadoes on 3 April 1974 that killed 47 people.

The tornado developed in a line of thunderstorms that rolled rapidly eastwards across the River Ohio valley during the morning. The National Weather Service posted severe thunderstorm warnings for sections of northern Ohio. The damage path through Newburgh, Indiana, 13 km east of Evansville, was about 1.2 km wide and extended for roughly 32 km. Emergency sirens sounded, but most people didn't hear them because it happened in the middle of the night. Some 25 000 homes were without power, mostly in Warrick County, Indiana. There also were reports of natural gas leaks.

Figure 3.11 Tornado fatalities in Indiana by decade

Tornadoes in Indiana

Indiana is in what is considered to be 'Tornado Alley', a swath of states extending from Texas to South Dakota. Although the state lacks the high frequency of tornadoes seen in places like Kansas and Oklahoma, it makes up for it in tornadic intensity. From 1950 to November 2001, 1024 tornadoes caused more than US$ 1.7 billion in damage and killed 223 people.

Tornadoes can occur in any month, but March to June is considered tornado season. Historically, the most destructive tornadoes strike in March and April. June holds the record for the most tornadoes in Indiana on any given day (37) and for the most in a single month (44). Both records were set in 1990, which is also the year for the most tornadoes in the state (49).

Some decades have seen greater fatalities than others (Figure 3.11). In the 1920s and 1960s over 130 people were killed by tornadoes in Indiana. In contrast, during the 1950s and 1980s less than 10 people were killed, and in the 1930s none.

Tri-State Tornado

Indiana was one of three midwestern states in the path of the deadliest tornado in US history. On 18 March 1925, the Tri-State Tornado travelled a record 352 km on the ground from Missouri through Illinois and into Indiana where it struck Posey, Gibson and Pike counties. The town of Griffin, Indiana, lost 150 homes, and 85 farms near Griffin and Princeton were devastated. About half of Princeton was destroyed with losses totalling nearly US$ 2 million. The funnel finally dissipated just outside Princeton, 3½ hours after it began. Nearly 700 people died, 74 of them in Indiana. Murphysboro, Illinois, lost 234 people, a record for a single community.

Activities

1. Why does Indiana experience a large number of tornadoes?
2. Describe and explain the seasonal nature of tornadoes in Indiana.
3. Is there a pattern over time during the twentieth century for changes in number of fatalities due to tornadoes (Figure 3.11)? Give reasons for your answer.

Discussion point

'Human activities may be making hurricanes and tornadoes more frequent and more damaging.' To what extent do you agree with this statement?

Theory into practice

Write an account of the impacts of tornadoes in the UK.

Take it further activity 3.1 on CD-ROM

3.2 How do atmospheric systems cause heavy snowfall, intense cold spells, heatwaves and drought and in what ways do they represent a hazard to people?

Subtropical and polar air masses converge in the mid-latitudes (Figure 3.12). This means that areas such as the UK are affected by warm air at certain times and cold air at other times, and frequently by the two meeting and producing a depression or cyclone.

The **polar front** refers to the boundary between the warm air mass and the cold air mass. At high altitude there is a polar front **jet stream**. This is a very fast meandering thermal wind which causes air to rise and sink. At low pressure systems air is rising, whereas at high pressure systems air is sinking.

> **Key terms**
>
> **Jet stream:** high altitude relatively strong wind concentrated in a narrow band.
>
> **Polar front:** the line which marks the meeting place of cold polar air and warm tropical air.

Air masses

Air masses are large bodies of air whose physical properties, notably temperature and humidity, vary only slightly over a large area. Air masses derive their temperature and humidity from the regions over which they form. These regions are known as source regions.

Principal air masses

The main air masses to affect the UK are polar (P) and tropical (T) air masses which meet at the polar front. Some polar air masses may originate in cool temperate or subarctic areas. Sometimes the UK is affected by Equatorial (E) and Arctic (A) air masses. These are then generally divided into maritime (m) or continental (c) depending upon the humidity characteristics of the air. Air masses which form over oceans are moist (m), whereas those that form over land are relatively dry (c).

Frontal weather

When two different air masses converge they form a front. When a Pm and a Tm meet, the temperature differences between them may be 13°C. Such a difference causes variations in density and allows the warm air mass to rise over the cool one. In any depression (low pressure system) there are a number of forces operating:

- the two air masses meet along a front
- the Coriolis force (spin of the Earth)
- the divergence of air aloft in the upper regions of the troposphere.

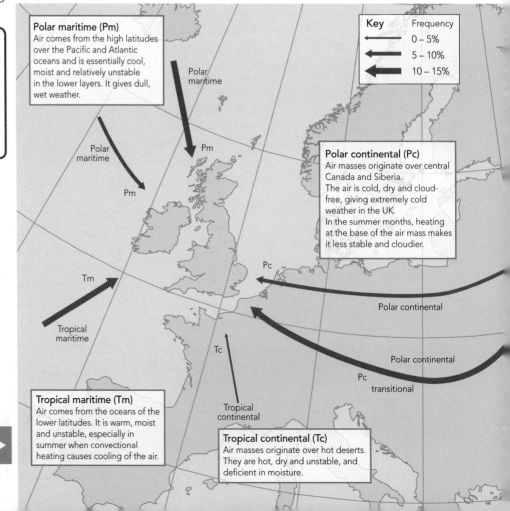

Figure 3.12 Principal air masses to affect the British Isles

The combination of these is to drag air inwards to the centre of low pressure. Warmer, lighter air invades the colder, denser air to form the warm sector, while warm air rises over the cold air at the warm front. Where the cold air pushes the warm air up, a cold front is formed. The rising air is removed at altitude by the jet stream.

In general, the appearance of a warm front is heralded by high cirrus clouds (Figure 3.13). Gradually, the cloud thickens and the base of cloud lowers. Altostratus clouds may produce some drizzle, while at the warm front nimbostratus clouds produce rain. A number of changes occur at the warm front: winds reach a peak, are gusty and come from another direction; temperatures suddenly rise; pressure which had been falling remains more constant. By contrast, the cold front is marked by a decrease in temperature; cumulonimbus clouds; heavy rain; increased wind speeds and gustiness; a further change in wind direction; and a gradual increase in pressure. After the cold front has passed, the clouds begin to break up and sunny periods are more frequent, although there may be isolated scattered showers associated with unstable polar maritime air.

The weather that is found in any depression depends on the air masses involved. The greater the temperature difference between the air masses involved, the more severe the weather. Depressions are divided into two main types – ana and kata depressions – depending on the vigour of the uplift of warm air. **Ana fronts** are formed when there is a great difference in the Tm and Pm air masses involved. This makes uplift more vigorous. By contrast, **kata fronts** are formed when the two air masses are fairly similar in temperature and density, making uplift less intense.

Figure 3.13 Weather associated with a 'typical' depression

Source: Guinness, P. and Nagle, G., *AS Geography: Concepts and Cases* (2000)

Key terms

Ana front: weather front in which there is strong vertical movement of air caused by the large temperature differences between the warm air mass and the cold air mass.

Kata front: weather front in which the temperature differences between warm air and cold air are small, so the vertical uplift of air is limited.

Activities

1 What are the main air masses to affect the British Isles?
2 Why is heavy snowfall not associated with very cold air masses?

Please refer to the CD-ROM for a case study about stability, instability and lapse rates.

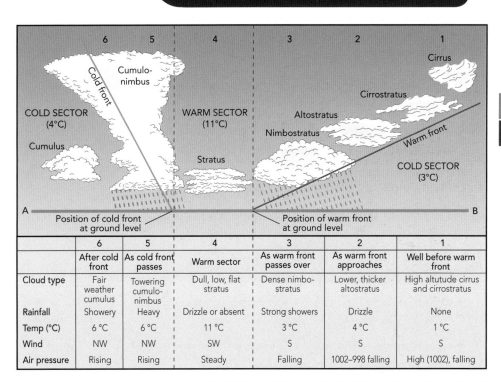

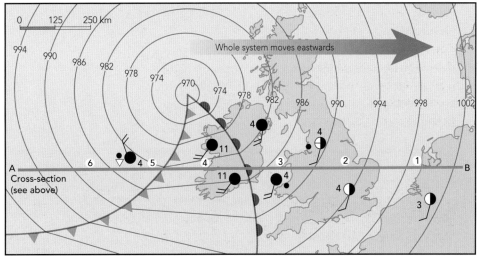

	6	5	4	3	2	1
	After cold front	As cold front passes	Warm sector	As warm front passes over	As warm front approaches	Well before warm front
Cloud type	Fair weather cumulus	Towering cumulo-nimbus	Dull, low, flat stratus	Dense nimbo-stratus	Lower, thicker altostratus	High altutude cirrus and cirrostratus
Rainfall	Showery	Heavy	Drizzle or absent	Strong showers	Drizzle	None
Temp (°C)	6 °C	6 °C	11 °C	3 °C	4 °C	1 °C
Wind	NW	NW	SW	S	S	S
Air pressure	Rising	Rising	Steady	Falling	1002–998 falling	High (1002), falling

Case study | Severe storms and gales in Britain

The Great Storm of 1987 was one of the most severe storms ever to hit southern Britain. It started as a low pressure system off the coast of Spain but was dragged north-eastwards to Britain by high level jet streams. Some of its energy was supplied by warm waters from the Bay of Biscay.

This was the strongest gale to affect the UK since 1703, bringing with it gusts of up to 167 km (104 mph) (Figure 3.14). The worst gusts were during the night – had they been during the day the effects would have been much greater. The total cost of insurance claims was over £860 million, but this figure could have been much higher.

> **Key term**
>
> **Gale:** strong winds with speeds of between 63 and 74 km/h; severe gales have wind speeds of between 76 and 87 km/h.

There are several elements that affect the nature of a disaster. These include:

- the time of day – it is more difficult to warn or evacuate people in the middle of the night
- areal extent of the storm – the wider the area affected the greater the potential damage
- the time of year – trees in leaf may intercept rain, but not deciduous trees in winter
- awareness of the event – people who are aware of the event can take precautionary measures
- precautions taken – such as leaving the area, avoiding driving during a storm, etc.
- the duration of the event – the longer it goes on the greater the risk of damage
- the strength of the storm or hazard – the more extreme the event the greater the damage inflicted.

There are many management problems to deal with: disruptions to transport and communications; electricity black-outs; localised flooding; damage to properties; injury and loss of life, and the psychological impacts these have; insurance claims; and preparation to withstand any future storms.

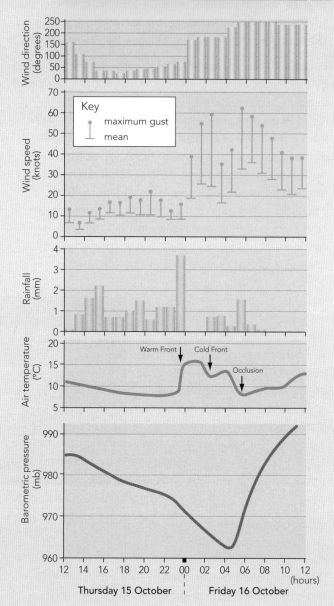

Figure 3.14 The Great Storm of 1987 – weather data collected in Oxford between midday on 15 October and midday on 16 October

Source: Coones, P. (1987), 'The Great Gale of 16 October, 1987', *Geography Review*, 1, 3, 6–10

Activities

1. Suggest why the Great Storm had such an impact on southern England.
2. Comment on the factors which affected the impact of the Great Storm.

Low pressure

A depression is a low pressure system in mid- and high-latitudes. It is also called a cyclone. It is caused by the mixing of cold and warm air when air masses of differing temperatures meet. The warm air mass, being lighter, rises above the denser, cold air mass and forms a centre of low pressure.

Hazards associated with cold spells

There are many hazards associated with cold spells. These include snowfalls and blizzards, wind chill, frostbite and hypothermia. These hazards can affect human health and mortality rates, and they can affect the economy, in particular the growth of crops, transport and heating bills.

Snowfalls and blizzards

The impact of snowfall depends on many factors, such as the amount of snow, the frequency of high winds, air temperature, and the length of time there is snow cover. The impacts affect transport, economic activities and human health, though how the authorities prepare for the event could have some bearing on its impacts.

The urban snow hazard depends on three variables – the snow, human activities and human attitudes. For example, the hazard depends on the amount of snow, temperature, wind speed and accumulation rate. The impacts vary on whether jobs are outside or inside, and whether transport and communications are disrupted. In addition, the hazard may be perceived differently if it is regular, rare or random, and if there is a programme to deal with the hazard.

Heavy snowfalls are associated with warm air masses (which can contain large amounts of water) meeting with very cold air masses, causing the rain to fall as snow. Some very cold winters, such as that of 1962–3, are associated with **blocking anticyclones**, resulting in long periods of stable conditions. If conditions remain very cold, any precipitation will fall as snow.

> **Key term**
>
> **Blocking anticyclone:** area of high pressure which remains relatively stationary in comparison to approaching low pressure systems (depressions) and so blocks their passage over an area.

Heavy snowfall can cause disruption to transport (Figure 3.15) – road casualties increase by 25 per cent on snowy days. Rail transport is also disrupted. Large accumulations of snow can lead to avalanches on slopes of 35–45 degrees; avalanches are more likely when there is a fall of at least 50 mm in one day.

Snowfall can affect human activities in many ways. Heavy snowfalls (25–30 cm), which may be expected only once every 15–20 years in the south of England, can block roads, paralyse communications, isolate rural settlements, cause power failures, lead to major absenteeism in schools and at places of work, and require rescue work undertaken by air. On the other hand, smaller snowfalls can be more of a nuisance or irritant rather than having a crippling or paralysing effect. Smaller falls, which may occur every winter, can cause traffic movement to be slowed and require gritting lorries rather than snowploughs. Generally, there is minimal press coverage with the smaller snowfalls.

The lambing season is especially affected by spring snows. High mortality rates for lambs and ewes occur during snowfall conditions (although much less so now because most lambing in the UK is done indoors). Commercial and residential fruit-bearing plants are also affected by late snowfalls (Figure 3.16).

In the winter of 1962–3, economic activity decreased by 7 per cent and unemployment increased. Up to 160 000 workers were laid off. Consumption of electricity and gas increased by 20 per cent. In the cold winter of 1979, insurance claims for weather-related events came to over £70 million.

The impact on the elderly is significant. Death by hypothermia, heart attacks, pneumonia and bronchitis may be triggered by extreme cold spells.

Figure 3.15 Heavy snowfall can cause major disruption to transport routeways

Activities

What are the hazards associated with heavy snowfalls?

▲ **Figure 3.16** Heavy snowfalls can damage fruit crop

Winter deaths related to ice and snow:
- about 70 per cent occur in cars
- about 25 per cent are due to people caught out in a storm
- the majority are males over 40 years old

Winter deaths related to exposure to cold:
- 50 per cent are people over 60 years old
- over 75 per cent are males
- about 20 per cent occur in the home.

Source: website of the United States Search and Rescue Task Force

Figure 3.17 Winter death statistics, USA ▲

Blizzards

Blizzards are extreme winter storms consisting of snow and wind causing poor visibility. They can create a variety of hazardous conditions (Figure 3.17). Travelling can become difficult or impossible due to poor visibility and drifting snow.

Another hazard is wind chill. This is the temperature 'felt' due to the combination of wind and temperature (Figure 3.18). For example, a strong wind (48 km/h) combined with a temperature of just below freezing (minus 1.1°C) can have the same effect as a still air temperature of minus 20°C.

Low wind chill can cause **frostbite** or **hypothermia**. Frostbite is a severe reaction to cold that can permanently damage a person. Hypothermia occurs when the body's temperature falls to very low temperatures.

Blizzards are also associated with other hazards and dangers, such as power cuts when strong winds and heavy snow bring power cables down. Liquids in pipes can freeze because of the cold weather; fuel supplies may be cut off because of disruptions to communications.

Key terms

Frostbite: the freezing of skin or flesh leading to tissue damage.

Hypothermia: condition in which the temperature of the body falls below 35°C, which can lead to a coma and death.

To access a website describing measures that can be taken to reduce the impact of blizzards, go to www.heinemann.co.uk/hotlinks, enter the express code 7627P and click on the relevant link.

Actual thermometer reading °C

Estimated wind speed km/h	10	4	−1	−6	−12	−18	−23	−29	−34	−40	−45
0	10	4	−1	−6	−12	−18	−23	−29	−34	−40	−45
8	9	3	−3	−9	−14	−21	−26	−32	−38	−44	−49
16	4	−2	−9	−16	−23	−29	−36	−43	−44	−57	−64
24	2	−6	−13	−21	−28	−38	−43	−50	−58	−65	−73
32	0	−8	−16	−23	−32	−39	−47	−55	−63	−71	−79
40	−1	−9	−18	−26	−34	−42	−51	−59	−67	−76	−83
48	−2	−11	−19	−28	−36	−44	−53	−62	−70	−78	−87
56	−3	−12	−20	−29	−37	−45	−55	−64	−72	−81	−89
64	−3	−12	−21	−29	−38	−47	−56	−65	−73	−82	−91

Equivalent temperature °C

Little danger for properly clothed person
Increasing danger
Great danger

Danger from freezing of exposed flesh

(Note: wind speeds greater than 64 km/h have little additional effect.)

▲ **Figure 3.18** Wind chill chart

Impacts of cold spells on transport and forestry

In February 2003 Britain was badly affected by heavy snow. It caused chaos. Arctic winds swept in, bringing gales and snow showers. In Scotland snow closed a number of roads in the Highlands and winds disrupted ferries. In Whitby, North Yorkshire, gales whipped the sea into a sticky foam which blew inland, causing hazardous conditions on the streets.

Freezing temperatures and heavy snow caused chaos on the road and on the railways. Parts of the M40 and M4 were blocked by accidents, while in parts of Scotland the A9 and A90 could only be passed with care. The worst conditions were said to be in northern Scotland and eastern England. In Humberside, police recorded snow falls of up to 23 cm near Hull.

Britain appeared to have been taken by surprise with trains, planes and motorways brought to a halt by 5 cm of snow causing motorists to spend up to 20 hours without food, warmth or water on the M11. Cambridgeshire police even sent out a helicopter to drop food parcels to the thousands of motorists who spent the night on the M11.

A total of 41 London underground stations were closed and 225 flights were cancelled at Heathrow. The greatest disruption was on the roads, particularly those in Essex, Cambridgeshire, Hertfordshire and Bedfordshire. Motoring organisations blamed the highways agency and local authorities for inadequate gritting.

Other effects:

- there was a 64-kilometres tailback on the M11
- there were severe delays and cancellations on WAGN, Anglia, First Great Eastern and Silverlink
- Stansted airport closed overnight during the worst of the storms

Impacts of snow storms and cold weather on forestry

In February 2008 snowstorms destroyed or damaged up to 10 per cent of China's forests. Gales and the weight of snow and ice tore down trees as well as bamboo across 16 million ha – an area larger than England. In recent years China had stepped up reforestation and had planned to plant 2.5 billion trees in 2008.

Ice storms are important meteorological disturbances affecting forests over a large portion of the USA from east Texas to New England. This area experiences major ice storms at least once a decade, and major events once or twice a century. They occur mostly in winter. Ice storms affect species composition, woodland structure, and condition over wide areas. Impacts of individual storms are highly patchy and variable, and depend on the nature of the storm. In January 1998 freezing rain produced ice accumulation of 40–100 mm in eastern Ontario. Much damage was done to the forests and tree-related industries, such as maple sugar production.

Case study | North American blizzard of 2003

The blizzard that occurred between 14 and 19 February 2003 on the East Coast of the USA and Canada was record-breaking. There were 27 deaths and over US$ 14 million of damage. All cities between Washington D.C. and Boston were brought to a standstill because of the thick covering of snow (between 38 and 76 cm) and the low temperatures. In Baltimore and Boston, this was the biggest snowstorm on record, with 71.6 and 69.9 cm of snow respectively.

The severity of the blizzard can be attributed to the unusually favourable conditions surrounding the storm. First, the moisture from the Atlantic Ocean enhanced precipitation totals along the Eastern Seaboard of the USA. Second, a high pressure system over eastern Canada allowed cold air to be brought down into coastal areas. This cold air ensured that many areas where storms typically produce mixed precipitation (rain and snow) received most or all precipitation from the storm in the form of snow.

The snowstorm paralysed much of the East Coast's transport infrastructure. Washington's Reagan National Airport, Baltimore–Washington International Airport and New York's LaGuardia Airport were shut down completely. Road travel was nearly impossible. In Baltimore, the roof of the historic B & O Railroad Museum collapsed.

Activities

1. What are the hazards associated with blizzards?
2. Explain why the north-east coast of the USA and Canada is subject to blizzards.

High pressure

Global high pressure

One of the most permanent features of the global distribution of high and low pressure is the subtropical high pressure belts, especially over ocean areas (Figure 3.19). In the southern hemisphere these are almost continuous at about 30 degrees latitude, although in summer over South Africa and Australia they are broken somewhat. These areas are associated with arid conditions on account of the subsiding air which prevents convectional uplift (and the associated cooling of air, condensation and formation of rain). In polar areas pressure is relatively high throughout the year, especially over Antarctica, owing to the coldness of the land mass.

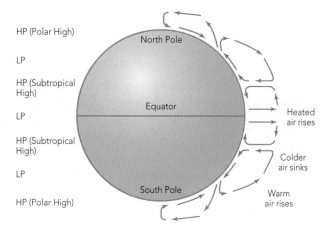

Subtropical high pressure and polar pressure affect the location of deserts

Figure 3.19 Global high pressure regions

High pressure systems

High pressure systems or anticyclones act very differently compared with low pressure systems or depressions. An anticyclone is a large mass of subsiding air, which produces high pressure at the surface. Air moving to the poles from tropical areas sinks to form the subtropical high pressure belt. Whereas depressions produce wet, windy weather, high pressure systems produce hot, sunny, dry calm days in summer and cold, sharp, crisp days in winter. Nights are cold as the lack of cloud cover allows heat to escape. Frost and fog are common in winter and autumn. Winds in a high pressure system blow out from the centre of high pressure in a clockwise direction (compared with a low pressure system where they blow into the centre of low pressure in an anticlockwise direction). As a result of the pressure gradient, the winds are light hence the isobars on a high pressure weather chart are circular and spaced far apart.

Sometimes large stationary anticyclones become established near the British Isles, and they 'block' the passage of depressions. These anticyclones are most common in spring, but can occur in all seasons; sometimes they can persist for a month or more and completely change the character of the weather.

In summer, the blocking anticyclones sometimes bring a prolonged spell of warm or hot weather to the British Isles; they are responsible for 'heatwaves'. In winter, too, dry weather prevails in anticyclones, but during cloudless nights the temperature falls and does not recover during the following day because there is only weak sunshine or persistent fog.

Anticyclonic weather in the UK

A **temperature inversion** is a small layer of the atmosphere where the decrease in temperature with height is much less than normal (or even increases with height). Winter anticyclones are associated with a range of hazards. In some cases they bring extremely

Winter weather (polar source)	Summer weather (tropical source)
• Cold daytime temperatures – usually below freezing to a maximum of 6°C.	• Hot daytime temperatures – over 23°C.
• Very cold night temperatures, with frosts (Figure 3.22).	• Warm night-time temperatures – may not fall below 15°C.
• Clear skies by day and night with low relative humidity.	• Generally clear, sometimes cloudless skies (Figure 3.21).
• Stable conditions so fogs may form, especially radiation fogs in low-lying areas.	• Some mists and fogs early morning – especially at the coast where they may be persistent. (These are advection fogs created by sea/land differences. Advection fog is formed where warm, moist air passes over a cold surface.)
• High levels of atmospheric pollution in urban areas, caused by subsiding air and lack of wind.	• Thunderstorms may result from convectional uplift, usually in late afternoon/early evening.
• Pollutants are trapped.	• High amounts of low-level ozone and formation of photochemical smog are a major pollution hazard.

Figure 3.20 Characteristics of anti-cyclonic winter and summer weather

Figure 3.21 Summer high pressure at Blenheim Palace, Oxfordshire

Figure 3.22 Winter high pressure at Blenheim Palace, Oxfordshire

cold weather conditions with days of continuous frost and temperatures well below freezing. This causes problems for those who may inadequate heating in their homes and/or find it difficult to keep warm, such as the elderly; in extreme cases there may be death by hypothermia. Widespread incidence of frost increases the risk of accidents on the roads and for pedestrians as a result of skidding. Winter anticyclones are bad news for the health service as they lead to an increase in emergency admissions to hospitals and bed shortages. Winter high pressure can have major implications for wildlife as they may find it difficult to find food.

A further hazard is winter fog that is associated with very stable air conditions. **Radiation fog** is a common winter hazard. This may develop into **smog** in industrial areas and large conurbations, which leads to health hazards as a result of the effects of pollutants such as sulphur dioxide (SO_2) or nitrogen dioxide (NO_2). It is the stability of the anticyclone systems that allows pollution levels to build up as pollutants are not dispersed by the wind. During summer anticyclones, high amounts of low-level ozone and the formation of **photochemical smog** are a major pollution hazard.

Activities

1 Suggest the types of hazards likely to be experienced in Figures 3.21 and 3.22.

2 Brainstorm the sorts of people this range of hazards might affect.

Key terms

Photochemical smog: smog is a mixture of smoke and fog. Photochemical smog forms when photons of sunlight hit molecules of different pollutants and cause chemical reactions. The pollutants include nitrogen oxides (NOx), volatile organic compounds (VOCs) and ozone.

Radiation fog: a shallow layer of fog formed usually in spring and autumn. The ground surface is cooled by radiation at night and this in turn chills the air above it. If the air is cooled to below dew-point (the temperature at which relative humidity is 100 per cent), condensation will occur and a fog is formed.

Smog: fog which combines with pollutants such as sulphur dioxide and oxides of nitrogen.

Temperature inversion: an increase in temperature with altitude, which may occur near the surface (low level inversion) or higher air temperatures encountered at a particular altitude (upper air inversion) – in each case a reversal of the normal decrease occurs.

Drought

Drought is an extended period of dry weather leading to conditions of extreme dryness. Absolute drought is a period of at least 15 consecutive days with less than 0.2 mm of rainfall. Partial drought is a period of at least 29 consecutive days during which the average daily rainfall does not exceed 0.2 mm. In large parts of the UK there were periods of absolute drought in 1988–92. There are some advantages of predictable drought (Figure 3.22) but there are also many dangers.

Case study | Drought in Britain and Europe, 2003

The effects of drought in the UK

For those involved in tourism, the heatwave of summer 2003 brought a rise in profits. The UK's tourist industry had slumped 10 per cent in 2001 with the effects of foot and mouth disease (FMD) and the terrorist attacks of September 11 (9/11), but crept up slightly in 2002. With the heatwave, it was expected to grow by another 3–4 per cent in 2003.

Nevertheless, dry conditions and heat affected many areas (Figure 3.23). Transport was disrupted by the heatwave, for example the roads melted in the Essex village of Great Leighs, and on 15 July temperatures on the London Underground reached 37°C – hotter than the legal limit for transporting animals!

Some retailers benefited from advance warnings of the warm spell and were able to increase stocks of home barbeques, sun cream, paddling pools, beer, ice cream and lettuce. In contrast, demand for umbrellas and bubble bath were well below the seasonal average!

The heatwave was also associated with an increase in low-level ozone, especially over southern England. In the two weeks at the peak of the heatwave, as many as 900 people may have died as a result of poor air quality, according to Hansard, the record of UK parliamentary discussion. (Nearly 10 per cent more people died than was normal for that time of year.)

Figures from the CBI suggested that the cost of people taking days off work (to enjoy the hot weather) was between £7.5 million and £10 million per day.

France

Estimates for the death toll from the French heatwave in 2003 were as high as 30 000. Harvests were down between 30 per cent and 50 per cent on 2002. France's electricity grid was also affected as demand soared while the population turned up air conditioning and fridges. However, nuclear power stations, which generate around 75 per cent of France's electricity, were operating at a much reduced capacity because there was less water available for cooling.

Portugal

Portugal declared a State of Emergency after the worst forest fires for 30 years. Temperatures reached 43°C in Lisbon in August 2003: 15°C hotter than the average for the month. Over 1300 deaths occurred in the first half of August, and up to 35 000 hectares of forest, farmland and scrub was burned. More than 80 families were forced to abandon their homes. Some fires were, in fact, deliberately started by arsonists seeking insurance or compensation money, and over 70 people were arrested.

Figure 3.23 Crowded beach in Bournemouth showing the effects of high pressure

The prolonged heatwave left some countries facing their worst harvests since the end of the Second World War. Some countries such as the Ukraine, Hungary, Bulgaria and Romania that usually export food were forced to import it for the first time in decades. Across the EU, wheat production was down 10 million tonnes – about 10 per cent.

Activities

1. Outline the hazards associated with drought.
2. Suggest reasons why the impact of drought may be increasing.
3. Suggest how human activities may have intensified the climatic hazards of 2003.
4. Use newspaper articles to illustrate types and scale of hazards affecting the UK in recent years.

Discussion point

Are people in the UK more capable, or less capable, of coping with natural hazards than they were 100 years ago? Give reasons for your answer.

Theory into practice

While you are studying this unit, monitor the news, newspapers and Internet sites to assess the impact of low pressure and high pressure on a named area.

 Take it further activity 3.2 on CD-ROM

3.3 Why do the impacts of climatic hazards vary over time and location?

As discussed in Chapter 1, the impact of hazards varies between MEDCs and LEDCs. Variations in impacts by location is largely related to the development continuum. For example, in MEDCs loss of life is generally lower than in LEDCs, but the economic losses are much greater. Impacts might be considered to be greater in large urban areas, due to higher population densities, although such areas are generally well protected. Moreover, impacts change over time – short-term hazards may be replaced by long-term secondary hazards. However, individual events may challenge conventional wisdom.

 Please refer to the CD-ROM for a case study on Hurricane Katrina.

Smog and climate change

The increasing incidence of **aerosols** has been explained by the 35 per cent increase in sulphur dioxide over the past decade and increase in soot emissions over south-east Asia. Aerosols affect the size of water droplets. The more pollution particles there are in the air, the smaller the water droplets will be. Smaller droplets are less likely to run into each other and coalesce into drops of rain, meaning clouds stay in the air longer. However, with more pollutants in the atmosphere, there is increased potential for rainfall, since the pollutants act as hydroscopic nuclei, attracting vapour to them. Once large enough, the water droplets fall as rain. It should be expected then that increased air pollution could both increase cloud cover and potentially also increase rainfall. Indeed, the number of such clouds in Asia increased by 20 to 50 per cent between 1994 and 2005, compared with the previous 10-year period. Increased rainfall has long-term impacts, as the following examples show.

Key term

Aerosol: tiny particles of liquids or solids suspended in the atmosphere.

Case study | Smog in south-east Asia

During 1997–8 fires burned across 4.5 million hectares of rainforest and plantation in Indonesia, especially Kalimantan and Sumatra. The smog that resulted covered parts of Indonesia, Malaysia, Brunei, Singapore, Philippines and Thailand (Figures 3.24 and 3.25), an area larger than western Europe. In places visibility was less than one metre. Up to 70 million people were affected and there were widespread

Case Study: Smog in south-east Asia

concerns that long-term climate disruption could result. The fires were mostly forest fires but also included the burning of peat, with peat burning up to 3 m deep.

The fires had many short-term effects. Over 60 000 Malaysians and Indonesians, mostly children and the elderly, were treated for smog-related illnesses. Schools in Sarawak were forced to close down; hospitals struggled to cope with the increase in the number of throat infections, diarrhoea, conjunctivitis and other eye problems. An Indonesian airliner crashed in dense forest as it was trying to land in the thick smog on the Indonesian island of Sumatra; all 234 passengers and crew on board were killed. Over 275 people died from starvation in the Indonesian province of Irian Jaya. Others died from cholera caused by a lack of clean water. The worst affected people in Sarawak were the indigenous communities (Figure 3.26). Living in remote forests, cut off from medical care and access to bottled water, they received very little help from the government and from aid organisations.

The effect of the smog has been likened to smoking three packets of cigarettes each day. Some doctors claim that the effects were even worse than this – the sulphide and carbon gases obstruct airways and destroy lungs. The young and the elderly were especially vulnerable. The problem was compounded by the El Niño effect, which was thought to have contributed to drought in Asia in 1997 and the late arrival of the monsoon rains. Reservoir levels dropped and some were contaminated by seawater.

Many of the Indonesian fires, which affected up to 4.5 million ha of rainforest and plantations, were started deliberately by plantation owners as a cheap way of clearing the land. The Indonesian government blamed 176 plantation companies for causing

Key term

El Niño effect: a flow of abnormally warm water across the eastern Pacific towards the coast of South America. It is associated with a reversal of airflow in the Pacific, and is capable of disturbing weather patterns in many parts of the world. It generally occurs about once every seven years.

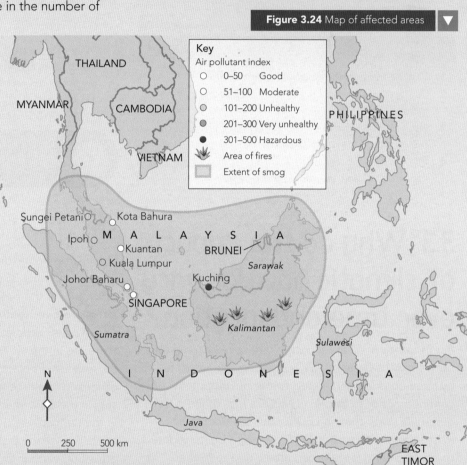

Figure 3.24 Map of affected areas

Figure 3.25 Satellite picture showing forest fires in south-east Asia

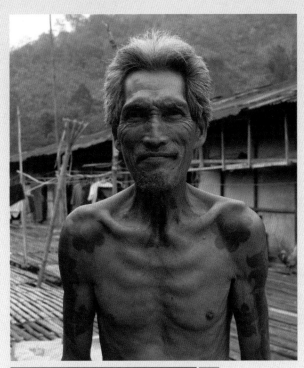

Figure 3.26 A member of the Iban – an indigenous group in Sarawak

the fires, but has since taken limited action against any of them. Many of these companies have western consumers. For example, in 1996 the UK imported over 200 000 cubic metres of tropical timber from Indonesia. In Indonesia alone, over 1 million ha of rainforest are lost to logging each year.

The long-term ecological effects of the fires are not well understood, although the effects on global climate are feared to be very significant. The lowland rainforests of Sumatra and Kalimantan are among the most diverse ecosystems on Earth. The fires could lead to a huge loss of biodiversity. The problem is increased because much of the forest lies on a bed of peat, up to 20 m thick in places. Scientists warned that up to 1 million ha of peat swamp forest could be destroyed by the fires, and that peat fires burning underground could take up to ten years before they go out. Seeds are unable to germinate in the burning soil, affecting the food chain and possibly leading to the decline of such species as tigers, orang-utans and elephants. The effect of the fire, especially if it takes hold in the peatlands, results in the release of carbon dioxide into the atmosphere, speeding up global warming.

In addition, the fires have had a negative effect on economic activities. Tourism dropped almost immediately. Following the Indonesian airliner disaster, many domestic flights were cancelled because of the thick smog. Ships without radar navigation were advised not to sail in the Strait of Malacca, which separates Malaysia from Indonesia. Crop yields were down and many food-exporting countries, such as Thailand and the Philippines, had to import staple crops such as coffee and rice rather than export them.

Key term

Staple crop: a food crop that forms the major part of a diet, e.g. potatoes, yams, cassava, rice and wheat. They are generally high in calories and inexpensive.

Activities

1. Suggest the likely causes of the South Asian smogs in the late 1990s.
2. What were the short-term impacts of the South Asian smogs?
3. In what ways did the causes of the smogs lead to long-term impacts of hazards in the region?

Case Study | Flash flooding and mudslides in Sumatra and the Philippines

At least 72 people were killed and scores were missing after a flash flood washed through Bukit Lawant, a tourist resort on the Indonesian island of Sumatra in November 2003. The disaster was sparked by several days of heavy rain in the Leuser national park. The district chief blamed man-made environmental devastation for the tragedy, he said 'The flood was caused by massive logging in the Leuser national park.' Illegal logging is rampant in most of Indonesia's national parks as few other sources of primary rainforest remain.

Similarly, mudslides in the Philippines resulted in hundreds of villagers being buried in mud as they slept. Floods and landslides swept through the province of Southern Leyte in December 2004. In 1991, landslides and flash floods occurred in an area south-east of Ormoc, also in Leyte province, killing more than 5000 people in a few hours. Heavy rains had fallen on hillsides denuded by uncontrolled logging, with debris sweeping down the hillsides devastating everything in their path. Metres-high piles of bodies were found in villages. Since 1991, the government has banned commercial logging of virgin forests, but problems of enforcement and corruption persist.

In March 2006 more mudslides buried the village of Guinsaugon on the Philippine island of Leyte. Rescuers found fewer than 125 bodies, with no trace of 1000 missing villagers.

The government's immediate response was to announce extra spending on a survey meant to identify settlements at risk from natural calamities. The hope is that these settlements would be evacuated either permanently or when threatened. But people are reluctant to abandon their livelihood and even if they agree to leave temporarily, they often return before the danger has passed for fear of looting.

Discussion point

Large natural hazards, such as Hurricane Katrina, are just as likely to devastate MEDCs as LEDCs. To what extent do you agree?

Theory into practice

Keep a 'hazards diary' and record the nature and impacts of climate hazards worldwide over a term. Comment on your findings.

Please refer to the CD-ROM for a case study on smog in Beijing.

3.4 What can humans do to reduce the impact of climatic hazards?

Hurricanes

Some hazards are easier to predict than others. The monsoon rains and floods occur most summers in specific areas, such as south-east Asia, especially in river valleys. In contrast, drought is a long-term hazard that may take years to materialise. Hurricane paths, on the other hand, are notoriously difficult to predict (Figure 3.2, page 87). In general, climatic hazards are difficult to predict as their mechanisms are complex, involving the interaction of atmospheric, oceanic and land-based systems.

There are many ways in which people can try to reduce the impact of hazards, for example hurricane shelters (Figure 3.27), hurricane screens (Figure 3.28) and avalanche breaks (Figure 3.29). There is evidence to suggest that protection schemes are working. For example, during 2007 over four-fifths of the state of Tabasco, in Mexico's south-east, was under water, damaging the homes of nearly a million people. The damage was estimated to be about US$ 5 billion. The hurricane was described as 'one of the worst natural disasters in Mexican history'. Nevertheless,

Figure 3.27 Hurricane shelter

Figure 3.28 Hurricane screens (open and closed)

despite the catastrophic scale of the flooding, the death toll has been relatively light (estimates of 25 people). Mexico and neighbouring countries have made major advances in their civil protection systems. The main improvement has been in evacuating residents when hurricanes approach.

Before Hurricane Felix struck in September 2007, Honduran officials managed to evacuate 25 000 people from the coast – a poor area. Emergency aid has also improved, thanks partly to better technology. However, in Nicaragua the same year, officials found it harder to convince the Miskito Indians to leave their scattered coastal villages. Consequently, 338 people were either dead or missing making it the worst loss of life that year (Figure 3.30). In Haiti and the Dominican Republic, where Hurricane Noel (2007) killed some 150 people, procedures were far from perfect.

While improvements in evacuation procedures and emergency aid can save many lives from the climatic

Figure 3.29 Avalanche breaks

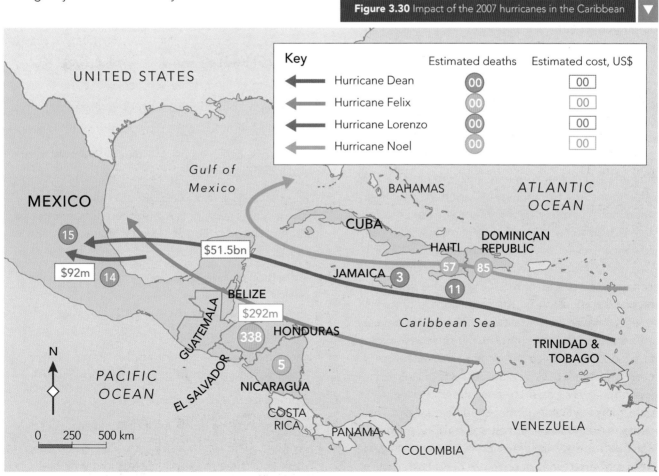

Figure 3.30 Impact of the 2007 hurricanes in the Caribbean

hazard, it is harder to mitigate the immense economic damage natural disasters inflict. Hurricane Felix deprived 200 000 people of their houses or land. Because the staple crop on the Mosquito Coast, Nicaragua, is coconut palm, it will take at least five years until villagers can return to their previous standard of living. Bananas, a staple across the Caribbean, grow back faster, but new plants still take almost a year to bear fruit.

One of the best ways of reducing the impact of hurricanes is to maintain a complete cover of coastal forest. This need only extend a few hundred metres from the shoreline. This has been successfully done in parts of Dover Beach, Barbados. The forest reduces the strength of the wind, and the vegetation slows down the force and height of the surge. In contrast, when the vegetation is removed, not only is there no protection from the hurricane, but the tidal surge and the heavy rains cause considerable soil erosion and flooding.

An excellent way to reduce the economic cost of hurricanes would therefore be through better planning of human settlement, avoiding the riskiest areas. In Jamaica, officials talk of creating 'no-build' zones in exposed coastal areas. But in many countries authorities have no powers to prevent building in vulnerable spots. The country best prepared for disaster with a good response plan is Cuba, one of the world's few remaining Communist countries. In contrast to Hispaniola (Haiti and Dominican Republic) where there were many deaths as a result of Hurricane Noel, in Cuba there was one.

There is also scope to mitigate the economic impact of disasters through making sure people have access to money to help them rebuild their lives and businesses. Even after Hurricane Katrina, most US citizens can still get hurricane insurance, thanks to government guarantees. In Central America such protection is in its early stages. The Community of Caribbean countries has made more progress: 16 of its members – or associated states – have joined a new Caribbean Catastrophe Risk Insurance Facility, a scheme organised by the World Bank after Hurricane Ivan caused severe economic damage in several islands in 2004. They include Haiti, the poorest country in North, South and Central America, as well as several very small island states such as Barbuda and Montserrat. The member governments have paid US$ 40 million in premiums and start-up fees. Payouts are based on the intensity of storms or earthquakes assessed by wind speed and Richter scale, rather than the number of buildings destroyed. It is hoped that this will discourage building in areas of high risk. The cover falls far short of replacement value but the scheme is designed to pay out quickly, acting as a bridge between emergency help and long-term rebuilding.

Almost a decade after Hurricane Mitch, in which 11 000 people were killed, largely in Honduras and Nicaragua, governments have learned how to save more lives. There are many practical tips for surviving hurricanes. The next task is to safeguard livelihoods, especially since many scientists predict that climate change is likely to increase the frequency and intensity of hurricanes.

Activities

Suggest ways in which people can reduce the impact of climatic hazards. Give examples to support your answer.

Dealing with drought – managing water supplies in dry areas

In many areas of the world, the problem is not too much water but too little. Approximately 25 per cent of the world's surface experiences a shortage of water. Water can be 'sourced' in many ways. The main ones are:

- extraction from rivers and lakes (including traditional forms of irrigation such as the shadoof and Archimedean screw)
- trapping behind dams and banks
- pumping from aquifers (water-bearing rocks)
- desalinisation (changing salt water to fresh water).

These can be achieved using either high-technology or low-technology methods.

'Water harvesting' is making use of available water before it drains away or is evaporated. Efficient use or storage of water can be achieved in many ways, for example:

- irrigation of individual plants rather than of whole fields
- covering expanses of water with plastic or chemicals to reduce evaporation
- storage of water underground in gravel-filled reservoirs (again to reduce evaporation losses).

Inappropriate use of water can lead to:

- salinisation (the build-up of toxic salts in the soil)
- waterlogging (which produces cold, oxygen-deficient conditions in a soil and is useless for farming).

Case study | Water development in Chad

Chad is a land-locked country in north-central Africa (Figure 3.31). It has been independent from France since 1960, but it has not been at peace. There are three distinct zones in Chad:

- the tropical, cotton-producing south (the most prosperous region in Chad)
- the central semi-arid Sahelian belt
- the northern desert and the Tibesti Mountains – here the nomadic Tubu tribe make use of wild plants and herd their camels.

Oil deposits have recently been discovered in the south. This may bring economic prosperity to the country or it could keep factions at war for even longer as they battle over the new-found riches.

Chad is a hot, dry country.

- The northern half of Chad has temperatures between 10°C and 20°C between November and April, and less than 250 mm of rain over this same period.
- By contrast, the southern half experiences temperatures of between 20°C and 29°C in the same period. It also has less than 250 mm of rain.
- During winter, the Harmattan, a hot, dry wind, blows from the Sahara raising daytime temperatures in the south.
- Between May and October much of the northern half has temperatures over 30°C, whereas the south has temperatures between 20°C and 30°C. Rainfall at this time in the north is low (< 250 mm) but is higher in the south (> 500 mm).
- The Tibesti Mountains contain reserves of groundwater, which could be used for irrigation.

Chad is dry on account of its location relative to the heat equator. The thermal equator refers to the zone of maximum insolation and largely reflects the migration of the overhead sun. During late March and late September, the sun is over the Equator. At that time trade winds blow in towards the Equator from the subtropics. As NE trades and SE trades are similar in character, they form a convergence zone rather than a front. The inter-tropical convergence zone (ITCZ) or inter-tropical discontinuity marks the position of the thermal equator.

Latitudinal variations in the ITCZ occur as a result of the movement of the overhead sun. In June, the ITCZ lies further north whereas in December it lies in the Southern Hemisphere. Winds at the ITCZ are generally light – the doldrums. These are broken by occasional strong westerlies, generally in the summer months.

Seasonal variations in the thermal equator have an impact on rainfall and on the distribution and type of ecosystems. In general, with increasing distance from the equator there is increasing seasonality of rainfall and an increasing dry season. The vegetation changes from rainforest to savanna to desert.

Lake Chad, which is widely regarded as the world's third largest lake and the world's largest freshwater lake, has become an endangered water source to the people in the country due to natural (drought, desertification, high evapotranspiration rates and possible river capture) and man-made causes (forest over-exploitation, bush burning, tree cutting, etc.). These natural phenomena cannot easily be

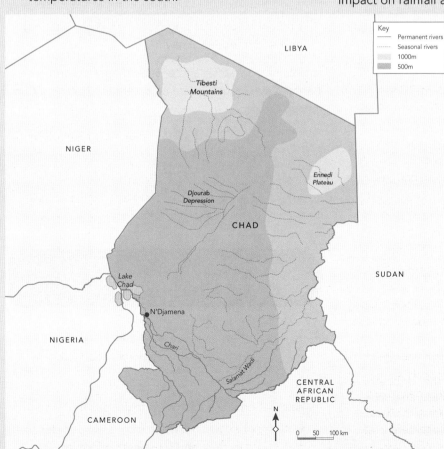

◀ **Figure 3.31** The location of Chad

predicted and contained, and new systems of water resources management and planning have to be devised through water conservation and management techniques. Such techniques would include the improvement of the efficiency of current water uses, particularly large-scale irrigation projects, as well as such water conservation measures as improving soil texture to reduce evaporation, adopting proper operation rules for upstream dams, and intra- and inter-basin water transfer options within the basin.

Options for water development in Chad

There are a number of options for water development in Chad. These include:

- long-distance transfers of water from wet areas
- inter-basin transfers (moving water from one part of a region to another; along a canal, for example)
- new dams
- desalinisation (though Chad is a land-locked country)
- using icebergs as a source of fresh water (unlikely to be a real option)
- developing groundwater supplies
- reusing effluent (water used for sewage treatment)
- building local, low-level earthen dams to catch and hold back water.

Management options include:

- the use of meters in people's homes (though most people would not have piped water)
- repairing leaking pipes (though piped water is rare)
- new technology such as desalinisation.

The problem of managing resources is becoming harder. Population growth and economic growth is increasing the demand for water and it is being used up at a faster rate than it is being renewed. There are, however, means of improving the situation. These include:

- use of dryland farming techniques; using drought-resistant fodder crops such as American Aloe (pastureland is especially fragile owing to a combination of drought, overgrazing, and population pressure)
- less water wastage
- more efficient pipes
- better irrigation methods
- desalinisation of seawater
- less government control of water resources.

The Integrated Plan for Water Development and Management (SDEA) is a multisectoral, strategic master plan providing guidelines for the sustainable development and management of water resources in Chad. It was developed by UNDP. The overall aim of the plan is to achieve the Millennium Development Goal – particularly the target to increase access to safe drinking water – and ensure integrated and participative management of water resources and their infrastructure.

Activities

1. Suggest why there is a water shortage problem in Chad.
2. Outline the ways in which it is possible to manage water in dry areas such as Chad.

Discussion point

Is it possible to manage climatic hazards?

 Take it further activity 3.3 on CD-ROM

Theory into practice

Compare the impacts of hurricanes in the Caribbean, for example, in the 1990s and 2000s. What does this tell you about the ability to manage climate hazards?

3.5 In what ways do human activities create climatic hazards?

Global warming

Enhanced greenhouse effect

The greenhouse effect is the process by which certain gases absorb outgoing long-wave radiation from the Earth and return some of it back to Earth (Figure 3.32). In all, greenhouse gases such as carbon dioxide, methane, CFCs, nitrous oxides and water vapour raise the Earth's temperatures by about 33°C (compared to the Moon, which has no atmosphere). Greenhouse gases vary in their abundance and contribution to global warming (Figure 3.33). The enhanced greenhouse effect is also referred to as global warming – it is the increase in greenhouse gases (and in global temperatures) as a result of human activity.

One of the concerns about global warming relates to the build up of greenhouse gases. CO_2 levels have risen from about 315 parts per million (ppm) in 1950 to 383 ppm in 2008 and are expected to reach 600 ppm by 2050. The increase is due to human activities – burning fossil fuel (coal, oil and natural gas) and deforestation. Deforestation of the tropical rainforest is a double blow – not only does it increase atmospheric CO_2 levels, it removes the trees that convert CO_2 into oxygen.

Methane is the second largest contributor to global warming, and is increasing at a rate of one per cent per annum. It is estimated that cattle convert up to 10 per cent of the food they eat into methane, and emit 100 million tonnes of methane into the atmosphere each year. Natural wetland and paddy fields are another important source – paddy fields emit up to 150 million tonnes of methane annually. As global warming increases, bogs trapped in permafrost will melt and release vast quantities of methane.

Chlorofluorocarbons (CFCs) are synthetic chemicals that destroy ozone as well as absorb long-wave radiation. CFCs are increasing at a rate of 4 per cent per annum, and are up to 10 000 times more efficient at trapping heat than CO_2.

Various models predict that effects of global warming are mixed:

- Sea levels will rise causing flooding in low-lying areas such as the Netherlands, Egypt and Bangladesh.
- Storm activity will increase owing to more atmospheric energy.
- Agricultural patterns will change, for example there will be a decline in the US grain belt but an increase in Canada's growing season.
- Rainfall will be reduced over the USA, southern Europe and former Soviet Union.
- Vulnerable wildlife species will become extinct.

Figure 3.32 Greenhouse effect

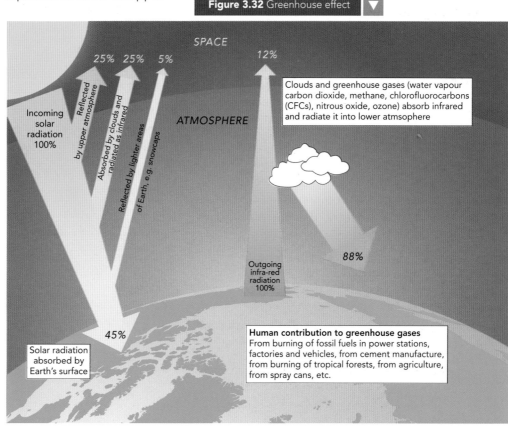

Greenhouse gas	Average atmospheric concentration (ppm)	Rate of increase (percentage per annum)	Direct global warming potential (GWP)
Carbon dioxide	355	0.5	1
Methane	1.72	0.6–0.75	11
Nitrous oxide	0.31	0.2–0.3	270
CFC 11	0.000255	4.0	3400
CFC 12	0.000453	4.0	7100

Figure 3.33 Properties of key greenhouse gases

Reducing the greenhouse effect

International policy to protect climate

The first world conference on climate change was held in Geneva in 1979. However, the 1987 Montreal Protocol was the crucial first step in limiting further damage to the ozone layer in the stratosphere. It was signed by many countries to greatly reduce the production and use of CFCs that had been shown to be responsible for damage to the ozone layer. The main CFCs have not been produced by any of the signatories since 1995, except for a limited amount for essential uses, such as for medical sprays.

The Toronto Conference of 1988 called for the reduction of CO_2 emissions by 20 per cent of the 1988 levels by 2005 (not achieved). Also in 1988, UNEP and the World Meteorological Organization established the Intergovernmental Panel on Climate Change (IPCC). 'The ultimate objective is to achieve... stabilisation of greenhouse gas concentrations in the atmosphere at a level that would prevent dangerous anthropogenic interference with the climate system.' (United Nations Framework Convention on Climate Change.) Anthropogenic inference here means interference caused by human activity.

The Kyoto Protocol (1997) gave all MEDCs legally binding targets for cuts in emissions from the 1990 level by 2008–2012. The EU agreed to cut emissions by 8 per cent, Japan by 7 per cent and the USA by 6 per cent. Some countries found it easier to make cuts than others. The UK, for example, had already reduced its use of coal and oil and changed to natural gas. Germany, too, had exhausted many of its traditional heavy industries – such as iron and steel that consume vast quantities of coal – and Japan was largely dependent on nuclear power and hydroelectric power. By contrast, the USA was experiencing an economic boom in the 1990s and was reluctant to make cuts. According to Kyoto, MEDCs were on average to make cuts by 5.2 per cent.

In the Bali Conference of 2007, all countries, including the USA, have made a commitment to make 'deep cuts' in greenhouse gas emissions. Two years of negotiations will now start. Rich countries have agreed that poor countries must be given money to help them adapt to climate change. Money may also go to countries not to cut down or degrade forests – one of the most serious sources of climate change emissions. Developing countries will be helped to cut emissions with a transfer of new technologies. Small-scale projects intended to cut emissions will be helped more. The least developed countries will be given extra help to adapt.

However, there are no clear goals or timetables set for emission reductions, only vague guidelines that 'deep cuts' should be made. These could be watered down. In addition:

- no significant extra money was pledged until after 2012 to help poor countries adapt
- no binding targets were set for future funding (it is expected to cost poor countries in the region of £25bn a year to adapt to climate change)
- there is a danger that climate change money will not go to benefit local people and could even be taken from existing development aid budgets
- extra forestry money could be hijacked by industry for plantations.

Activities

1. What is the greenhouse effect?
2. To what extent is global warming the responsibility of the global community?

Global dimming

Researchers have confirmed a suggestion put forward in 1985 that the amount of sunlight falling on the Earth's surface is declining – a phenomenon dubbed 'global dimming' (Figure 3.34). Now, two groups of researchers report that global dimming, which is believed to have started in the 1960s, came to a halt in the 1990s. In other words, it is now getting sunnier.

Global dimming was discovered by Atsumu Ohmura, a geographer at the Swiss Federal Institute of Technology in Zurich. He found that the level in the early 1980s was 10 per cent lower than in the 1960s. Further analysis suggested that the global fall was actually slightly less than this (between 7 per cent and 9 per cent), while in some places, such as the then Soviet Union, the drop was higher (in that particular case, 20 per cent).

Rachel Pinker, at the University of Maryland, has analysed satellite records of the amount of sunlight reaching the Earth's surface between 1983 and 2001. It declined until about 1990, and then started rising. Overall, it increased by about 2 per cent over the study period.

The reason is thought to be a decrease in the Earth's albedo (how much sunlight it reflects back into space), a finding that contradicts earlier results. Much of the sunlight reflected into space by the Earth is turned away by clouds and dust in the atmosphere.

In 2004, a group of researchers suggested that global dimming was caused by clouds getting clogged up with particulate matter such as soot. These particles act as nuclei on which water vapour condenses to form clouds. When such nuclei are rare, clouds tend to be formed of relatively few, large droplets. When they are common, many small droplets form – and lots of small droplets are more reflective than a few large ones.

The change from dimming to brightening occurs when the atmosphere is cloud-free. This shift was due to a decrease in the amount of particulate material in the atmosphere. And that, in turn, was due to more effective clean-air regulations and a decline in the economies of east European countries in the late 1980s. The pattern of brightening was also found in other parts of the world, such as North America, Antarctica, Japan and Australia. In rapidly industrialising (and dusty) India, however, dimming continued. Not surprising, then, that during the period of global dimming the dirty heavy industry of the Soviet Union dimmed that country most of all.

It is nevertheless ironic to note that while particles such as soot are bad for human health, they may yet turn out to be good for the planet if they cause global cooling. The 'global dimming' effect could have implications for everything from the effectiveness of solar power to the growth of plants and trees.

Figure 3.34 Industrial atmospheric pollution linked to 'global dimming'

> **Activities**
>
> 1. What is meant by global dimming?
> 2. What are the causes of global dimming?
> 3. What are the effects of global dimming?

Acid rain

The major causes of acid rain (acid deposition) are sulphur dioxide and nitrogen oxides produced when fossil fuels such as coal, oil and gas are burned. Sulphur dioxide and nitrogen oxides are released into the atmosphere where they can be absorbed by the moisture and become sulphuric and nitric acids, sometimes with a pH of around 3. Most natural gas contains little or no sulphur and causes less pollution.

Coal-fire power stations are the major producers of sulphur dioxide, although all processes that burn coal and oil contribute. Vehicles, especially cars, are responsible for most of the nitrogen oxides in the atmosphere. Some come from the vehicle exhaust itself, but others form when the exhaust gases react with the air. Exhaust gases also react with strong sunlight to produce poisonous ozone gas that damages plant growth and in some cases human health.

For example, coal and oil burned in the UK and Germany produces SO_2 and NOx which are carried by the prevailing south-westerly winds towards Norway and Sweden. Freshwater ecosystems, including fish stocks, have been badly affected by acidification, especially in the southern half of Norway where soils are thin and the bedrock is acidic granite. In contrast, trees in Norway appear relatively healthy.

Dry deposition typically occurs close to the source of emission and causes damage to buildings and structures (Figure 3.35). Wet deposition, by contrast, occurs when the acids are dissolved in precipitation, and may fall at great distances from the sources. **Wet deposition** has been called a 'trans-frontier' pollution, as it crosses international boundaries.

> **Key terms**
>
> **Dry deposition:** materials transported in suspension in the air that eventually settle out as dry matter.
>
> **Wet deposition:** materials transported in suspension in the air that eventually settle out as precipitation.

Acidification has a number of effects:
- buildings are weathered (Figure 3.35)
- metals, especially iron and aluminium, are mobilised by acidic water and flushed into streams and lakes
- aluminium damages fish gills
- forest growth is severely affected
- soil acidity increases.

The effects of acid deposition are greatest in those areas which have high levels of precipitation (causing more acidity to be transferred to the ground) and those that have base poor (acidic) rocks that cannot neutralise the deposited acidity.

Figure 3.35 Weathering of buildings due to acid deposition

The solutions

Various methods are used to try to reduce the damaging effects of acid deposition. One of these is to add powdered limestone to lakes to increase their pH values. However, the only really effective and practical long-term treatment is to curb the emissions of the offending gases. This can be achieved in a variety of ways:

- by reducing the amount of fossil fuel combustion
- by using less sulphur-rich fossil fuels
- by using alternative energy sources that do not produce nitrate or sulphate gases (e.g. hydropower or nuclear power)
- by removing the pollutants before they reach the atmosphere.

However, while victims and environmentalists stress the risks of acidification, industrialists stress the uncertainties. For example:

- rainfall is naturally acidic
- no single industry/country is the sole emitter of SO_2/NOx
- more cars have catalytic convertors
- different types of coal have variable sulphur content.

Activities

1. What are the causes of acid deposition?
2. What are the impacts of acid deposition?
3. What can be done to reduce the effects of acid deposition?

Case study | Acid rain in China

Location

In 2003 acid rain fell on more than 250 cities and caused direct annual economic losses of 110 billion yuan (US$ 13.3 billion), equal to nearly 3 per cent of the country's gross domestic product. The regional acid rain pollution is still out of control in some southern cities. The acid rain pollution in the south-western areas is very serious. With the exception of Chongqing, the average pH value of the central districts was lower than 5.0 and the acid rain frequency was 70 per cent. The acid rain in southern China was mainly distributed in the Pearl River delta and central and eastern areas of Guangxi.

Causes

Two major causes were the rapidly growing number of cars and increasing consumption of cheap, abundant coal as the country struggles to cope with energy shortages and meet power demand. China is the world's largest source of soot and sulphur dioxide (SO_2) emissions from coal, which fires three-quarters of the country's power plants. More than 21 tonnes of SO_2 were discharged in China in 2003, up 12 per cent from the year earlier. It is estimated that the country will consume more than 1.8 billion tonnes of coal in 2005, emitting an additional 6 million tonnes of SO_2. The growth of nitrates, due to a swift rise of automobile and coal consumption plus overuse of fertilisers, is playing an increasing role in the country's acid rain pollution. In short, China's explosive economic growth is outpacing environmental protection efforts.

Possible solutions

The Chinese government has made efforts and made significant progress in energy saving and consumption reduction. The energy consumption amount has gone down year by year over the past two decades. The rate of smoke and dust removal from industrial waste gas has been reduced. In industry, the government has taken technical measures like the introduction of levying charges for pollution emissions and issuing license for discharging air pollutants. To mitigate the acid rain, the government adopted a series of measures to help with the prevention of acid rain, such as the adoption of clean coal, energy conservation and desulphurisation technologies.

By 2010, the Chinese government hopes that the total amount of discharged pollutants in industrial waste water and the total amount of industrial waste discharged shall be lower than 1995. This may be achieved by setting quotas for SO_2 emissions from thermal power plants and urging them to install desulphurisation facilities. China has already banned the use of coal in some areas most severely affected by SO_2 emissions, but sulphur is not the only enemy in the fight against acid rain.

Activities

1. What are the main causes of acid rain in China?
2. Outline the potential solutions of the problems of acid rain in China.

Discussion point

The problem of acid rain will only disappear once coal reserves are all gone. Discuss the validity of this statement.

Photochemical smog

Poor air quality affects 50 per cent of the world's urban population, a total of about 1.6 billion people. Each year several hundred thousand people die due to poor air quality. The problem is increasing due to increasing population growth in urban areas, industrial development, and an increase in the number of vehicles worldwide.

Los Angeles, for example, is sometimes known as 'car city' or 'smog city'. There are over 9 million vehicles in Los Angeles and it has experienced photochemical smog since the 1940s. The effects of pollution are serious: children growing up in Los Angeles have a 10–15 per cent reduction in lung capacity, and there are 1600 premature deaths related to chronic respiratory disorders each year. Road transport accounts for 44 per cent of VOCs (volatile organic compounds such as the cancer-producing benzene), 55 per cent of NOx and 87 per cent of carbon monoxide emissions.

Air pollution and health

Air pollution is associated with high pressure. This is because winds in a high pressure system are usually weak, meaning that pollutants remain in the area and are not dispersed. Air pollution has been linked with health problems for many decades. People most at risk include asthmatics, those with heart and lung disease, infants and pregnant women. This accounts for 20 per cent of the population in the developed world and an even higher proportion in the developing world. The death rate from asthma has increased 40–60 per cent in recent decades and it is now one of the most common causes of hospital admissions for children in MEDCs.

The main pollutants which aggravate asthma are sulphur dioxide, ozone, acid aerosols, PM10s (very fine particulates, less than ten microns in diameter, which can flow into the lung), nitrogen dioxide and dust. The number of people diagnosed with asthma is rising steadily and about one in seven children in the UK now suffer from asthma. In the USA it is estimated that the costs related to air pollution are about US$ 40 billion each year. This is made up of US$ 16 billion in healthcare costs and a further US$ 24 billion in lost productivity and absenteeism. The main urban air pollutants are sulphur dioxide, suspended particles (smoke, dust, PM10s), nitrogen oxides, carbon monoxide, VOCs, ozone, lead and other metals.

Summer and winter smog

Under anticyclonic conditions, poor air quality often persists for many days. In some climates, notably Mediterranean climates, stable high pressure conditions persist all season, hence poor air quality can remain for months in cities like Athens. Although smog occurs under certain natural high pressure atmospheric conditions, human activity – the emission of pollutants – is responsible for the environmental hazard. Summer smog occurs on calm sunny days when photochemical activity leads to low-level ozone formation.

At ground level ozone is a pollutant and is formed when nitrogen oxides and volatile organic compounds react in the sunlight and other compounds are formed, including acid aerosols (sulphates, sulphuric acid, nitrates and nitric acid), aldehydes, hydrogen peroxide and PAN (peroxacetyl nitrate). This process may take a number of hours to occur, by which time the air has drifted into surrounding suburban and rural areas. This means that ozone pollution may be greater outside the city centre. Concentrations can rise substantially above background levels in summer heatwaves when there are continuous periods of bright sunlight with temperatures above 20°C and light winds.

The effect of ozone pollution is to cause stinging eyes, coughing, headaches and chest pains. Above 20°C reactions are accelerated. Ozone can harm lung tissues, impair the body's defence mechanism, increase respiratory tract infections, and aggravate asthma, bronchitis and pneumonia. The long-term effects include the premature ageing of the lung. Children born and raised in areas where there are high levels of ozone can experience up to a 15 per cent reduction in their lung capacity.

Background levels of ground-level ozone have risen substantially over the last century. There is evidence that the pre-industrial near ground-level concentrations of ozone were typically 10–15 ppb. In the past 100 years the current annual mean concentrations are approximately 30 ppb over the UK. The number of hours of 'poor' ozone concentrations tends to increase from the north to the south of the UK. Once formed, ozone can persist for several days and can be transported long distances; pollution transported with continental air masses plays a significant role in UK ozone episodes. Note that ground level ozone found in the lower atmosphere (troposphere) is different from the ozone layer found in the upper atmosphere.

In city centres it is nitrogen dioxide rather than ozone that is likely to be the main pollutant. Vehicles emit two forms of nitrogen oxide – nitric oxide and nitrogen

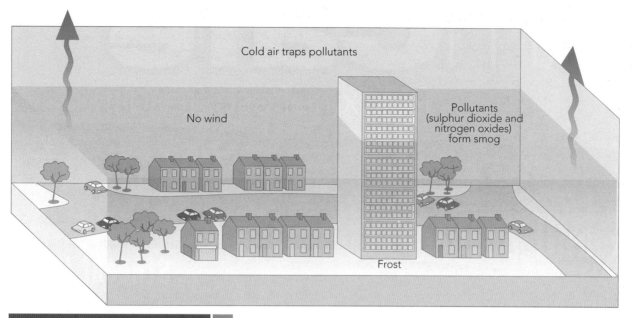

Figure 3.36 Formation of winter smog

Figure 3.37 Smog over Seoul

dioxide. The nitric oxide is converted (oxidised) into nitrogen dioxide by reactions with oxygen and ozone. It again leads to health problems for asthma and bronchitis sufferers. The good point about NO is that it leads to the rapid removal of ozone in city centres. Winter smogs are associated with temperature inversions and high rates of sulphur dioxide and other pollutants, due to increased heating of homes, offices and industries. Under cold conditions vehicles operate less efficiently and need more time to 'warm up'. This longer time for warming up releases larger amounts of carbon monoxide, nitrogen dioxide and hydrocarbons (Figure 3.36). These can cause coughs and sore throats, and can aggravate asthma and bronchitis.

Urban areas surrounded by high ground are especially at risk from winter smogs. This is because cold air sinks in from the surrounding hills, reinforcing the temperature inversion. This occurs on a range of scales from relatively minor albeit noticeable, such as Oxford, to major, such as Seoul (Figure 3.37) and has an impact on people's daily lives.

Discussion point

As the United Nations suggests, is climate change more dangerous than international terrorism (page 85)?

Is acid rain 'yesterday's problem'?

Take it further activity 3.4 on CD-ROM

Activities

1 What are the causes of photochemical smog?
2 What are the effects of photochemical smog?
3 Suggest ways in which it is possible to manage photochemical smog.

Knowledge check

1 How and why does the impact of climatic hazards vary between MEDCs and LEDCs?
2 Outline how the impacts of climatic hazards, and the ability to prepare for them, vary between urban and rural areas.
3 In what ways are human activities intensifying climatic hazards?

Exam Café
Relax, refresh, result!

Relax and prepare

Student tips

What I wish I had known at the start of the year…

Keda

"I don't always read all of the issue questions carefully as I expect to do the three from the three options I have studied. Often I do find I can do one of the other issues questions as the application of logic and the use of my knowledge from AS means I can do it. This does take time so I always start with the expected three – if they seem straightforward I don't read any further."

Ned

"There is always an example of some climatic disaster that I can use to illustrate my answer. As it is topical there is usually lots of information in the national papers, on television and on the Internet. Being topical impresses the examiner and demonstrates that I am willing to study around the subject. My teacher says it is a good way of demonstrating synopticity."

Bethany

"I find A2 examinations more stressful than AS since the larger number of essays required puts pressure on me as I am a slow writer. I used to panic when my friends were well into their essays when I was still doing my plan. I have found out that whereas they might write more my answers scored more highly as they were better structured and clearly focused. The time spent planning paid off and quality does beat quantity."

Common mistakes – Jezel

"A common error is to assume that in LEDCs populations suffer more from climatic hazards as they lack the resources and technology to cope with their impacts. It is worth remembering that many of these communities have been coping with difficult or challenging climatic conditions for centuries and have adapted their lifestyles and activities to the conditions. Often low-level technology or different crops can be used to cope with the conditions. No better example exists than the way that people have adapted their farming rhythm to the monsoon rains in south-east Asia. Remember, you can cope with or adapt to climatic conditions."

Refresh your memory

3.1. Tropical storms form and develop under particular conditions

Definitions	Tropical storm – winds 64–118 km/h; hurricane over 118 km/h
Form	High humidity – releases latent heat, sea temp 26–27°C – at least for 60m depth. More than 5° N or S of Equator for Coriolos force to impart spin. Almost constant vertical conditions. Divergent airflow with height to draw air upwards. Unstable air – surface winds converge.
Develop	Move westwards due to Earth's rotation – 15–30 km/h
	Break up over land (friction and little moisture) or 35° N/S as sea cooler

3.2 Tropical storms have serious environmental, social and economic impacts

Impacts	Primary – wind, storm surge, high rainfall, lightning
	Secondary – flooding, pollution (sewage, drains, etc.), disease, hunger
	Tertiary – long-term economic impacts, e.g. cost, loss of jobs, etc.
Environmental	Relief – landslides, mudflows. Drainage – floods, waterlogging. Vegetation – trees destroyed/uprooted, habitats destroyed. Pollution – water, disease.
Social	Health – injuries and deaths, disease, depression
	Housing – destroyed, temporary shelter, forced to migrate
	Social unrest – looting, family break up, tension, e.g. Katrina
Economic	Infrastructure – destroyed, e.g. roads, power, schools
	Agriculture – cash and food crops lost, pollution, tree crops hard hit
	Transport – bridges destroyed, road and rail damage, loss of aeroplanes
	Trade – loss of exports, need to import, cost of aid

3.3 Atmospheric systems can produce extreme weather

Anticyclone	Large slow moving area of high pressure – clear skies. Temperature inversion – fog gets trapped; smog. Clear skies – prolonged cold spells and frost in winter, heatwaves in summer; no clouds. Drought – fires, water shortages. Often ends in heavy thunderstorm in summer or snow in winter.
Depression	Fast-moving area of low pressure (uplift) with warm and cold fronts
	Heavy cloud cover – heavy rain at cold front, hail, lightning – snow in winter
	Strong winds – can produce blizzards in winter
	Heavy rain – flash floods

3.4 These hazards have serious environmental, social and economic impacts

Impacts	Anticyclone	Depression
Environmental	Wildlife – drought, fires, heat, frost	Flooding
	Vegetation – drought, fires, heat, frost	Lack of sun
	Drainage – drought and flash floods	Flooding, waterlogging
	Soils – shrink, subsidence	Swell – landslides
Social	Health – poor as heat/cold/fog; high pollen	Damp, depression
	Accidents – frost and fog	
Economic	Agriculture – need to irrigate; frost kills	Lack of sun
	Forestry – fires, drought	
	Transport – accidents in fog and on frost; heat buckles rail, melts roads	Wet surfaces
	Industry – water shortage	

Refresh your memory

3.5 The impacts of climatic hazards vary over time and location	
Time	Short versus long; duration; time of year, season
Location	Rural versus urban, coastal versus inland. LEDC versus MEDC; level of development. Nature of the hazard. West versus east (depressions move west to east in northern hemisphere).
3.6 There are a variety of ways to reduce or manage the impacts of climatic hazards	
Prediction	Based on past events, monitoring of pressure, satellites
Risk assessment	Calculate size and extent of risk – inform population, building design, location of vital buildings/facilities, e.g. power stations
Planning	Individual, e.g. store water; local authority, e.g. emergency centres; state or central, e.g. mobilisation of rescue services, building controls
Prevention	Build reservoirs and transfer schemes, river channel modification
Warnings	Use of media, planned evacuations
Recovery	Emergency aid, insurance, state aid for reconstruction
3.7 Human activities may impact on the global climate to create climatic hazards	
Activities	Industry, transport, housing, power production, farming – pollution
Causes	
Global warming	Burning fossil fuels produces carbon dioxide. Farming produces methane and chemicals releasing greenhouse gases. Deforestation – less carbon absorbed and cleared by burning. Waste tips produce methane. Industry produces hydrofluorocarbons.
Global dimming	Last 50 years average sunlight reaching the surface fallen by 3 per cent a decade. Heavy industry produces sulphate pollution that reflects light. Traffic fumes. Could be a side of global warming producing more clouds.
Acid rain	Burning fossil fuels. Industry – especially heavy as release sulphur. Traffic – releases nitrogen oxides. Farming – livestock produce ammonia.
Photochemical smog	Traffic fumes and/or industrial fumes react with sunlight. Often trapped by relief or avenues of tall buildings. Trapped by temperature inversion, domestic fires

Top tips...

If you have selected both climatic and earth hazards, some of the basic ideas of hazard perception (MEDC v LEDC types of mitigation) are very similar, so this can help save study time but has its own dangers. Candidates may get confused and include incorrect examples in their answer. Again, the emphasis is on reading the question carefully and many do not do this in the pressure of the exam room. The misreading or partial reading (so often candidates seem to stop reading before the end of the sentence) of questions is the single biggest cause of underperformance by candidates. It is vital to keep practising the analysis of questions to avoid this pitfall.

Get the result!

Example question

'Coastal areas are more vulnerable to climatic hazards than areas further inland.' How far do you agree with this statement? [30 marks]

Student answer

Coastal areas are more vulnerable to some types of climatic hazard. Hurricanes and tropical storms originate over the warm sea and bring flooding and 100 km/h winds to low-lying coastal areas, such as New Orleans when Hurricane Katrina hit in August 2005. Damage was much less inland as there was no warm water to power the storm. However, inland areas have their own climatic hazards, such as tornadoes. The USA gets over 1000 of these each year especially in 'tornado alley' stretching from Texas to Nebraska. Also, inland areas are away from the modifying effect of the sea so suffer more droughts, snowstorms, thunderstorms, etc. Coastal areas may be more vulnerable as more people live near coasts and there is only one direction to flee.

Examiner says

Clear attempt at evaluation. More knowledge of the example with a better explanation of why inland areas are less hazardous. Does argue that it is not that simple.

Examiner says

Recognises other climatic hazards with example and explanation. Pity they didn't make more of the population issue as hazards are related to there being a population at risk.

Examiner's tips

When you use examples to illustrate or support a point, try to ensure that there is a variety. It is not a good idea to use just one detailed case study as an example as all too often detail is given that is not strictly relevant to the question being answered. Many candidates regurgitate detailed case studies whether they fit the topic or not. Try to get at least two contrasting examples – contrasting in type of location, e.g. LEDC versus MEDC or upland versus lowland; in scale, e.g. local versus regional; and in level or type of impact.

Climatic hazards examples well illustrate these points. The nature and location of hurricanes, for example, forces a non-European focus. If looking at the impact or damage done by a hurricane or tropical storm, it is ideal to compare two areas at either end of the development spectrum, such as the USA and Honduras. Provided the hurricane was of the same strength then the impact can be directly related to the comparison in development as well as the differences in geography.

»» Exam Café

Chapter 4
Population and resources

The opening years of the twenty-first century have reawakened major concerns about the sustainability of global resources. As more countries have industrialised the competition for resources has intensified. This has been led by the phenomenal economic growth of China. As incomes rise in newly industrialised countries their populations aspire to Western lifestyles. Is the planet's resource base sufficient to give everyone a high standard of living? The current high prices for energy and food suggest not.

Questions for investigation

- How and why does the number and rate of growth of population vary over time and space?
- How can resources be defined and classified?
- What factors affect the supply and use of resources?
- Why does the demand for resources vary with time and location?
- In what ways does human activity attempt to manage the demand and supply of resources and development?

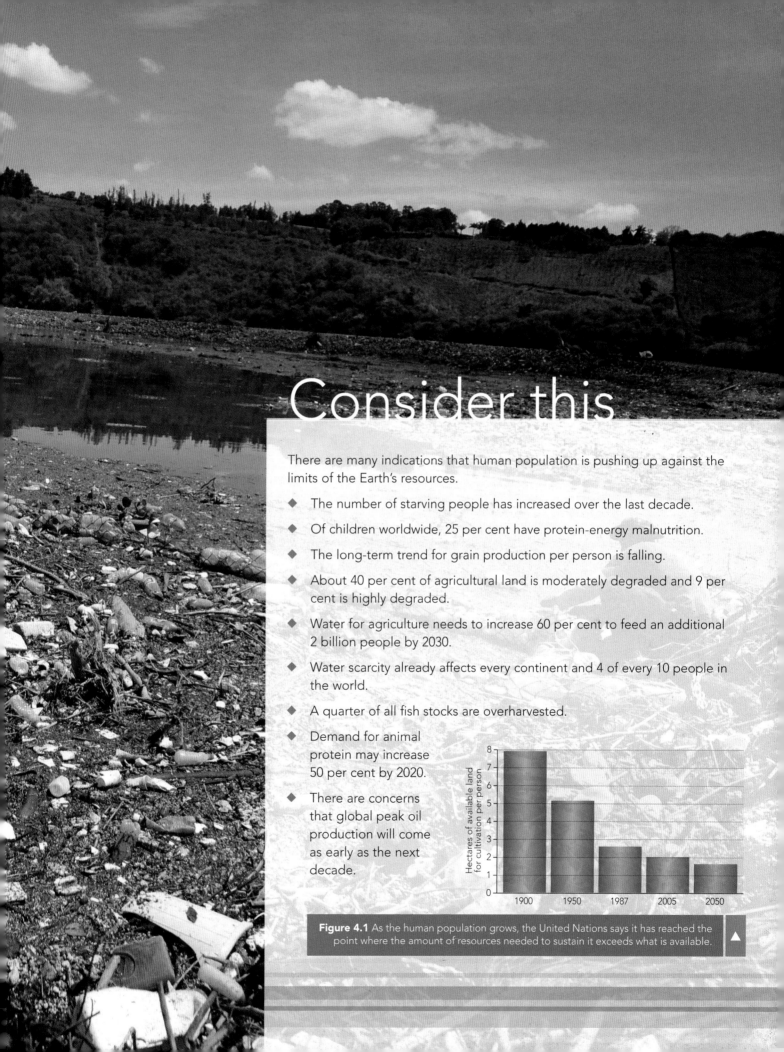

Consider this

There are many indications that human population is pushing up against the limits of the Earth's resources.

- The number of starving people has increased over the last decade.
- Of children worldwide, 25 per cent have protein-energy malnutrition.
- The long-term trend for grain production per person is falling.
- About 40 per cent of agricultural land is moderately degraded and 9 per cent is highly degraded.
- Water for agriculture needs to increase 60 per cent to feed an additional 2 billion people by 2030.
- Water scarcity already affects every continent and 4 of every 10 people in the world.
- A quarter of all fish stocks are overharvested.
- Demand for animal protein may increase 50 per cent by 2020.
- There are concerns that global peak oil production will come as early as the next decade.

Figure 4.1 As the human population grows, the United Nations says it has reached the point where the amount of resources needed to sustain it exceeds what is available.

4.1 How and why does the number and rate of growth of population vary over time and space?

Early humankind: growth and diffusion

The first hominids appeared in Africa around 5 million years ago, on a planet which was 4600 million years old. They differed from other apes by being able to walk on two legs and did not use their hands for weight-bearing. Other uses were soon found for these now liberated hands, with the acquisition of new skills charted in the evolutionary record as an increase in brain size.

The evolution of humankind was matched by its geographical diffusion. Whereas the locational evidence for *Australopithecus*, the first hominid, living 3–4 million years ago, is confined to Africa, remains of *Homo erectus* have been found stretching from Europe to south-east Asia. *Homo sapiens* roamed even further, making the first incursions into the cold environments of high latitudes.

During most of the period since *Homo sapiens* first appeared, global population was very low, reaching perhaps some 125 000 people a million years ago. It has been estimated that 10 000 years ago, when people first began to domesticate animals and cultivate crops, world population was no more than 5 million. Known as the Neolithic Revolution, this period of economic and cultural change significantly altered the relationship between people and their environments. But even then, the average annual growth rate was less than 0.1 per cent per year, extremely low compared with contemporary trends.

However, as a result of technological advance the **carrying capacity** of the land improved and population increased. By 3500 BC, global population reached 30 million. Over the next 4000 years or so this rose to about 250 million (Figure 4.2).

Demographers estimate that world population reached 500 million by about 1650. From this time population

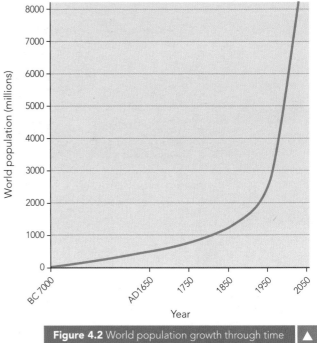

Figure 4.2 World population growth through time

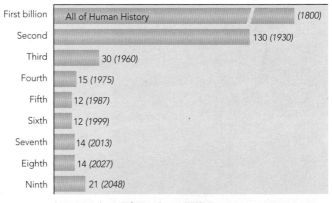

Source: Population Reference Bureau (2005)

Figure 4.3 World population growth, in billions

grew at an increasing rate. By 1830 global population had doubled to reach 1 billion. Figure 4.3 shows the time taken for each subsequent billion to be reached, while Figure 4.4 shows population change in 2005.

Demographic transition: how the rate of growth changes over time

Although no two countries have had exactly the same kind of population change, some broad generalisations can be made about population growth since the

Natural increase per	World	More developed countries	Less developed countries	Less developed countries (less China)
Year	80,794,218	1,234,907	79,559,311	71,906,587
Day	221,354	3,383	217,971	197,004
Minute	154	2	151	137

Source: *2005 World Population Data Sheet*, Population Reference Bureau

Figure 4.4 World population clock 2005

middle of the eighteenth century. These trends are illustrated by the model of **demographic transition** (Figure 4.5).

Key terms

Carrying capacity: the number of people who can be adequately supported by the productive capacity of the land.

Demographic transition: the historical shift of birth and death rates from high to low levels in a population.

No country as a whole retains the characteristics of Stage 1, which only applies to the most remote societies on Earth such as isolated tribes in New Guinea and the Amazon Basin which have little or no contact with the outside world. All the developed countries of the world are now in Stage 4 or Stage 5, most having experienced all of the previous stages at different times. The poorest of the developing countries (for example, Bangladesh, Niger, Bolivia) are in Stage 2 but are joined in this stage by the oil-rich Middle East states where increasing affluence has not been accompanied by a significant fall in fertility. Most developing countries which have registered significant social and economic advances are in Stage 3 (for example, Brazil and India) while some of the Newly Industrialised Countries (NICs) such as South Korea and Taiwan have just entered Stage 4. With the passage of time there can be little doubt that more countries will attain the demographic characteristics of the fourth stage of the model. Stage 5, natural decrease, is mainly confined to eastern and southern Europe at present. The basic characteristics of each stage are described below.

The High Fluctuating Stage (Stage 1)

During this stage the **birth rate** is high and stable while the **death rate** is high and fluctuating due to the sporadic incidence of famine, disease and war. Population growth is very slow and there may be periods of considerable decline. Infant mortality is high and life expectancy low. A high proportion of the population is under the age of 15 years. Society is pre-industrial with most people living in rural areas and dependent on subsistence agriculture.

The Early Expanding Stage (Stage 2)

The death rate declines to levels never before witnessed. The birth rate remains at its previous level as the social norms governing fertility take time to change. As the gap between the two vital rates widens, the **rate of natural change** increases to a peak at the end of this stage. The **infant mortality rate** falls and **life expectancy** increases, as does the proportion of the population under 15 years. Although the reasons for the decline in mortality vary in intensity and sequence from one country to another, the essential causal factors are: better nutrition; improved public health particularly in terms of clean water supply and efficient sewage systems; and medical advance. Considerable rural to urban migration occurs

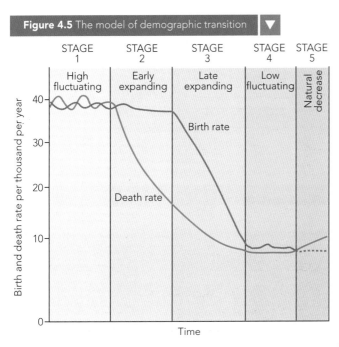

Figure 4.5 The model of demographic transition

during this stage. However, for developing countries in recent decades, urbanisation has often not been accompanied by the industrialisation which was characteristic of the developed nations during the nineteenth century.

The Late Expanding Stage (Stage 3)

After a period of time, social norms adjust to the lower level of mortality and the birth rate begins to decline. Urbanisation generally slows and average age increases. Life expectancy continues to increase and infant mortality to decrease. Countries in this stage usually experience lower death rates than nations in the final stage, due to their relatively young population structures.

The Low Fluctuating Stage (Stage 4)

During this stage, both birth and death rates are low. The former is generally slightly higher, fluctuating somewhat due to changing economic conditions. Population growth is slow. Death rates rise slightly as the average age of the population increases. However, life expectancy still improves as age-specific mortality rates continue to fall.

The Natural Decrease Stage (Stage 5)

In a limited but increasing number of countries, mainly European, the birth rate has fallen below the death rate. In the absence of net migration inflows these populations are declining.

Key terms

Crude birth rate (generally referred to as the **birth rate**): the number of births per thousand population in a given year. It is only a very broad indicator as it does not take into account the age and sex distribution of the population.

Crude death rate (generally referred to as the **death rate**): the number of deaths per thousand population in a given year. This is only a broad indicator as it is heavily influenced by the age structure of the population.

Infant mortality rate: the number of deaths of infants under one year of age per thousand live births in a given year.

Life expectancy (at birth): the average number of years a person may expect to live when born, assuming past trends continue.

Rate of natural change: the difference between the birth rate and the death rate.

Demographic transition in the developing world

There are a number of important differences in the way that developing countries have undergone population change compared to the experiences of most developed nations before them. In the developing world:

- birth rates in Stages 1 and 2 were generally higher
- the death rate fell much more steeply and for different reasons
- some countries had much larger base populations and thus the impact of high growth in Stage 2 and the early part of Stage 3 has been far greater
- for those countries in Stage 3, the fall in fertility has also been steeper
- the relationship between population change and economic development has been much more tenuous.

Figure 4.6 contrasts fertility decline in the USA and Bangladesh. In Bangladesh, fertility decline was much later than in the USA, but also much steeper when it occurred.

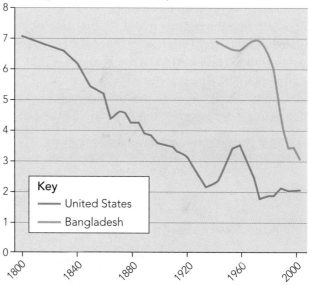

Source: Population Reference Bureau, *Population Bulletin* (2005)

Figure 4.6 Fertility decline in the USA and Bangladesh

Discussion point

What do you think are the main factors responsible for the UK's relatively low birth rate of approximately 12 per 1000 births?

Take it further activity 4.1 on CD-ROM

Recent demographic change and forecasts

Figure 4.7 shows that both total population and the rate of population growth are much higher in the less developed world compared to the more developed world. The highest ever global population growth rate was reached in the early- to mid-1960s when population growth in the less developed world reached a peak of 2.4 per cent a year. However, **demographic momentum** meant that the numbers being added each year did not peak until the late 1980s (Figure 4.8). Figure 4.9 shows a recent forecast for population change around the world between 2005 and 2050.

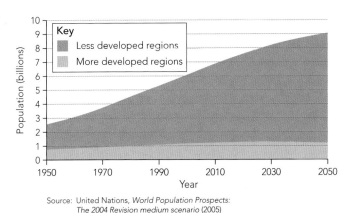

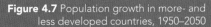

Figure 4.7 Population growth in more- and less developed countries, 1950–2050

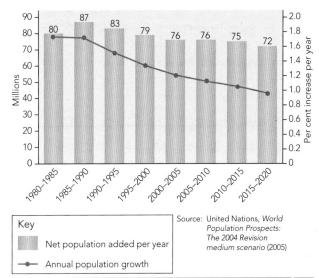

Figure 4.8 Population increase and growth rate over five-year periods, 1980–2020

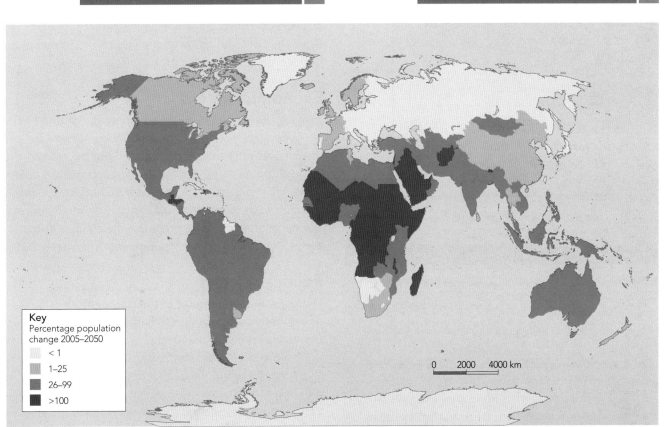

Source: Population Reference Bureau, *2005 World Population Data Sheet*

Figure 4.9 Percentage population change by country, 2005–2050

The components of population change

The relationship between births and deaths (natural change) is not the only factor in population change. The balance between immigration and emigration – **net migration** – must also be taken into account as the input–output model of population change shows (Figure 4.10). The corrugated divide on the diagram indicates that the relative contributions of natural change and net migration can vary over time within a particular country, as well as varying between countries at any one point in time. The model is a simple graphical alternative to the population equation $P = (B-D) +/- M$, the letters standing for population, births, deaths and migration respectively.

Key terms

Demographic momentum: the phenomenon of continued population increase despite reduced fertility rates. The population continues to grow due to a large proportion of its population entering its reproductive years.

Net migration (rate of): the balance between the immigration rate and the emigration rate.

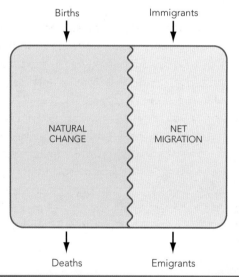

Figure 4.10 The input–output model of population change

Activities

1. Briefly describe the growth of human population over time.
2. Discuss the merits and limitations of the model of demographic transition.
3. Describe and explain the trends shown in Figure 4.8.
4. Identify and suggest reasons for the spatial patterns shown in Figure 4.9.

Factors affecting population change

The factors affecting population change can be grouped into four categories. These can vary in their relative importance from place to place.

Demographic factors

Other demographic factors, particularly mortality rates, influence the social norms regarding fertility. One study of sub-Saharan Africa, where the average infant mortality is over 80 per 1000, calculated that a woman must have an average of ten children to be 95 per cent certain of a surviving adult son.

Social/cultural factors

In some societies, particularly in Africa, tradition demands high rates of reproduction. What women of reproductive age may want for their lives may have little influence because of the intense weight of cultural expectations that they should marry and have large families. Education, especially female literacy, is the key to lower fertility. With education comes a knowledge of birth control, greater social awareness, more opportunity for employment and a wider choice of action generally. Most countries exhibit different fertility levels according to social class with fertility decline occurring in the highest social classes first. In some countries religion is an important factor. For example, some religions oppose certain methods of birth control, such as the oral contraceptive pill.

Economic factors

In many of the least developed countries, children are seen as an economic asset. They are viewed as producers rather than consumers. In the developed world the general perception is reversed and the cost of the child-dependency years is a major factor in the decision to begin or extend a family. Economic growth allows greater spending on health, housing, nutrition and education which is important in lowering mortality.

Political factors

There are many examples in the past century of governments attempting to change the rate of population growth for economic and strategic reasons. During the late 1930s, Germany, Italy and Japan all offered inducements and concessions to those with large families. In more recent years Malaysia has adopted a similar policy. However, today, most governments that are interventionist in terms of fertility want to reduce population growth.

The population and resource relationship

The relationship between population and resources has concerned philosophers and writers for thousands of years. However, the assumptions made by earlier writers were based on very limited evidence as few statistical records existed more than two centuries ago. For example, Confucius, the ancient Chinese philosopher, said that excessive population growth reduced output per worker, depressed the level of living and produced strife. He discussed the concept of optimum numbers, arguing that an ideal proportion between land and numbers existed and any major deviation from this created poverty. When imbalance occurred he believed the government should move people from overpopulated to underpopulated areas.

The Rev. Thomas Malthus (1766–1834) said that the crux of the population problem was 'the existence of a tendency in mankind to increase, if unchecked, beyond the possibility of an adequate supply of food in a limited territory'. Malthus thought that an increased food supply was achieved mainly by bringing more land into arable production, and maintained that while the supply of food could, at best, only be increased by a constant amount in arithmetical progression (1 – 2 – 3 – 4 – 5 – 6), the human population tends to increase in geometrical progression (1 – 2 – 4 – 8 – 16 – 32), multiplying itself by a constant amount each time. In time, population would outstrip food supply until a catastrophe occurred in the form of famine, disease or war. The latter would occur as human groups fought over increasingly scarce resources. These limiting factors maintained a balance between population and resources in the long term. In a later paper Malthus placed significant emphasis on 'moral restraint' as an important factor in controlling population.

Clearly Malthus was influenced by events in and before the eighteenth century and could not have foreseen the great technological advances that were to unfold in the following two centuries. Such technological developments have allowed resources to be produced at levels well beyond Malthus' concept of arithmetical progression.

Underpopulation, overpopulation and optimum population

The concept of **optimum population** has been mainly understood in an economic sense (Figure 4.11). At first, an increasing population allows for a fuller exploitation of a country's resource base, causing living standards to rise. However, beyond a certain level, rising numbers place increasing pressure on resources and living standards begin to decline. The highest average living standard marks the optimum population, or more accurately the **economic optimum**. Before that population is reached, the country or region can be said to be **underpopulated**. As the population rises beyond the optimum, the country or region can be said to be **overpopulated**.

> **Key terms**
>
> **Economic optimum:** the level of population which, through the production of goods and services, provides the highest average standard of living.
>
> **Optimum population:** the population that achieves a given aim in the most satisfactory way
>
> **Overpopulated:** when there are too many people in an area relative to the resources and the level of technology available.
>
> **Underpopulated:** when there are too few people in an area to use the resources available efficiently.

There is no historical example of a stationary population having achieved appreciable economic progress, although this may not be so in the future. In the past it is not coincidental that periods of rapid population growth have paralleled eras of technological advance which have increased the carrying capacity of countries and regions. Thus we are led from the idea of optimum population as a static concept to the dynamic concept of **optimum rhythm of growth** (Figure 4.12) in which population growth responds to substantial technological advances. For example, Abbé Raynal (Revolution de l'Amerique,

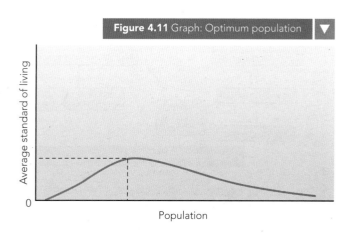

Figure 4.11 Graph: Optimum population

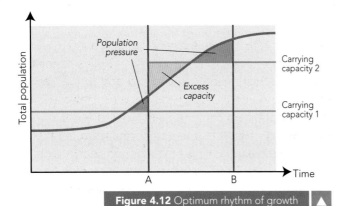

Figure 4.12 Optimum rhythm of growth

Figure 4.13 summarises the opposing views of the **neo-Malthusians** and the **resource optimists** such as Boserup. Neo-Malthusians argue that an expanding population will lead to unsustainable pressure on food and other resources. Resource optimists believe that human ingenuity will continue to conquer resource problems, pointing to so many examples in human history where, through innovation or intensification, humans have responded to increased numbers.

Key terms

Anti-Malthusians/resource optimists: those who argue that either population growth will slow well before the limits of resources are reached or that the ingenuity of humankind will solve resource problems when they arise.

Malthusians (or neo-Malthusians): the pessimistic lobby who fear that population growth will outstrip resources leading to the consequences predicted by Thomas Malthus.

1781) said of America 'If 10 million men ever manage to support themselves in these provinces it will be a great deal.' Yet today the population of the USA is over 300 million and hardly anyone would consider the country to be over-populated.

The most obvious examples of **population pressure** are in the developing world, but the question here is – are these cases of absolute overpopulation (**excess capacity**) or the results of underdevelopment that can be rectified by adopting remedial strategies over time?

Key terms

Excess capacity: when carrying capacity exceeds population number (per unit area).

Optimum rhythm of growth: the level of population growth that best utilises the resources and technology available. Improvements in the resource situation or/and technology are paralleled by more rapid population growth.

Population pressure: when population per unit area exceeds the carrying capacity.

Activities

1. Discuss the (a) demographic (b) socio/cultural (c) economic and (d) political factors which affect population change.
2. Explain the following:
   a underpopulation
   b overpopulation
   c optimum population.
3. Study Figure 4.12.
   a Suggest why the population initially started to increase.
   b What could account for the increases in carrying capacity at times A and B?
   c Why can Figure 4.12 be described as a dynamic model while Figure 4.11 is a static model?
4. With the aid of Figure 4.13, explain the opposing views of the neo-Malthusians and the resource optimists.

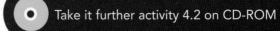
Take it further activity 4.2 on CD-ROM

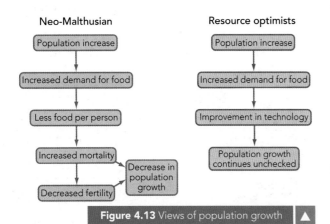

Figure 4.13 Views of population growth

4.2 How can resources be defined and classified?

Defining resources

Resources can be classed as either natural or human. Natural resources are naturally occurring substances, such as oil, forests and fish, which are considered to have value by human populations. They can be used directly or they can be processed to make other products. Human resources are the number of people in a population and its abilities and skills. Some countries, such as Japan, which have limited natural resources, have built up their wealth over time through the high-level skills of their populations. In contrast, some countries which are well endowed with natural resources have relatively low average incomes. Recent economic research shows a negative correlation between abundant natural resources and economic growth. Human and natural resources are closely linked because of the importance of the former in the development and exploitation of the latter.

For natural resources the traditional distinction is between renewable and non-renewable or stock resources (Figure 4.14). However, some resources can be classed as semi-renewable. Many renewable resources naturally regenerate within a human-defined time span to provide new supplies of these resources. Such resources, like soil and forests, are often connected in ecological systems. However, if renewable resources are used up at a level greater than their natural rate of replacement, the standing stock of such a resource will decline and may eventually become exhausted. Examples are the destruction of rainforest and the overfishing of marine species in high demand. Sustainable development policies are required to redress the balance between supply and demand for these renewable resources.

— Key term —

Resource: any aspect of the environment which can be used to meet human needs.

As shown, the distinction between non-renewable and renewable is not absolutely clear cut for all resources. In some cases it is useful to think in terms of the extent of renewability (Figure 4.14). This applies to resources such as forests and fish, where renewal depends on the rate at which they are used and the **resource management** techniques employed.

Flow renewable resources are renewable resources which do not need regeneration. Such resources are in constant supply. The latter include renewable energy sources such as solar, wind and tidal power. Such resources have been utilised to only a limited extent so far due to technology problems and cost. However, the

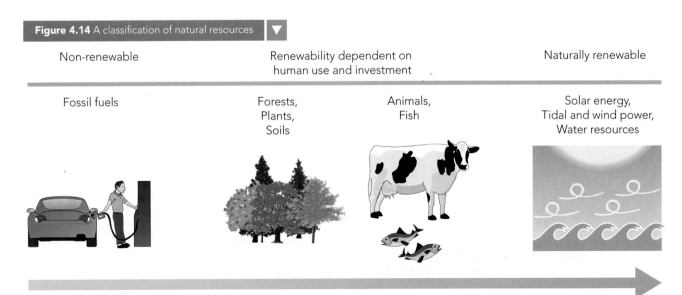

Figure 4.14 A classification of natural resources

Figure 4.15 Solar power – a flow resource

use of flow resources will increase significantly in the future as fossil fuel deposits become further depleted and environmental concerns about nuclear power remain.

In contrast, non-renewable resources such as coal and oil take millions of years to form and are therefore, in human terms, fixed in the supply available. **Resource depletion** can occur relatively quickly and eventually resources become exhausted with no viable further production possible. In the EU, many coalfields which were mining large quantities of coal in the nineteenth and early-twentieth centuries no longer produce any coal at all.

Key terms

Resource depletion: the consumption of non-renewable, finite resources which will eventually lead to their exhaustion.

Resource management: the control of exploitation and use of resources in relation to economic and environmental costs.

Figure 4.16 Iron ore – a non-renewable resource

Resources can also be classified as biotic or abiotic. Biotic resources such as wood are derived from living organisms, whereas abiotic resources such as copper and iron ore are not. A distinction can also be made between surface and sub-surface resources. In general, biotic resources are found at the surface of the Earth while abiotic resources are mainly below the surface of the Earth.

Blunden has classified minerals into the following groups:

- ubiquitous non-metallic minerals such as gravel
- localised non-metallic minerals such as salt
- non-ferrous metals including copper and tin
- ferrous metals
- carbon and hydro-carbon fuels.

The importance of aesthetic resources such as landscape is being increasingly recognised. High landscape quality is a major asset for both the quality of life of the resident population and for tourism, a major contribution to the economy of many areas.

Resources, reserves and supply

Figure 4.17 shows the United States Geological Survey's classification of mineral resources. The resources which a country has of any mineral is the total amount of that mineral contained in the Earth's crust within its borders. The reserve is that fraction of the total resources of the mineral that can be calculated to exist at commercially exploitable values under known technology. The supply of the mineral is the amount that can be produced and delivered to customers. The level of resources, reserves and supply can change very rapidly, each being partly dependent and partly independent of the other levels.

Resources can be divided into those which have been discovered and those which are as yet undiscovered. Identified resources which are not classified as reserves are classed as being paramarginal or submarginal depending on the estimated cost of extraction. Paramarginal resources are those recoverable at costs as much as one and a half times those that can be borne now. Submarginal resources are more costly still to extract.

Classification by source

The distribution of natural resources varies at the global, continental and national scales. Some resources are very geographically concentrated (point resources) while others occur widely (diffuse resources). Gold and iron ore are examples of point resources while sand and gravel are diffuse resources. Spatial concentration usually increases the scarcity of a resource, resulting in a relatively high monetary value. This is certainly the case with regard to gold,

Figure 4.17 USGS classification of mineral resources

	Identified resources			Undiscovered resources	
	Proved	Probable	Possible	In known districts	In undiscovered districts
Recoverable	RESERVES				
Paramarginal	IDENTIFIED PARA- AND SUBMARGINAL RESOURCES			HYPOTHETICAL RESOURCES	SPECULATIVE RESOURCES
Submarginal					

← Degree of certainty →

Feasibility of economic recovery ↑

Source: Cameron, E.N. (ed) *The Mineral Position of the United States 1975–2000*, University of Wisconsin Press (1972)

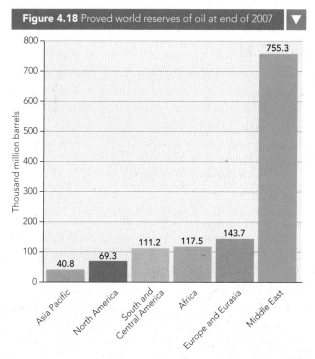

Figure 4.18 Proved world reserves of oil at end of 2007

Asia Pacific: 40.8
North America: 69.3
South and Central America: 111.2
Africa: 117.5
Europe and Eurasia: 143.7
Middle East: 755.3

(Thousand million barrels)

Source: BP Statistical Review of World Energy, 2008

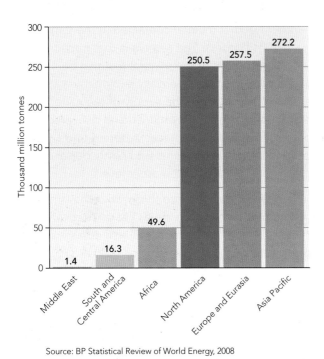

Source: BP Statistical Review of World Energy, 2008

Figure 4.19 Proved world reserves of coal at end of 2007

Activities

1. **a** Explain the distinction between natural and human resources.
   b What is the relationship between these two types of resources?
2. Use examples to explain the difference between point and diffuse resources.
3. Present a brief analysis of Figure 4.17.
4. Compare the data provided in Figures 4.18 and 4.19.

Theory into practice

Make a list of all the resources you think were used to construct and equip your school. Assess the impact of the school in terms of the balance between different types of resources.

---

diamonds, oil and copper. In contrast, resources such as sand and gravel occur widely and as a result their value is relatively low.

Figure 4.18 shows the distribution of proved reserves of oil at the end of 2006. The spatial pattern is dominated by the Middle East. In contrast, the Middle East plays an insignificant role in the proved reserves of coal (Figure 4.19).

As traditional deposits of fuel and non-fuel minerals have been depleted, mining companies have been forced to look increasingly to 'frontier' areas such as the extreme north of Canada and Siberia. There has also been a considerable increase in the development of 'offshore' deposits of oil and natural gas. Experience gained in drilling for oil and gas in the deep waters of the North Sea is being applied to other offshore areas around the world.

Classification by use

There is a huge gap in the rate of use of natural resources between MEDCs and LEDCs. However, the fastest rates of growth in natural resource consumption are in NICs. The availability of natural resources is vital for economic development. For example, the USA consumed 24.1 per cent of the world's oil in 2006 compared to Africa's 3.4 per cent and South and Central America's 6.1 per cent.

How changes in technology and society may result in changes in the definition of resources

The global usage of resources has changed dramatically over time. Such changes can be illustrated with reference to the UK. Here, technological advance has been the key to:

- the development of new resources
- the replacement of less efficient with more efficient resources.

The combination of the many advances in technology during the Industrial Revolution in Britain brought about a huge increase in resource use and considerable changes in the demand for different resources.

- The replacement of water power by steam power resulted in the rapid development of the UK's coalfields from the mid-1700s.
- The invention of the Gilchrist-Thomas process in the iron and steel industry in 1878 made it possible to smelt iron from phosphoric iron ores, leading to the development of the Jurassic iron ore fields of Lincolnshire and the East Midlands. Before 1878 these ores had no economic use. The mining of the Jurassic iron ores led to the construction of steelworks in the region and the expansion of urban settlements.

More recent developments have included the following:

- The intense pressure on food supplies during the Second World War resulted in the ploughing up of large areas of chalk downland for the first time. Advances in agricultural science made it possible to obtain reasonable crop yields with the application of the correct fertiliser mix. Prior to this understanding, the economic use of the chalk downlands was almost totally for sheep farming.

- The development of the nuclear power industry in Britain and other countries found a new use for uranium which significantly increased its price.

- The electrification of the railway system transformed a network which was once totally steam-driven.

- The location of oil and gas in the deeper parts of the North Sea was established some time before production from these areas commenced. What was required was (a) higher oil prices to justify the costs of deep-water production and (b) advances in deep-sea oil production technology.

- Renewable energy technology, particularly the construction of offshore wind farms, is now beginning to utilise flow resources in a significant way.

- Recycling has increased considerably in importance over the last decade, involving a much wider range of materials and products.

As society has changed, attitudes to certain resources and their use have also changed. For example, the demand for organic food is much higher today than even ten years ago and this upward trend is expected to continue strongly. Some power companies provide 'green energy' options which are attracting an increasing number of customers. Attitudes to plastic bags and the packaging of goods in general are changing. More and more people are questioning the environmental credentials of the companies from which they purchase goods and services.

> **Activities**
>
> 1. Explain three ways in which technological development has changed resource use in the UK.
> 2. Discuss one way in which major technological advance might change resource use in the future.

> **Discussion point**
>
> Suggest how changes in society can affect the demand for resources.

4.3 What factors affect the supply and use of resources?

The supply and use of resources is determined by a combination of physical and socio-economic factors (Figure 4.20). Figure 4.21 examines some of the major influences for variations in energy supply.

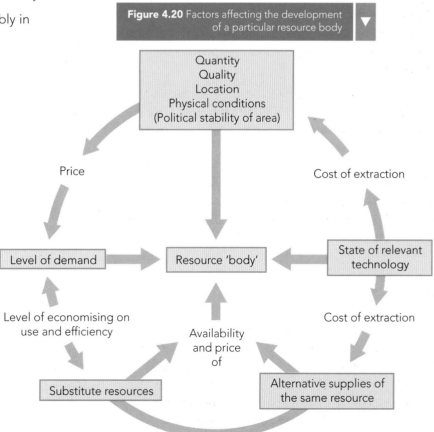

Figure 4.20 Factors affecting the development of a particular resource body

Source: Carr, M. *New Patterns: Process and Changes in Human Geography*, Nelson (1999)

PHYSICAL	• Deposits of fossil fuels are only found in a limited number of locations.
• Large-scale hydroelectric development requires high precipitation, major steep-sided valleys and impermeable rock.
• Large power stations require flat land and geologically stable foundations.
• Solar power needs a large number of days a year with strong sunlight.
• Wind power needs high average wind speeds throughout the year.
• Tidal power stations require a very large tidal range.
• The availability of biomass varies widely due to climatic conditions.
ECONOMIC	• The most accessible, and lowest cost, deposits of fossil fuels are invariably developed first.
• Onshore deposits of oil and gas are usually cheaper to develop than offshore deposits.
• Potential hydroelectric sites close to major transport routes and existing electricity transmission corridors are more economical to build than those in very inaccessible locations.
• In poor countries foreign direct investment is often essential for the development of energy resources.
• When energy prices rise significantly, companies increase spending on exploration and development.
POLITICAL	• Countries wanting to develop nuclear electricity require permission from the International Atomic Energy Agency.
• International agreements such as the Kyoto Protocol can have a considerable influence on the energy decisions of individual countries.
• Potential HEP schemes on 'international rivers' may require the agreement of other countries that share the river.
• Governments may insist on energy companies producing a certain proportion of their energy from renewable sources.
• Legislation regarding emissions from power stations will favour the use, for example, of low-sulphur coal as opposed to coal with a high sulphur content.

Figure 4.21 The physical, economic and political reasons for variations in energy supply

Case study | Problems of water supply in western USA

The USA is a huge consumer of water and supply would appear not to be a problem. However, the western states of the USA, covering 60 per cent of the land area with 40 per cent of the total population, receive only 25 per cent of the country's mean annual precipitation. Yet each day the west uses as much water as the east. The west has prospered this century. Agriculture, industry and settlement have flourished due to a huge investment in water transfer schemes. Hundreds of aqueducts take water from areas of surplus to areas of shortage. The federal government has paid most of the bill but now the demand for water is greater than the supply. If the west is to continue to expand, a solution to the water problem must be found.

Although much of the western states are desert or semi-desert, large areas of dry land have been transformed into fertile farms and sprawling cities. It all began with the Reclamation Act of 1902 which allowed the building of canals, dams and hydroelectric power systems in the states that lie, all or in part, west of the 100th meridian. Water supply was to be the key to economic development in general, benefiting not only the west but the USA as a whole.

California has benefited most from this investment in water supply. A great imbalance exists between the distributions of precipitation and population as 70 per cent of runoff originates in the northern one-third of the state but 80 per cent of the demand for water is in the southern two-thirds. While irrigation is the prime water user, the sprawling urban areas have also greatly increased demand. The 3.5 million hectares of irrigated land in California are situated mainly in the Imperial, Coachella, San Joaquin and the lower Sacramento valleys. Figure 4.23 shows the major component parts of water transfer and storage in the state.

Agriculture uses more than 80 per cent of the state's water, though it accounts for less than a tenth of the economy. Water development, largely financed by the federal government, has been a huge subsidy to California in general and to big water-users in particular. However, recently there has been a start to bringing the price mechanism to bear on water resources as farmers are being made to pay more for the water they use.

The Colorado: a river under pressure

The 2333 km-long Colorado River is an important source of water in the south-west (Figure 4.24). The river rises 4250 metres up in the Rocky Mountains of northern Colorado and flows south-west through Colorado, Utah, Arizona and between Nevada and Arizona, and Arizona and California before crossing the border into Mexico. The river drains an area of about 632 000 km².

Figure 4.22 Desert region in south-western USA

Figure 4.23 Water management schemes in California

Source: P. Guinness and G. Nagle, *Advanced Geography: Concepts and Cases*, Hodder and Stoughton

Case Study: Problems of water supply in western USA

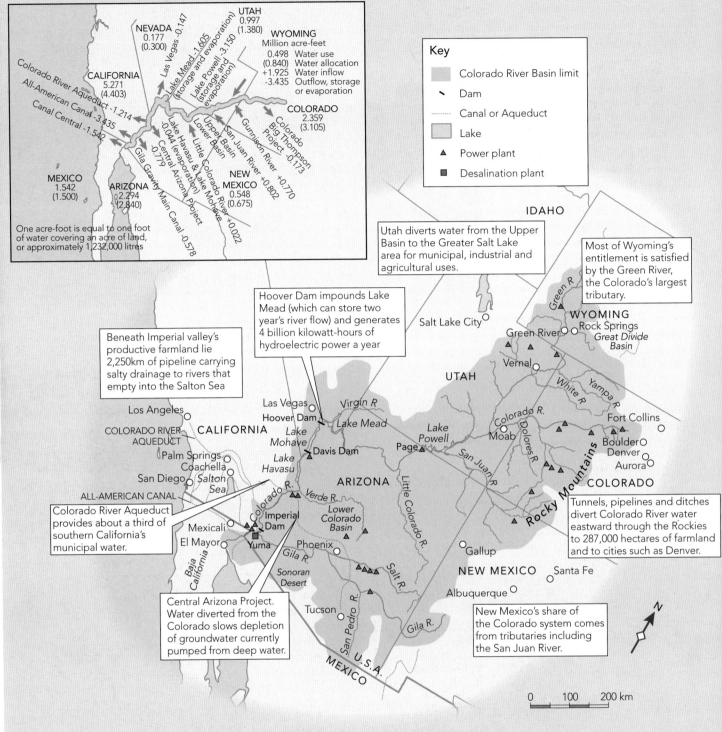

Figure 4.24 The Colorado River Basin

Source: P. Guinness and G. Nagle, *Advanced Geography: Concepts and Cases*, Hodder and Stoughton

In 1912 Joseph Lippincott, seeking future water supplies for the growing city of Los Angeles, described the Colorado as 'an American Nile awaiting regulation'. The Colorado was the first river system in which the concept of multiple use of water was attempted by the US Bureau of Reclamation. In 1922 the Colorado River Compact divided the seven states of the basin into two groups: Upper Basin and Lower Basin. Each group was allocated 9.25 trillion litres of water annually, while a 1944 treaty guaranteed a further 1.85 trillion litres to Mexico. Completed in 1936, the Hoover Dam and Lake Mead marked the beginning of the era of artificial control of the Colorado.

Despite the inter-state and international agreements, major problems have arisen over the river's resources.

- Although the river was committed to deliver 20.35 trillion litres every year, its annual flow has averaged only 17.25 trillion litres since 1930. Evaporation from artificial lakes and reservoirs has removed 2.45 trillion litres, and in drought periods this shortfall is accentuated.

- Demand has escalated. Between 1970 and 1990 the population of the seven Compact states increased from 22.8 million to 36.1 million. The river now sustains around 25 million people and 820 000 ha of irrigated farmland in the USA and Mexico.

The US$ 4 billion Central Arizona Project (CAP) is the latest, and probably the last, big money scheme to divert water from the river (Figures 4.24 and 4.25). Before CAP, Arizona had taken much less than its legal entitlement from the Colorado: it could not afford to build a water transfer system from the Colorado to its main cities and at the time the federal government did not feel that national funding was justified. Most of the state's water came from aquifers but it was overdrawing this supply by about two million acre-feet a year. If the cities of Phoenix and Tucson were to remain prosperous, something had to be done. The answer was CAP, which the federal government agreed to part fund. Since CAP was completed in 1992, 1.85 trillion litres of water a year has been distributed to farms, Indian reservations, industries and fast-growing towns and cities along its 570 km route between Lake Havasu and Tucson. However, providing more water for Arizona has meant that less is available for California. In 1997, the federal government told California that the state would have to learn to live with the 4.4 million acre-feet of water from the Colorado it is entitled to under the 1922 Compact, instead of taking 5.2 million acre-feet a year.

Resource management strategies

Implementation of the following strategies would conserve considerable quantities of water.

- The adoption of measures to reduce leakage and evaporation losses; up to 25 per cent of all water moved is currently lost by leakage and evaporation.

- Recycling water in industry where, for example, it takes 225 000 litres to make one tonne of steel.

- Recycling municipal sewage for watering lawns, gardens and golf courses could be implemented or extended. In Los Angeles such measures are already implemented.

- Introducing more efficient toilet systems which use only 6.5 litres for each flush instead of the conventional 26 litres.

- Charging more realistic prices for irrigation water. Many farmers pay only one-tenth of the true cost of water pumped to them; the rest is subsidised by the federal government. When long-term water contracts are eventually renewed, prices could be raised to more economic levels.

- Extending the use of drip irrigation systems which allocate specific quantities of water to individual plants and which are 100 times more efficient than the open-ditch system still used by many farmers; or sprinkler systems, which are up to ten times more efficient than open-ditch irrigation.

- Changing from highly water-dependent crops such as rice and alfalfa to ones needing less water.

- Changing the law to permit farmers to sell surplus water to the highest bidders. Since 1992 this has

Figure 4.25 Central Arizona Project (CAP)

been allowed in California. An emerging network of specialist brokers sells 'agricultural water' to cities for less than they already pay but at a profit for the farmers.

- The requirement that both cities and rural areas identify the source of water to be used before new developments can commence. This proposal, first mooted in southern California in 1994, proved to be politically unacceptable.

Future options

- Developing new groundwater resources. Although groundwater has been heavily depleted in many areas, in regions of water surplus such as northern California they remain virtually untapped. However, the transfer of even more water from such areas would probably prove politically unacceptable.

- It has been claimed that various techniques of weather modification, especially cloud seeding, can provide water at reasonable cost. However, environmental and political considerations cannot be ignored.

- In 1991, after several years of drought, the city of Santa Barbara approved the construction of a US$ 37.4 million desalination plant. Although much too expensive for irrigation water, it is likely that more plants will be built for urban use.

- Exploiting the frozen reserves of Antarctic water. Serious proposals have been made to find a 100 million-tonne iceberg (1.5 km long, 300 m wide, 270 m deep) off Antarctica, wrap it in sailcloth or thick plastic, and tow it to southern California. The critical questions here are cost, evaporation loss and the environmental effects of anchoring such a huge block of ice off an arid coast.

There is now general agreement that planning for the future water supply of south-west USA should embrace all practicable options. Sensible management of this vital resource shouldn't rule out any feasible strategy if this important region is to sustain its economic viability and growing population.

Activities

1. Using a precipitation map of the USA, describe the spatial distribution.
2. Using Figure 4.23, and with additional research, describe the main systems for water transfer in California.
3. Discuss the factors that have caused increasing pressure on the Colorado River over time.

Take it further activity 4.3 on CD-ROM

Case study | Wheat: growing concerns about global supplies

In December 2007 the price of wheat broke through the US$ 10 a bushel level, sparking protests in Pakistan and other countries as rising wheat prices were passed on to consumers. In 2005 the price of wheat had averaged US$ 3.5 a bushel.

There can be little doubt that the days of easy grain surpluses are a thing of the past. For example, India, the second largest consumer of wheat, became a large net importer in 2006 after a six-year period as a net exporter. In 2007, India tried to buy 50 per cent more grain than suppliers were offering. At about the same time Russia was considering curbs on wheat exports to prevent domestic prices rising too rapidly. In addition, Australia, the third biggest exporter of wheat, warned its output might be 18 per cent less than a previous government estimate due to a second year of drought. A late season drought in the Ukraine had a considerable impact on its wheat production. In Morocco the crop was down 76 per cent. Argentina temporarily halted wheat exports to assess damage caused by cold weather.

Demand for grain and other crops is rising because of the following.

- Global population continues to increase significantly and will reach 7 billion by 2010.
- Living standards are improving in many countries, especially in highly populated China and India. Higher incomes result in the increasing demand for meat. However, it takes 7 kg of grain to produce 1 kg of beef. Consumers in NICs are following the lifestyles developed in MEDCs.

Figure 4.27 Large-scale wheat production

- The diversion of agricultural resources from food production to bio-fuel manufacture because of concerns about energy security. This is reducing food production significantly in some areas.

The main problems with the supply of grain are that:

- most good-quality farmland is already being used
- about a third of this has been significantly degraded by intensive farming in the last half century (the worst predictions are that 30 per cent of all agricultural land could be unusable by 2025)
- the world's deserts expanded by 160 million hectares between 1970 and 2000
- the global land area used for the cultivation of wheat and barley has been falling for 25 years
- ocean freight transportation rates are at record highs
- drought and other adverse environmental factors are significantly reducing yields in key producing countries. (More and more countries are becoming concerned about the impact of climate variability on food production.)

In terms of global 'ending stocks-to-use ratio', the recent estimate for 2007–8 is 17 per cent, a 30-year low (Figure 4.26). (The ending stocks-to-use ratio is the amount of wheat in storage at the end of the year as a percentage of annual consumption.) This compares to 36 per cent between 1998–2001. A global stock of 25 per cent is considered to be a 'safe' level – anything below causes concern.

Steep price increases are not confined to wheat. Corn prices doubled in 2006–7, with the price of soybeans 50 per cent higher. Agricultural economists are concerned as historically prices have tended to move in long cycles driven by extended periods of mismatched supply and demand. The main impact of rising food prices is on the low income populations of LEDCs.

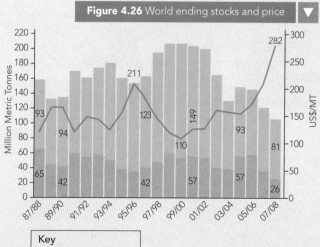

Figure 4.26 World ending stocks and price

Key
- Top 5 exporters
- Rest of World
- Hard Red Winter Wheat price

(Top 5 exporters include USA, Canada, Australia, Argentina and EU-25)

Source: 'World agricultural supply and demand estimates', United States Department of Agriculture (2007)

Activities

1 Describe the scene shown in Figure 4.27.
2 Explain why there is growing concern about global stocks of wheat.
3 Analyse the trends shown in Figure 4.26.

Case Study: Wheat: growing concerns about global supplies

4.4 Why does the demand for resources vary with time and location?

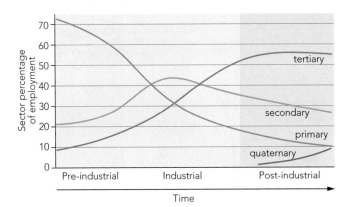

Figure 4.28 The Clark-Fisher Sector Model

Different parts of the world have differing demands in terms of quantities and types of resources, and these change with time and development. The overall global demand for resources increases with population growth and rising living standards. Thus, more affluent countries and regions have a much higher demand for most resources than poorer countries and regions. Over time the volume of demand increases in most countries which are not subject to major setbacks such as civil war or economic recession. Population size and density are major influences on demand. For example, in Brazil average population density increased from 6.1 per km² in 1950 to 19.95 per km² in 2000, with living standards also rising significantly during this time period.

As economic and social development occurs the structure of demand also changes. The sector model (Figure 4.28) is a useful aid in helping to explain how this process occurs. As an economy advances, the proportion of people employed in each sector changes. Countries such as the USA, Japan and the UK are 'post-industrial societies' where the majority of people are employed in the tertiary sector with the quaternary sector growing rapidly. Yet in 1900, 40 per cent of employment in the USA was in the primary sector. However, the mechanisation of farming, mining, forestry and fishing drastically reduced the demand for labour in these industries. Less than 4 per cent of employment in the USA is now in the primary sector. Major resource inputs were required as these industries became more capital intensive. In turn, such investment has significantly increased the production of many resources.

As primary jobs disappeared, people moved to urban areas where most secondary and tertiary employment is located. As the USA became the world's major manufacturer in the early-twentieth century, the range

Figure 4.29 Deindustrialisation – a closed steelworks in a MEDC

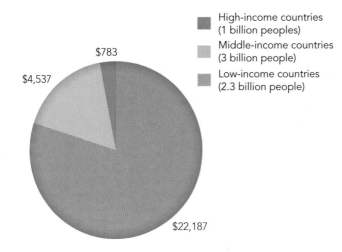

Source: Cassara, A. *Ask Earth Trends: How much of the world's resource consumption occurs in rich countries*, Earth Trends (2007)

Figure 4.30 Global consumption 2004 (billion US dollars)

and total amount of resources required to meet the demands of its industries rose dramatically. Human labour is steadily being replaced in manufacturing by robots and other advanced machinery. In 1950, the number of Americans employed in manufacturing was the same as in services. By 1980, two-thirds were working in services and this trend has continued.

As the USA and the other current MEDCs moved through the stages of the sector model, average incomes rose, triggering new resource demands. In such countries, the demand for resources is higher than it has ever been.

The process of globalisation has seen the loss of heavy industry in MEDCs (see Figure 4.29) as these industries have filtered down to NICs. Thus, the resources for manufacturing have fallen in many MEDCs and risen in NICs. The dramatic increase in demand for resources in NICs in particular has led to significant price increases for commodities.

Figure 4.30 shows global consumption in terms of monetary expenditure. Here, the 2.3 billion people of low-income countries accounted for less than 3 per cent of public and private consumption in 2004. In contrast, the 1 billion people in high-income countries consumed more than 80 per cent of the world total. The USA, with 4.6 per cent of the world's population, accounted for 33 per cent of global consumption.

Location is a major influence on the demand for resources. The environmental conditions in countries and regions can also have a significant impact on demand for resources. For example:

- countries in cold latitudes such as Russia will have a demand for energy well above the average
- countries with hot climates will have a higher than average demand for water.

Ecological footprints

The concept of **ecological footprints** has been used to measure natural resource consumption, how it varies from country to country, and how it has changed over time. The ecological footprint for a country has been defined as 'the total area required to produce the food and fibre that it consumes, absorb the waste from its energy consumption, and provide space for its infrastructure' (Living planet report 2004, World Wildlife Fund).

Figure 4.31 shows how the global ecological footprint has increased since 1961. In 2001, the total human ecological footprint exceeded **global biocapacity** by 0.4 **global hectares** per person, or 21 per cent. This means that world population is using natural resources faster than they can be regenerated. Figure 4.32 shows how the ecological footprint varies by world region.

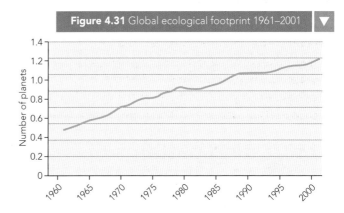

Source: Holmes, D. 'Eco-footprints' *Geofactsheet no. 183*, Hodder Education (2005)

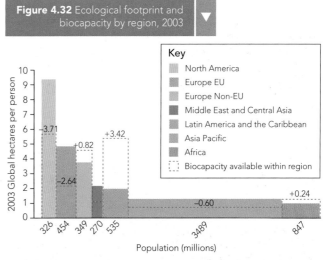

Figure 4.32 Ecological footprint and biocapacity by region, 2003

Source: The Living Planet Report 2006, page 18, World Wildlife Fund

Key terms

Ecological footprint: a sustainability indicator which expresses the relationship between population and the natural environment. It accounts for the use of natural resources by a country's population.

Global biocapacity: the capacity of global ecosystems to produce useful biological materials and to absorb waste materials generated by humans using current management schemes and extraction technologies.

Global hectares (gha): one global hectare is equivalent to one hectare of biologically productive space with world average productivity.

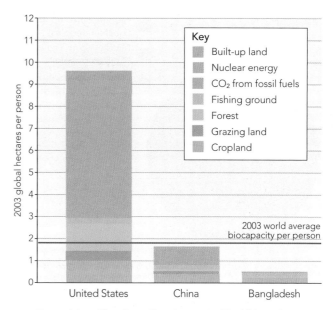

Source: Adapted from 'Living Planet Report', World Wildlife Fund (2006)

Figure 4.33 Ecological footprint of the USA, China and Bangladesh

Activities

1. Why is the Clark-Fisher Sector Model (Figure 4.28) a useful aid in explaining how the demand for resources changes over time?
2. **a** What is meant by the term ecological footprint?
   b Present an analysis of spatial and temporal variations in the ecological footprint using Figures 4.31 and 4.32.

Theory into practice

To access a website that allows you to calculate your own ecological footprint, go to www.heinemann.co.uk, enter the express code 7627P and click on the relevant link.

Contrasting patterns of demand in MEDCs, NICs and LEDCs

Figure 4.33 compares the ecological footprint of the USA (a MEDC), China (a NIC) and Bangladesh (a LEDC) in terms of built-up land, food and fibre, and energy. Figure 4.34 shows actual figures for a number of individual resources for the three countries. Per capita oil consumption is a good indicator of contrasts in economic development: oil consumption in the USA is almost 14 times higher than in China; in turn, per capita oil consumption in China is almost nine times that of Bangladesh. The contrast in meat, water and electricity consumption is also considerable, although, as with oil, consumption in China of all these resources is rising rapidly.

Figure 4.35 is a simplified model showing the contrast in (a) the level of resource use and (b) the rate of change in resource use between LEDCs, NICs and MEDCs. Figure 4.35 can be compared with the rates of change in consumption of four commodities between 1990 and 2005 (Figure 4.36).

Bangladesh is one of the most densely populated countries in the world. Although the rate of population growth has fallen, it was still estimated at 1.6 per cent a year in 2006. This compares to 0.5 per cent in China and 0.6 per cent in the USA. Almost half of its population lives on less than one dollar a day. The majority of people are employed in agriculture but there is simply not enough work in this sector to go round (Figure 4.37). The government is trying to attract foreign investment into the manufacturing and energy

Figure 4.34 Individual resource use for USA, China and Bangladesh

	Oil consumption per capita 2005. Barrels per day per 1000 people	Meat consumption 2002. Kilograms per person	Average water use per person per day in litres 2006	Electricity consumption per capita (KWh/year) 2006
USA	68.838	124.8	576	12187
China	4.943	52.4	86	2140
Bangladesh	0.565	3.1	46	120

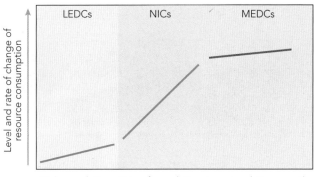

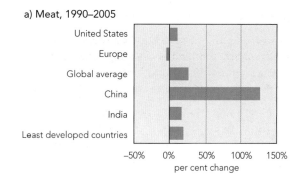

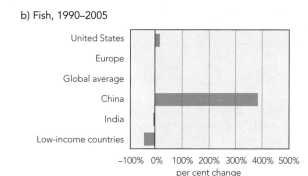

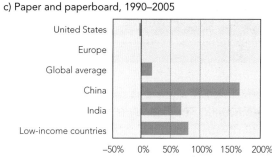

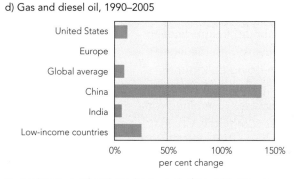

Figure 4.35 Model of the relationship between resource consumption and the level of economic development

Source: Cassara, A. *Ask Earth Trends: How much of the world's resource consumption occurs in rich countries*, Earth Trends (2007)

Figure 4.36 Trends in consumption of selected commodities

sectors and has achieved some degree of success. A large pool of low-cost labour is certainly an attraction to foreign TNCs, particularly when labour costs in other Asian countries are rising. If Bangladesh can transform itself into one of the next generation of newly industrialised countries, its demand for a wide range of resources will increase rapidly.

In the energy sector, onshore and offshore gas reserves could generate significant income but there is some debate about whether these reserves should be exported or kept for domestic use. However, a plan to develop major opencast coal reserves in northern Bangladesh has sparked off huge controversy because of the large number of people who would be displaced from their homes to make the project viable.

The scale of economic growth in China in recent years has been so significant that the country has been dubbed 'the new workshop of the world', a phrase first applied to Britain during the height of its Industrial Revolution in the nineteenth century. This rapid industrialisation has not only had a phenomenal effect on China itself but it has also impacted significantly on the rest of the world. The business magazine *Fortune* recently quoted CLSA Asia economist Jim Walker as saying 'We're at the early stages of one of the greatest industrial revolutions in world history'.

China has required huge raw material inputs to feed its rapidly expanding industries. A high percentage of these raw materials is imported (Figure 4.38). Rapidly increasing Chinese demand is one of the main reasons for rises in commodity prices of, for example, oil and copper.

China is the world's largest producer of coal, steel and cement. At 220 million tons in 2003, China produced more steel than the USA and Japan combined. In addition, China had to import an extra 40 million tons to satisfy domestic demand. Although China is the world's largest coal producer it may well become a net

Figure 4.37 Labour intensive agriculture in Bangladesh

importer within a few years as demand continues to grow significantly. The country is the second largest consumer of energy and the second largest importer of oil after the USA, surpassing Japan in 2003.

The huge demand for raw materials in China has been a major factor in pushing up shipping freight rates to new levels, with the world's ocean-going bulk carrier fleet operating at almost full capacity. Combined with higher commodity prices, there is concern that this might filter through to higher global inflation.

In the USA the demand for resources has changed over time as the economy has developed. The USA is now classed as a post-industrial society, having lost a significant part of its traditional heavy industrial base as these industries have filtered down to NICs, such as China. Changes in the rates of consumption are significantly lower than other parts of the world for some commodities. However, the overall use of resources remains colossal.

Figure 4.38 Imported raw materials arriving at a Chinese dock

Figure 4.39 Mass consumption: a traffic jam on a US motorway

Activities

1. Use Figures 4.33 and 4.34 to describe the differences in the ecological footprints of the USA, China and Bangladesh.
2. Describe and explain the three stages shown in Figure 4.35.
3. Identify and explain the trends shown in Figure 4.36.

4.5 In what ways does human activity attempt to manage the demand and supply of resources and development?

Both the demand for and supply of resources need to be planned and managed to achieve a sustainable system. In addition, there is growing pressure from individual governments and international organisations for a greater degree of equity in the use of the world's resources. Alongside this is a general recognition of the need for greater conservation of the environment.

Processes that can help achieve a sustainable future include the following:

- *Substitution*: the use of common and thus less valuable resources in place of rare, more expensive resources. An example is the replacement of copper by aluminium in the manufacture of a variety of products.

- *Beneficiation*: the upgrading of a resource that was previously considered too costly to develop. This process usually depends on technological advance, for example, the concentration of a dispersed mineral so that it can be transported and processed more easily.

- *Maximisation*: the use of a variety of methods that avoid waste and increase the production of a resource.

- *Recycling*: the concentration of used or waste materials, their reprocessing and their subsequent use in place of new materials. If organised efficiently, recycling can reduce demand considerably on fresh deposits of a resource. Recycling also involves the recovery of waste. New technology makes it possible to recover mineral content from the waste of earlier mining operations. However, the proportion of a material recycled is strongly related to the cost in proportion to the price of the original raw material, although governments are doing more and more to weaken this relationship.

- *Quotas* may change on an annual or longer time period basis.
- *Rationing*: this is very much a last-resort management strategy, which is adopted when demand is massively out of proportion to supply. Individuals may only be allowed a very small amount of fuel and food per week, for example rationing during and after the Second World War.

> **Key term**
>
> **Quota:** the limit set in an agreement between countries to take only a predetermined amount of a resource in a particular area over a given time, usually a year. The quota is usually divided among a number of different nations as a result of negotiation.

Figure 4.40 A recycling plant

Case study | Contrasting resource management in the EU: fishing and farming

The European Union's Common Fisheries Policy (CFP) and Common Agricultural Policy (CAP) are arguably the most sophisticated attempts to manage agriculture and fishing over a very large land area encompassing a steadily increasing number of countries. With the addition of ten new member countries in 2004 and another two in 2007, EU membership now stands at 27.

Both the CFP and the CAP have used various strategies in an attempt to balance supply and demand, with varying degrees of success over time, and both policies would claim sustainability as a major objective. However, most critics would argue that the CFP has done better in terms of sustainability than the CAP.

While decline in resources has been the main concern for fishing in the EU, a significant problem with EU agriculture has been overproduction and the environmental and economic consequences of this situation. The main concern of the CAP at first was growing enough food, but the generous grants and subsidies provided encouraged farmers to grow too much, resulting in 'food mountains' and 'wine lakes'. As a result, the cost of storing such surpluses became very high indeed.

A major management and planning strategy problem with both policies has been the difficulty in reaching an agreement, as each country has its own objectives and perceptions.

The Common Fisheries Policy (CFP): attempting to make resource development sustainable

Many of the world's fishing grounds are in crisis because of **overfishing**. In the worst affected areas it is feared that fish stocks may not recover for many years, if at all. The EU's Common Fisheries Policy is a significant example of resource management, although perceptions vary widely as to its effectiveness.

> **Key term**
>
> **Overfishing:** a level of fishing resulting in the depletion of the fish stock.

The 'tragedy of the commons' is a term used to explain what has happened in many fishing grounds including European waters. Because the seas and oceans have historically been viewed as common areas, open to everyone, the capacity of fishing vessels operating has exceeded the amount of fish available. The result has been resource depletion (Figure 4.41). To try to solve this problem countries have extended their territorial waters to 200 miles from their shores and instigated a variety of management techniques within these waters. Outside of these limits nations have sought to come to agreements with varying degrees of success.

Figure 4.42 'Factory' fishing vessels docked in Ravenna, Italy

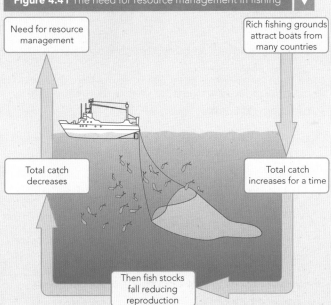

Figure 4.41 The need for resource management in fishing

> **Key term**
>
> **The tragedy of the commons:** the idea that common ownership of a resource leads to over-exploitation as some nations will always want to take more than other nations.

Technological advance has played a major role in overfishing. The impact of the modern generation of huge, hi-tech fishing vessels (Figure 4.42) has been so great that environmentalists have labelled them 'the strip miners of the sea'. Electronic navigation aids such as LORAN (Long-Range Navigation) and GPS (satellite positioning systems) have turned the Earth's water into a grid enabling vessels to locate within 10 metres fish breeding grounds. There is no comparison between modern day fishing and fishing pre-1950.

Large-scale commercial fishing can be extremely wasteful. Many fish are thrown back into the sea because they are damaged, unsaleable, the wrong species or too small. In many cases fish are thrown back because regulations demand it. If a vessel has a licence to catch only haddock then any other species caught in the nets must be thrown back. The rejected fish – called **bycatch** – amount to an estimated 27 million tonnes a year, more than 25 per cent of the total caught worldwide.

> **Key term**
>
> **Bycatch:** unwanted fish which are caught alongside the desired catch but thrown overboard after sorting.

The evolution of the Common Fisheries Policy

The first common measures relating to fishing date from 1970, setting regulations for access to fishing grounds, markets and structures. In 1976, member states followed the international convention and agreed to extend their rights to marine resources from 12 to 200 miles offshore. Then, after years of negotiation, the Common Fisheries Policy was introduced in 1983. It is the primary instrument for managing the fishing industry in the EU. The CFP regulates commercial, social and environmental aspects of the fishing industry within EU territorial waters. The policy has provisions for the following:

- subsidies
- marketing
- industry organisation
- external fisheries treaties
- enforcement of regulations.

The 1992 and 2002 CFP Reviews: new approaches to sustainable development

The 1992 reforms provided the basis for modernising the policy on the conservation and management of fishery resources. The regulation on the common organisation of the market was completely overhauled and a genuine structural policy introduced in the fisheries sector with the creation of the Financial Instrument for Fisheries Guidance (FIFG). This organisation, established in 1993, provides financial assistance for the restructuring of the EU fleet by providing aid for scrapping, exporting and converting fishing vessels, and for modernisation. Figure 4.43 shows the allocation of FIFG funding by measure.

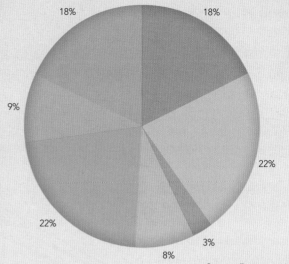

Permanent withdrawal of vessels
Renewal and modernisation of the fleet
Socio-economic measures
Fishing port facilities
Processing and marketing
Aquaculture
Other measures (inland fisheries, collective measures by professionals, promotion of products)

Figure 4.43 Allocation of FIFG funding 2000–06

In 2001, the EU Council on Fisheries produced a Green Paper which identified problems with both the content and enforcement of the CFP. It sought advice from interested parties in order to revise the policy. Following detailed discussion, new measures were adopted in December 2002 which came into force on 1 January 2003: The new measures included:

- taking a long-term approach by fixing **total allowable catches** (TACs) on the basis of fish stocks
- introducing accompanying conservation measures
- setting recovery plans for stocks below the safe biological limits
- managing the introduction of new vessels and the scrapping of old vessels in such a way as to reduce the overall capacity of the EU fleet
- EU aid for the modernisaton of vessels finishing at the end of 2004
- neutralising the socio-economic consequences of fleet reduction
- encouraging the development of sustainable aquaculture.

Total Allowable Catches (TACs)

Fish stocks need to renew themselves due to losses from both fishing and natural causes. It is therefore important that small fish be allowed to reach maturity so that they can reproduce. With this in mind the CFP sets the maximum quantity of fish or total allowable catch that can be caught each year. This is divided among the member states to give the national quota for each country. The number of small fish caught is limited by:

- minimum mesh sizes
- closure of certain areas to protect fish stocks
- the banning of certain types of fishing
- recording catches and landings in special log books.

Supporters of the CFP argue that it makes a strong contribution to the quest for sustainable fishing. However, environmentalists believe that short-term economic and political concerns override the objective of sustainability.

The Community Fisheries Control Agency

In March 2005, proposals were agreed for the establishment of a Community Fisheries Control Agency (CFCA). The new agency, which became operational in 2006, is located in Vigo, Spain. Its remit is to coordinate fisheries enforcement activity across EU member states, in particular the monitoring of and compliance with regulations on threatened stocks. The establishment of the CFCA stems from the European Commission's Road Map on the reform of the CFP published in May 2002.

Figure 4.44 shows the decline in the stock of North Sea cod. In contrast, the situation for herring is significantly better although stock levels have varied considerably over the last four decades. The impact of declining fish stocks on the UK fishing industry has been a fall in the number of fishermen from 18 000 in 1997 to 13 000 in 2006.

The CFP remains controversial. Most of the interested parties seem dissatisfied. Fishing operators frequently view the TACs as being too severe. On the other hand, scientists and environmentalists see many TACs as being far too high and way above sustainable levels. The balance of the catch allowed to individual countries is also often a source of friction. The next decade will be a crucial one not only for EU fishing but also for the global industry. It is possible that in ten years' time many types of fish will be viewed as a luxury because of their scarcity and resultant high price.

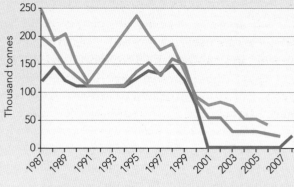

Source: Hickman, M. 'How do we balance conservation with the interests of the fishing industry?', *Independent* (2007)

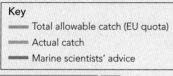

Figure 4.44 The decline in North Sea cod

Activities

1. Explain the main aspects of the CFP.
2. How has the CFP changed over time?
3. Describe the decline in North Sea cod shown in Figure 4.44.

The Common Agricultural Policy

The need to ensure a reliable and adequate supply of food in the post-Second World War period was one of the main reasons for the introduction of the Common Agricultural Policy in 1960 (Figure 4.45). The CAP is a set of rules and regulations governing agricultural activities in the EU. It is expensive to run with each EU taxpayer contributing about £80 a year to its existence.

In the late 1970s, agricultural subsidies, a major aspect of the CAP, accounted for more than 75 per cent of the entire EU budget. Although agriculture

Original aims	Evaluation
Improve production yields to guarantee farm supplies	The CAP is no longer needed to achieve this aim. Freely competitive agricultural markets and technological innovation in agriculture guarantee increased farm production and higher yields. In a competitive market system, farmers would have to produce efficiently to remain profitable in the long term.
Ensure a fair standard of living for EU farmers	There are major doubts as to the equity of the CAP in delivering this objective. There is wide and growing division between large-scale and small-scale farmers within the EU. The accession of new countries into the EU as part of enlargement has exposed the extent to which farm support affects farmers of different-sized land areas.
Stabilise agricultural markets	This original objective has been largely achieved, but at great economic and environmental cost. There is a limit to which any form of government intervention can and should seek market stability in terms of prices and incomes.
Ensure availability of farm supplies	Variations in farm output as a result of climatic differences are now much reduced by developments in agricultural technology and biotechnology. The globalisation of agricultural mass markets makes fears of food shortages less of an issue. Geo-political factors are not particularly relevant any more to the question of maintaining food supplies within the EU.
Ensure food supplies are available to consumers at reasonable prices	The CAP system has increased European food prices, not reduced them, leading to a long-term loss in economic welfare for consumers. Many economists believe that competitive market disciplines are the best route to achieve lower prices in the long term.

Figure 4.45 Original aims of the CAP

accounted for no more than 5 per cent of total employment of the 15 member EU countries at the beginning of 2004, it still used up 40 per cent of the EU's total budget. However, recent reforms have marked an important change in direction for the CAP as more and more people have questioned the relevance of its original aims to modern-day Europe.

Changes have been made to the CAP a number of times since 1960. Prior to the 2003 reforms the most important changes have been:

- the McSharry Reforms of 1992
- Agenda 2000 Reforms.

The 2003 CAP Reforms: a more sustainable approach

The latest set of reforms to the Common Agricultural Policy was agreed in June 2003.

- The most important change has been to break the link between subsidy and the level of production ('decoupling') by linking payment to the area of land farmed as opposed to the quantity produced. A major objective is to encourage farmers to produce food for the market (that is, to respond to real demand from consumers) rather than produce for the subsidies available.

- Decoupled payments will be made where there is 'cross-compliance' – when certain conditions have been met with regard to the environment, food safety and animal welfare.

- By removing the incentive for farmers to maximise production, the environmental impact of farming should be reduced.

- More subsidy will also be diverted from production to wider rural development and environmental initiatives.

- The reforms simplify the CAP and so cut the paperwork farmers face by rolling a range of subsidies into a new single payment.

- Member countries have been given some discretion in how they implement these reforms.

It has always been difficult to get agreement about the CAP between the different EU countries. In general, those countries where farms are larger and farming is more efficient have wanted to reduce CAP subsidies.

Britain, Sweden and the Netherlands have led the reform camp in the EU. On the other hand, countries with a less efficient agriculture have received a great deal of money from the CAP. To date, France has been the main beneficiary from the CAP and so it is not surprising that this country was least in favour of the changes.

Of CAP funds, 70 per cent goes to only 20 per cent of EU farms. These are mainly large farms. Nearly three-quarters of EU farmers survive on total revenue of less than £5000 a year. Small farmers make up about 40 per cent of EU farms but receive only 8 per cent of EU subsidies. The new reforms should reduce payments to large farms as they usually achieve the highest yields.

Ten new members join EU in 2004

The expansion of the EU into Eastern Europe posed considerable problems for the CAP. The new member countries have a greater dependency on farming as a source of income and employment. Agriculture in the new member countries accounts for over 13 per cent of employment compared to just over 4.2 per cent in the original member nations.

Farmers in the new EU nations had hoped that the level of existing subsidies would apply to them also. However, it was clear that the existing system of subsidies would be unaffordable with the enlargement of the EU from 15 to 25 countries. As a result, the CAP budget was frozen at 40 billion euros and it is set to remain at this level until 2013. The new EU nations qualified for 25 per cent of the level of subsidies given to established member farmers in 2004, 30 per cent in 2005, 35 per cent in 2006, reaching 100 per cent in 2013.

Under the terms of the Copenhagen agreement, the new EU members can pay their new CAP money to their farmers either as conventional arable area aid or livestock herdage premiums, or they may operate a 'single area payment scheme' for the first three years of membership. This involves dividing all of a country's aid over its entire agricultural area to produce a single flat-rate payment. Eight of the ten new countries are adopting this approach.

Figure 4.46 Large-scale farming in the EU

The CAP and the World Trade Organization

The CAP has come under severe criticism in the World Trade Organization from two main sources.

- LEDCs, who complain that (a) the EU's Common External Tariff denies them fair access to the large EU food market and (b) excess production in the EU is 'dumped' in LEDCs undermining their fragile food markets.

- The Cairn Group of agricultural exporting nations who are incensed by the dumping of food surpluses in their markets at below cost price, causing significant problems for their farmers. The Cairn Group was formed in 1986 to lobby for freer trade in agricultural products. Its members include Argentina, Brazil, Canada, New Zealand, Australia, the Philippines and South Africa.

Figure 4.47 Small-scale farming in the EU

Activities

1. With the help of Figure 4.45, discuss the extent to which the CAP has met its original aims.
2. a Discuss the main reforms introduced in June 2003.
   b Why has it always been difficult to get the EU countries to agree on reforms to the CAP?
   c Why are LEDCs in particular very critical of the CAP?
3. a Make a list of the countries which joined the EU in 2004.
   b Why will the new member countries receive lower CAP subsidies than those which have been available in the past?

Take it further activity 4.4 on CD-ROM

Knowledge check

1. Describe how the rate of global population growth has changed over time.
2. Identify and explain current global contrasts in population growth.
3. Discuss the ways in which resources can be defined and classified.
4. Suggest how changes in society may result in changes in the definition of resources.
5. With reference to examples, outline the physical factors that determine the supply and use of resources.
6. Examine the socio-economic factors that influence the supply and use of resources.
7. Why does the demand for resources vary over time?
8. Explain why the demand for resources varies with location.
9. Discuss the strategies that can be used to manage the demand and supply of resources.
10. To what extent can the EU's Common Fisheries Policy be regarded as a sustainable development policy?

Exam Café
Relax, refresh, result!

Relax and prepare

What I wish I had known at the start of the year…

Archie

"Remember that the demographic transition model is based on the European experience of mortality and fertility. It is often used to forecast future population trends in LEDCs. This is risky as cultures differ, not all European countries followed the model, there was less government influence in the past and accurate data was limited. Above all – why should a model that worked in the 1920s and 30s be applicable in the very different world of today?"

Pramod

"Sustainability is a difficult concept when looking at most resources, especially minerals, as there is a finite supply so true sustainability is difficult. This raises the issue of recycling which increases sustainability, but it is rare for recycling to be 100 per cent and some is always lost in the recycling process. Perfect sustainability is probably impossible but it is more a question of how we make them more sustainable than at present."

Student tips

Elena

"Try to study a range of resources. The specification asks for 'two including one non-energy resource', which means one can be an energy resource such as oil or coal. The other is best chosen from the semi-renewable resources, such as fish or forests, but it is also useful to have a renewable one such as water or the landscape scenery. This will ensure you can develop a range of discussion points across the variety of types of resources."

Common mistakes – Korin

It is easy to make the error of seeing resources as finite or fixed. The whole concept of resources is dynamic with the pressure of changing demand and prices encouraging the development of new sources or alternatives. Who knows what will be harnessed next as a resource. It is too easy to take a gloomy Malthusian view of resources seeing them as a limiting factor on population numbers and standards of living.

Don't just think of population as the demand side of the population/resource relationship. Population is itself a resource – it could be in terms of quantity or in terms of quality, its skill, education, culture, etc. Some countries have based prosperous economies on little else than their hardworking and skilled populations, such as Singapore.

Refresh your memory

4.1 Population number and rate of growth varies over time and space	
Natural increase	Birth rate – death rate
Net migration	In-migration – out-migration, push versus pull factors
Factors:	
Demographic	Age structure, sex ratio, infant death rate, ethnicity
Social	Health, socio-economic status, culture, status of women, religion, education and literacy, level of overcrowding, social services, age of marriage, alternate attractions
Economic	Income, employment type, infrastructure, standard of living, cost of children, income from children, housing, transport availability
Political	Policies, e.g. birth control; laws, e.g. women's status; tax policy, e.g. family allowance; indirect, e.g. education, old age pensions, inheritance
DTM model	High stationary – high BR/DR; Early expanding – high BR, DR falling; Late expanding – BR falling, DR low; Low stationary – low BR/DR; Low declining – low BR, rising DR (ageing population)

4.2 How can resources be classified?	
By source	Organic versus inorganic; extraterrestrial versus terrestrial; domestic versus imports
By form	Liquid, solid, gaseous, visual, etc.
By use	High value versus low value, stores versus flows, vital versus non-vital
By level of renewability	Renewable, non-renewable, semi-renewable
Resources change as:	Technology changes, e.g. uranium not a resource in eighteenth century Population grows in size, e.g. land Price or cost changes Education increases, e.g. scenery now seen as a resource

4.3 Factors affecting the supply of a resource	
Physical	Climate – type, extremes; Geology – rock and mineral type, structure, faults, hardness; Water – quantity and quality, seasonal variation; Soil – type, depth, fertility, drainage; Biotics – plants, animals, etc.; Distance – remoteness
Human	Technology – equipment, processes, detection, sustainability; Capital – quantity, type, e.g. fixed versus operating, long versus short term; Transport – type, bulk carrying capacity, cost, direction, time; Demography – workforce, technical know-how; Energy – type, volume, cost, consistency; Alternatives – competition, bi-products, role of imports, recycling; Agriculture – type, productivity, intensity
Others	Conservation and protection, cost of making good, pollution

Exam Café

Refresh your memory

4.4 Factors affecting use of/demand for resources

The nature of the resource	Cost/price – key factor Quality of the resource – purity, waste content, is it scattered? Quantity of the resource – volume, weight, life expectancy Bi-products – does it come with valuable bi-products, e.g. cotton comes with fibre and an oil seed? Importance – is it vital, how dependent are we on it, e.g. oil? Location – is it near or remote, e.g. Alaskan oil? Substitutes – are their possible substitutes, e.g. aluminium can replace copper? Flexibility – how many uses does it have? Ease of transport – bulk, perishability, hazards, cost
The nature of the area	Technology – ease of extraction, processing, etc. Conditions – do the physical conditions aid or hinder extraction? Demography – is there a local labour force? Standard of living – basics or necessities are demanded first. Existing type of agriculture and its type/level of output. Culture and tradition – may conflict with extraction. Its accessibility – core versus periphery

4.5 What is the link between population and resources?

Population versus resources	Overpopulation, under population – dynamic concepts as technology changes
Malthus	Population will increase until it exceeds resources then positive checks, e.g. famine, disease cause population to crash to below resources
Boserup	Population increase puts pressure on resources so triggering innovation and resource development
Carrying capacity	The number of people that can live at a given standard of living in any given environment and beyond which any increase in population causes a decrease in that standard

4.6 In what ways does human activity attempt to manage demand and supply to achieve a sustainable system?

Sustainability	Can this be achieved with growing demand and finite resources?
Management	May be governmental, individual or corporate, e.g. TNCs. Free markets find such controls difficult but totalitarian states have few problems
Why bother?	Help control prices, for health or security, avoid exhaustion of resource, reduce overdependence, protect trade balance
Demand	Rationing, quotas – on basis of need, own criteria Pricing – raise the price to reduce demand, people can choose Substitutes – cheaper often poorer quality or from different source Bans – difficult to enforce Advertising – encourages voluntary actions, types of media
Supply	Import controls – via import tax, quotas but may face retaliation Subsidies – reduces costs but paid by tax payer Price/cost controls – price fixing agreements but legal problems Rationing/quotas – on basis of need, own criteria

Exam Café

Get the result!

Example question

Study Figure 1, showing population structure in a LEDC. Suggest the possible issue(s) indicated and appropriate strategies that could be used to manage its impact. [10 marks]

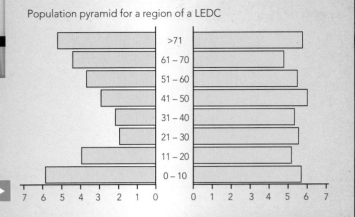

Figure 1

Student answer

This is an unstable pyramid. There is a lack of males, especially in the active age group, probably due to out-migration to find work. This leaves a predominantly female population and a high dependency ratio. This suggests a stagnant community with little economic growth, probably reliant on money sent back by migrant workers. Clearly the region needs some form of economic activity to reduce out-migration — possibly tourism or development of the local infrastructure such as a dam for power and irrigation. Such schemes are expensive so some form of assisted self-help might be more effective, such as a credit union, craft school, etc. Increased economic activity would increase the tax base so helping to support the dependent population, and increased wealth and jobs should reduce the birth rate.

Examiner says

Good use of the data with an attempt at interpretation. A number of issues are identified and well linked in cause-effect.

Examiner says

Appropriate strategies offered with some detail. Some evaluation and an appreciation of the inter-linkages lift this answer to an A grade.

Examiner's tips

A very useful approach to any evaluation is to adopt a cost–benefit approach. As long as the benefits outweigh the costs then the scheme should go ahead. It is not quite as simple as that and resource development is a good example of the complexity. Crucial is who decides what benefits and costs are included and how a value is put on them, e.g. how to quantify the loss of a lovely view or the impact and noise of heavy lorries on habitats. All too often the term 'in the national interest' is used as an indication of overwhelming benefit to the nation as a whole, so cost/benefit balances may look very different when local, regional and national views are compared. Even individuals will vary in their estimation of cost/benefit as shown by NIMBYISM (not in my back yard). So when you use this approach, remember its limitations and most of the evaluation becomes straightforward.

Exam Café

Chapter 5
Globalisation

Globalisation has brought about a restructuring of economics and politics on a huge scale which may prove to be the most historically significant event since the Industrial Revolution. However, critics argue that this process is occurring with little regard to the environment, local economics, democracy and human welfare.

Questions for investigation

- What is meant by the term 'globalisation' and why is it occurring?
- What are the issues associated with globalisation?
- What are transnational corporations (TNCs) and what is their contribution to the countries in which they operate?
- How far do international trade and aid influence global patterns of production?
- How can governments evaluate and manage the impact of globalisation?

Consider this

Globalisation has the potential to bring considerable benefits to all parts of the world. However, the benefits have so far been very unequally shared, with poor people in both developing and developed countries largely excluded from the advantages brought about by globalisation.

5.1 What is meant by the term 'globalisation' and why is this occurring?

What is globalisation?

Globalisation is the increasing interconnection and interdependence of the world's economic, cultural and political systems. There are many aspects of globalisation which are summarised in Figure 5.1.

Figure 5.2 shows Peter Dicken's view of the global economy. Transnational corporations and nation states are the two major decision makers. Nation states individually and collectively set the rules for the global economy but the bulk of investment is through TNCs, who are the main drivers of global shift. It is this process which has resulted in the emergence of an increasing number of Newly Industrialised Countries (NICs) since the 1960s.

In Asia, four generations of NICs can be recognised in terms of the timing of industrial development and their current economic characteristics. Within this region, only Japan is at a higher economic level than the NICs (Figure 5.4) but there are a number of countries at much lower levels of economic development. The latter form the least developed countries in the region.

Figure 5.1 The dimensions of globalisation

Dimension	Characteristics
Economic	Under the auspices of first GATT (General Agreement on Tariffs and Trade) and latterly WTO (World Trade Organization), world trade has expanded rapidly. Transnational corporations have been the major force in the process of increasing economic interdependence, and the emergence of different generations of newly industrialised countries has been the main evidence of success in the global economy. However, the frequency of 'anti-capitalist' demonstrations in recent years shows that many people have grave concerns about the direction the global economy is taking. Many LEDCs and a significant number of regions within MEDCs feel excluded from the benefits of globalisation.
Urban	A hierarchy of global cities has emerged to act as the command centres of the global economy. New York, London and Tokyo are at the highest level of this hierarchy. Competition within and between the different levels of the global urban hierarchy is intensifying.
Social/cultural	Western culture has diffused to all parts of the world to a considerable degree through TV, cinema, the Internet, newspapers and magazines. The international interest in brand name clothes and shoes, fast food and branded soft drinks and beers, pop music and major sports stars has never been greater. However, cultural transmission is not a one-way process. The popularity of Islam has increased considerably in many Western countries as has Asian, Latin American and African cuisine.
Linguistic	English has clearly emerged as the working language of the 'global village'. Of the 1.9 billion English speakers, some 1.5 billion people around the world speak English as a second language. In a number of countries there is great concern about the future of the native language.
Political	The power of nation states has been diminished in many parts of the world as countries organise themselves into trade blocs. The European Union is the most advanced model for this process of integration, taking on many of the powers that were once the sole preserve of its member nation states. The United Nations has intervened militarily in an increasing number of countries in recent times, leading some writers to talk about the gradual movement to 'world government'. On the other side of the coin is the growth of global terrorism.
Demographic	The desire of people to move across international borders has increased considerably in recent decades. More and more communities are becoming multicultural in nature.
Environmental	Increasingly, economic activity in one country has impacted on the environment in other nations. The long-range transportation of airborne pollutants is the most obvious evidence of this process. The global environmental conferences in Rio de Janeiro (1992) and Johannesburg (2002) are evidence that most countries see the scale of the problems as so large that only coordinated international action can bring realistic solutions.

> **Key terms**
>
> **Global shift:** the large-scale filter-down of economic activity from MEDCs to NICs and LEDCs.
>
> **Newly Industrialised Countries:** nations that have undergone rapid and successful industrialisation since the 1960s.
>
> **Transnational corporation:** corporation which has the ability to organise and control operations in more than one country, even if it does not own them.

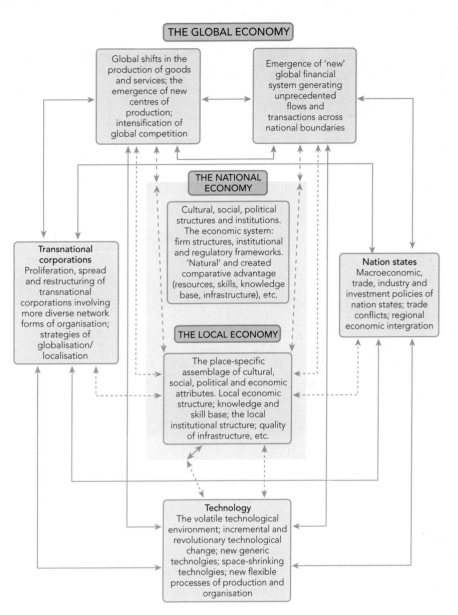

Figure 5.2 The global economy

Nowhere else in the world is the filter-down concept of industrial location better illustrated than in Asia. When Japanese companies first decided to locate abroad in the quest for cheap labour, they looked to the most developed of their neighbouring countries, particularly South Korea and Taiwan. Most other countries in the region lacked the physical infrastructure and skill levels required by Japanese companies. Companies from elsewhere in the developed world, especially the USA, also recognised the advantages of locating branch plants in such countries. As the economies of the first generation NICs developed, the level of wages increased resulting in:

- Japanese and Western TNCs seeking locations in second generation NICs where improvements in physical and human infrastructures now satisfied their demands but where wages remained low
- indigenous companies from the first generation NICs also moving routine tasks to their cheaper labour neighbours, such as Malaysia and Thailand.

With time, the process also included the third generation NICs, a significant factor in the recent very high growth rates in China and India. The least developed countries in the region, nearly all hindered by recent conflict of one sort or another, are now beginning to be drawn into the system. The recent high level of FDI (foreign direct investment) into Vietnam makes it reasonable to think of the country as an example of a fourth generation Asian NIC.

Globalisation is about worldwide markets in all economic sectors, not just raw materials and manufacturing but also the tertiary and quaternary sectors, for example call centres and tourism.

Figure 5.3 Car assembly plant in South Korea, a NIC

Figure 5.4 Levels of economic development in Asia

Level	Countries	GNP Per Capita 2005 (US$)
1	Japan: a MEDC	38 984
2	First generation NICs, e.g. Taiwan	16 764
3	Second generation NICs, e.g. Malaysia	4963
4	Third generation NICs, e.g. China	1735
5	Fourth generation NICs, e.g. Vietnam	623
6	Least developed countries, e.g. Mongolia	380

The development of globalisation

Globalisation is a recent phenomenon (post-1960) and very different from anything the world had previously experienced. It developed out of internationalisation.

Key terms

Foreign direct investment: overseas investments in physical capital by transnational corporations.

Internationalisation: the extension of economic activities across national boundaries. It is essentially a quantitative process which leads to a more extensive geographical pattern of economic activity. The phase precedes globalisation.

A key period in the process of internationalisation occurred between 1870 and 1914 when:

- transport and communications networks expanded rapidly
- world trade grew significantly with a considerable increase in the level of interdependence between rich and poor nations
- there were very large flows of capital from European companies to other parts of the world.

International trade tripled between 1870 and 1913. At this time the world trading system was dominated and organised by four nations: Britain, France, Germany and the USA. However, the global shocks of the First World War and the Great Depression put a stop to this period of phenomenal economic growth. It was not until the 1950s that international interdependence was back on track.

Since the 1950s, world trade has grown consistently faster than world GDP. However, even by 1990 the level was unremarkable compared with the late-nineteenth and early-twentieth centuries. It is therefore not surprising that some writers argue that the level of integration before 1914 was similar to that of today.

However, today's globalisation is very different from the global relationships of 50 or a 100 years ago. Peter Dicken makes the distinction between the 'shallow integration' of the pre-1914 period and the 'deep integration' of the present period. The global economy is more extensive and complicated than it has ever been before.

Discussion point

What is the difference between the phases of internationalisation and globalisation?

Economic globalisation

The factors responsible for economic globalisation

These include:

- until the post-1950 period the production process itself was mainly organised within national economies; this has changed rapidly in the last fifty years or so with the emergence of a new international division of labour (NIDL) reflecting a change in the geographical pattern of specialisation, with the fragmentation of many production processes across national boundaries
- the increasing complexity of international trade flows as this process has developed
- major advances in trade liberalisation under the World Trade Organization – economic and legal barriers to world trade (tariffs, quotas and regulations) are much lower today than in the past
- the emergence of fundamentalist free-market governments in the USA (Ronald Reagan) and Britain (Margaret Thatcher) around 1980 – the economic policies developed by these governments influenced policy making in many other countries
- the emergence of an increasing number of Newly Industrialised Countries
- the integration of the old Soviet Union and its Eastern European communist satellites into the capitalist system (Figure 5.6); today, no significant group of countries remains outside the free-market global system
- the opening up of other economies, particularly those of China and India
- the deregulation of world financial markets allowing a much greater level of international competition in financial services
- the 'transport and communications revolution' that has made possible the management of the complicated networks of production and trade that exist today.

Figure 5.5 Financial district of New York – one of the big three global financial centres

Key terms

New international division of labour: this divides production into different skills and tasks that are spread across regions and countries rather than within a single company.

World Trade Organization: established in 1995 as a permanent organisation with far greater powers to arbitrate trade disputes than its predecessor (GATT).

Figure 5.6 The fall of the Berlin Wall – the beginning of the integration of Eastern Europe into the free market system

The advantages for economic activity in working at the global scale

These include:

- sourcing of raw materials and components on a global basis reduces costs
- TNCs can seek out the lowest cost locations for labour and other factors, for example energy
- high-volume production at low cost in countries such as China helps to reduce the rate of inflation in other countries and raise living standards
- collaborative arrangements with international partners can increase the efficiency of operations considerably
- selling goods and services to a global market allows TNCs to achieve very significant economies of scale
- global marketing helps to establish brands with huge appeal all around the world.

Which are the most globalised countries?

Figure 5.7 shows the top 20 globalised countries in 2004 according to the Centre for the Study of Globalisation and Regionalisation (CSGR), based at the University of Warwick. The economic, social and political aspects of globalisation are aggregated to produce the final list of 'overall globalisation'. Singapore tops the list with the UK in fourth place. Figure 5.8 shows the criteria used by the CSGR to produce their rankings. Four factors are used to measure economic globalisation, nine for social globalisation and three for political globalisation.

According to the CSGR's criteria, globalisation increased significantly from 1982 to 2004 (Figure 5.9), although there was considerable variation between the three aspects.

In terms of the strength of overall globalisation by world region, North America ranks first followed by Western Europe, then East Asia and the Pacific, then South Asia.

Ranking	Overall globalisation	Economic globalisation	Social globalisation	Political globalisation
1	Singapore	Luxembourg	Bermuda	France
2	Belgium	Netherlands Antilles	Singapore	USA
3	Canada	Singapore	Hong Kong, China	Russian Federation
4	UK	Hong Kong, China	Switzerland	China
5	US	Ireland	New Zealand	UK
6	Austria	Malaysia	Austria	Canada
7	Sweden	Belgium	Canada	Belgium
8	Switzerland	Guyana	Netherlands Antilles	Egypt, Arab Rep.
9	France	Swaziland	Sweden	Germany
10	Denmark	Thailand	Denmark	Italy
11	Ireland	Angola	UK	Sweden
12	Germany	Bahrain	Malta	Austria
13	Italy	Hungary	Iceland	India
14	Malaysia	Malta	Belgium	Poland
15	Finland	Philippines	Australia	Malaysia
16	Australia	Moldova	Finland	Pakistan
17	Netherlands	Estonia	Netherlands	Denmark
18	New Zealand	Ukraine	USA	Nigeria
19	Russian Federation	Congo, Rep.	Barbados	Jordan
20	Korea, Rep.	Cambodia	Ireland	Turkey

Source: Lockwood, B. and Redoano, M. *The CSGR Globalisation Index: An introductory guide*, Centre for the Study of Globalisation and Regionalisation Working Paper 155/04 (2005)

Figure 5.7 The top 20 globalised countries in 2004

Figure 5.8 The CSGR measures of globalisation

Sub-index	Variable	Definition
Economic globalisation	Trade	Exports plus imports of goods and services as a proportion of GDP
	Foreign Direct Investment (FDI)	Inflows plus outflows of foreign direct investments as a proportion of GDP
	Portfolio investment	Inflows plus outflows of portfolio investments as a proportion of GDP
	Income	Employee compensation paid to non-resident workers and investment income from foreign assets owned by domestic residents plus employee compensation paid to resident workers working abroad and investment income from domestic assets owned by foreign residents, as a proportion of GDP
Social globalisation	*People*:	
	Foreign stock	Stock of foreign population as proportion of total population
	Foreign flow	Inflows of foreign population as proportion of total population
	Worker remittances	Worker remittances (receipts) as a proportion of GDP
	Tourists	Number of tourists (arrivals plus departures) as proportion of total population
	Ideas:	
	Phone calls	International outgoing telephone traffic (minutes) per capita
	Internet users	Internet users as a percentage of population
	Films	Number of films imported and exported
	Books and newspapers	Sum of value of books and newspapers imported and exported per capita (US dollars)
	Mail	Number of international letters delivered and sent per capita
Political globalisation	Embassies	Number of foreign embassies in country
	UN missions	Number of UN peacekeeping operations in which country participates
	Organisations	Number of memberships of international organisations

Source: Lockwood, B. and Redoano, M. *The CSGR Globalisation Index: An introductory guide*, Centre for the Study of Globalisation and Regionalisation Working Paper 155/04 (2005)

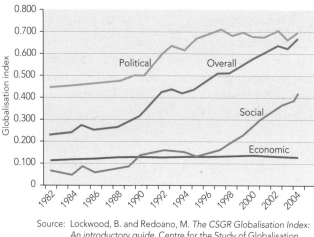

Source: Lockwood, B. and Redoano, M. *The CSGR Globalisation Index: An introductory guide*, Centre for the Study of Globalisation and Regionalisation Working Paper 155/04 (2005)

Figure 5.9 The development of globalisation 1982–2004

Which are the major global economies?

Figure 5.10 shows the relative size of the top 20 global economies according to the traditional measure of gross domestic product (GDP). However, organisations such as the UN and the International Monetary Fund are tending to publish GDP data at purchasing power parity (PPP). Once differences in the local purchasing power of currencies are taken into account, China's economy moves firmly into second place behind the USA (Figure 5.11). The other major emerging economy whose relative importance increases significantly once output is measured on a PPP basis is India.

Cultural globalisation

Culture can be defined as the total of the inherited ideas, beliefs, values and knowledge which constitutes the shared bases of social action. Culture is a major aspect of peoples' lives, and outside influences may have a very significant impact on traditions that have developed over many years.

The mixing of cultures is a major dimension of globalisation. This has occurred through:

- migration, which circulates ideas, values and beliefs around the world
- the rapid spread of news, ideas and fashions through the mass media, trade and travel
- the growth of global brands such as Coca-Cola and McDonald's, that serve as common

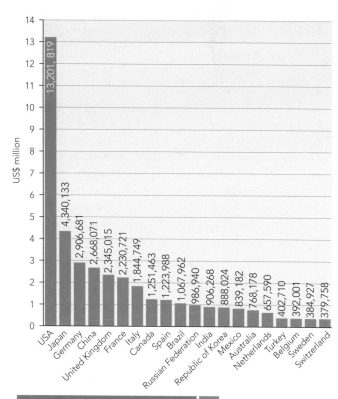

Figure 5.10 Total GDP for the top 20 global economies, 2006

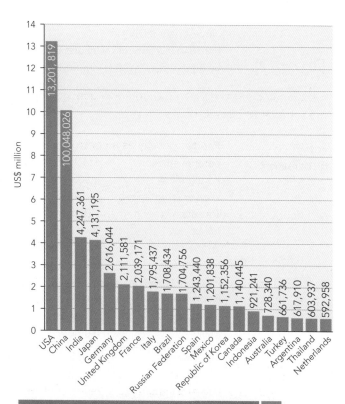

Figure 5.11 Purchasing power parity (PPP) GDP for the top 20 global economies, 2006

reference points (the terms 'Americanisation' and 'McDonaldisation' are often used to describe global consumer culture)

- the Internet has allowed unparalleled individual and mass communication
- the transport revolution.

Cultural hybridity is the term used to describe the extent to which cultures are intermixed. This process has been important to the success of many TNCs. The power of brands and their global marketing strategies cannot be underestimated. The increasing knowledge of Western consumer culture in the former Soviet Union and Eastern Europe in the Cold War was an important factor in the eventual disintegration of the Eastern Bloc.

However, in spite of the significant impact of globalisation on cultures around the world, it is important not to exaggerate the changes that have taken place. Considerable cultural differences remain, with some writers using terms such as 'culture wars' and 'clashes of civilisations'.

Discussion point

What is the evidence of the mixing of different cultures in the area where you live?

Possible future trends

Frequently cited are the following consequences of advanced globalisation:

- the elimination of geography as a controlling variable in the global economy
- the reduction in influence of the nation state
- economic synchronisation across the globe
- companies with no specific territorial location or national identity
- the disappearance of distinctions between LEDCs and MEDCs as structures of wealth and poverty become detached from territory
- English as the common public language of the globalised system.

Most people feel that it is inevitable that globalisation will continue to develop, but this cannot be taken for granted. Although there has been a strong movement towards a 'single system' in recent decades, the degree of conflict around the world emphasises that there is a lack of agreement on what shape the single system should take in the future. Fewer people are now prepared to leave it to governments and international economic organisations to decide. Many see the growing influence of global civil society as a major factor in countering the negative aspects of globalisation. The general message coming from this disparate array of individuals and organisations is that the starting point is to fundamentally change the way in which the global economy is organised.

The increasing pace of globalisation in recent decades has been paralleled by the rise of religious fundamentalism. Many writers see the latter as a reaction to the former. In the West the term 'fundamentalism' invariably means Islamic fundamentalism. The fundamentalist revival in Islam that began in the 1970s is, rightly or wrongly, a major concern to many people and governments in the West. This is because this 'aggressive' faction of Islam largely rejects Western modernisation and secularism. The call is for 'Islamisation' whereby:

- Sharia law replaces secular law
- education centres on the Koran
- the economic system is oriented to redistribution rather than the accumulation of wealth by individuals
- cultural products (TV, music, etc.) are tightly controlled.

However, the influence of fundamentalism on other religions should not be underestimated. The development of the New Christian Right in the USA has had a significant influence on attitudes and policy-making in that country, for example presidents Reagan, Bush and Bush.

Key terms

Global civil society: encompasses a range of organisations and individuals who are challenging the way globalisation operates. Their aim is to 'civilise' globalisation, making the process more democratic at all levels.

Religious fundamentalism: movements favouring strict observance of religious teaching, for example Christianity, Hinduism and Islam.

Activities

1. Define the terms (a) globalisation and (b) global shift.
2. Study Figure 5.2.
   a List the main components of the global economy.
   b Explain the relationships between them.
3. Briefly outline the main elements of:
   a economic globalisation
   b social globalisation
   c political globalisation.
4. Why are NICs considered to be the success story of globalisation?
5. What is:
   a the new international division of labour
   b cultural hybridity?

 Take it further activity 5.1 on CD-ROM

5.2 What are the issues associated with globalisation?

Globalisation of economic activity may bring advantages and disadvantages to various areas. These impacts may be environmental, economic, social or political.

Those who have major concerns about globalisation include:

- trade unionists and others worried about jobs being lost in their own country and moved to lower wage economies
- environmentalists who say that TNCs are disregarding the environment in the rush for profits and market share
- those fearful of the erosion of national sovereignty and culture
- people who are concerned about the power of big business
- small businesses afraid that they will become the victims of global economies of scale
- poverty campaigners who say that in the globalisation process the West's gain has been at the expense of developing countries
- those against the apparent control that a relatively small number of countries have over the major international organisations.

Globalisation and the environment

Virtually all aspects of globalisation impact on the environment. While the emergence of NICs is generally accepted as the success story of globalisation, the rapid industrialisation of China and India in particular is putting a huge strain on the Earth/atmosphere system. Environmentalists argue that the economics of globalisation is concerned primarily with internal costs, largely ignoring external costs such as environmental impact. While companies make profits, society has to pay the bill. There is a great deal of evidence that the planet's ecological health is in trouble. Between 1950 and 2000,

Figure 5.12 Destruction of Indonesian rainforests to satisfy huge demand for timber, in particular from China

humankind consumed more of the world's natural capital than during its entire previous history.

In many countries there is a shortage of good agricultural land and increasing demands are being placed on limited water supplies. In some nations forests are disappearing at a rapid rate for timber (Figure 5.12) and in an attempt to extend agricultural land, though far too often forest soils quickly degrade with such a change in use. Globalisation has resulted in rapidly increasing volumes of freight and passenger transport, with all modes of movement having an adverse impact on the environment. Spending cuts enforced by the International Monetary Fund have reduced spending on the environment in a number of countries.

International migration

The world's population is more mobile than ever before. The recent high level of movement has reinforced some long-established trends but also created some new ones. Migration is a major economic, social and political issue in both countries of origin and destination. One in every 35 people around the world is living outside the country of their birth. This amounts to about 175 million people, higher than ever before.

Globalisation in all its aspects has led to an increased awareness of opportunities in other countries. With advances in transportation and communication and a reduction in the real cost of both, the world's population has never had a higher level of potential mobility. Also, in various ways, economic and social development has made people more mobile and created the conditions for emigration.

Financial speculation

Prior to the relatively recent deregulation of global financial markets, the activities of banks, insurance companies and investment dealers had been confined largely within national boundaries. With the removal of regulations the financial sector scanned the globe for the best returns. In this new relaxed atmosphere, finance capital became a destabilising influence on the global economy with an increasing level of speculation (short term) as opposed to 'firm' (long term) investment. Nervous short-term investors who withdraw capital at the first hint of a problem can cause a vicious downward spiral to a country's economy which can be

Figure 5.13 Trading floor of a stock market

very difficult for a government to counter (Figure 5.14). Critics of globalisation often use the term 'global casino' to refer to rapid movement of speculative money around the world.

The rise of global civil society

Another relatively recent phenomenon has been the growing importance of cross-border pressure groups such as Friends of the Earth and Amnesty International. These are elements of the global civil society. The combined actions of these organisations and others have resulted in a growing framework of international rules on trade, the environment, human rights, war and other aspects of international relationships. However, the process is not always easy.

Figure 5.14 East Asian financial crisis

In 1996, capital was flowing into East Asia at almost US$ 100 billion a year

↓

The main targets were risky real estate ventures and local stock markets

↓

This inflated share and property prices well beyond the value of their underlying assets

↓

As the perception of the problem widened, investigators panicked and rushed to withdraw money

↓

Having abandoned capital controls in the early 1990s, Asian governments were unable to halt the massive withdrawal of funds

↓

Bankruptcies, unemployment and political crises followed

Case study | The impact of globalisation on the UK

Globalisation has had a major impact on the UK (Figure 5.15) as its international trading links have spread and become more complex in recent decades. The UK is not only one of the leading countries in attracting foreign direct investment (FDI) but is also one of the largest investors in other countries. The changes brought about by globalisation have extended the nature and scope of FDI both into and out of the UK.

Globalisation has been a major factor in the changing nature of employment, speeding up the filter-down of traditional manufacturing industries to NICs and LEDCs. The resulting deindustrialisation has seen considerable job losses in the traditional industrial areas in particular. Only 15 per cent of employment remains in manufacturing although output has increased by 13 per cent since the early 1990s. According to the Local Government Association, in the UK more people now work in Indian restaurants than in shipbuilding, steel manufacturing and coal-mining combined.

The UK has had to concentrate its efforts on the service sector and specialised high-value manufacturing to maintain a strong economy. Economic activity and the valuable employment that goes with it can now be moved around the world faster than ever before. The UK government is trying to ensure that the country's population has the skills to meet this challenge as international competition intensifies.

The business and financial sector of the economy has benefited hugely from the deregulation of global financial markets. This has enhanced London's position as a global city in economic terms. The success of the business and financial sector has resulted in very significant increases in wages at the 'top end'. In contrast, with high immigration and a resultant large pool of unskilled and semi-skilled labour, wages at the 'bottom end' have been kept down, resulting in widening social division.

The UK has become a global hub for migration. Record numbers of people are moving in to and out of the country and this process is bringing about significant social and cultural change. Many communities face considerable challenges in absorbing large numbers of migrants.

Key terms

Deindustrialisation: the long-term absolute decline of employment in manufacturing.

Global city: major world city supplying financial, business and other significant services to all parts of the world. The world's major stock markets and the headquarters of large TNCs are located in global cities.

Perspective	Benefits	Problems
Economic	• As one of the world's most 'open' economies, the UK attracts a very high level of foreign direct investment creating significant employment and contributing to GDP. • A high level of investment abroad by UK companies also increases national income. • Financial deregulation has enhanced the position of the City as one of the world's top three financial centres. • Low-cost manufactured products from China and elsewhere have helped keep inflation low.	• High job losses in traditional industries have resulted from global shift and deindustrialisation. • These job losses are concentrated geographically. • TNCs can move investment away from the UK as quickly as they can bring it in, causing loss of jobs and corporation tax. • Speculative investment (e.g. hedge funds), causing economic uncertainty, has increased with financial deregulation. • There is a widening gap between the highest and lowest paid workers.
Social	• Economic growth has facilitated high levels of spending on education and health in particular. • Globalisation is a large factor in the increasingly cosmopolitan nature of UK society. • The transport and communications revolution has transformed lifestyles.	• A strong economy has attracted a very high level of immigration in recent years, with increasing concerns that this is unsustainable.
Political	• Strong trading relationships with a large number of other countries brings political influence. • As a member of the EU, Britain can extend its influence to areas where it was not previously well represented.	• Voter apathy is linked to many people seeing a loss of political power to the EU and major TNCs. • International terrorism is a growing threat with increasing ethnic diversity, rapid transportation and more open borders.
Environmental	• Deindustrialisation has improved environmental conditions in many areas. • Increasing international cooperation to solve cross-border environmental issues gives a better chance of such problems being addressed.	• Population growth impacts on the environment with the increasing demand for land, water and other resources. • Increasing wealth has resulted in higher levels of transportation, particularly car ownership and air travel, leading to rising levels of pollution. • Rapid industrial growth in China and elsewhere impacts on the global environment, including the UK.

Figure 5.15 The UK: examples of the benefits and problems of globalisation

Consumer choice in terms of primary products, manufactured goods and services has never been greater. However, there is growing concern about the impact of such choice on the environment. The issue of 'food miles' (the distance food is transported to reach markets) has been a major discussion point in the UK recently, with supermarkets being urged to stock more local produce.

The transport revolution has had a major impact on the relative proximity of places. For example, Eurostar can take passengers 293 miles from London to Paris in 2 hours and 15 minutes, an hour less than the journey to Carlisle. The geographical diffusion of British tourists has extended considerably in recent decades as has the origin of tourists coming to the UK.

So many aspects of life in the UK have been influenced by globalisation. For example, no football league in the world illustrates the globalisation of sport better than the English premier league. An estimated 1 billion worldwide audience watched Arsenal play Manchester United in November 2007.

Activities

1. Discuss the positive impact of globalisation on the UK.
2. What problems has globalisation created for the UK?

Theory into practice

Identify the countries of origin of the manufactured goods in your house. Pool the results of your class to illustrate on a world map.

Case study | The impact of globalisation on China

The rapid growth of economies in the 1960s and 1970s as globalisation gained momentum was of major concern to China, which operated at arm's length to the wider global economy. However, in the late 1970s, China decided to end its relative economic isolation and try to imitate the export-led success of countries such as South Korea.

The global economy has provided huge opportunities for Chinese companies. The Chinese economy has been growing at an average of 9.2 per cent a year for three decades. Figure 5.17 shows how China narrowed the gap with the UK, and then overtook the UK in terms of total GDP.

Figure 5.16 illustrates some of the opportunities and problems associated with China's rapid integration into the global economy. The current scale of Chinese manufacturing is astounding. For example, China produces:

- half of the world's microwaves
- two-thirds of all shoes
- two-thirds of the world's photocopiers
- half of the world's clothes
- one-third of all mobile phones.

However, China is also growing rapidly as a market. In 2006 it overtook Japan as the world's second biggest car market. Volkswagen now sells more cars in China than in Germany.

However, rapid growth and increasing wages have also brought some economic concerns (Figure 5.16). China's success has led to increasing inflation and rising prices.

Figure 5.16 China: examples of the benefits and problems of globalisation

Perspective	Benefits	Problems
Economic	◆ A high level of foreign direct investment ◆ Huge increase in employment opportunities in manufacturing but also in services ◆ Increasing GDP and rising average wages ◆ Large surplus in trade balance ◆ Greater influence in global and regional economic forums	◆ Rising wages and other costs are causing some companies to look for locations that are lower cost than China ◆ The appreciating value of the currency increases the cost of living for the lower paid in particular ◆ International concerns over the safety of some Chinese-made products has affected demand for some companies
Social	◆ Economic growth, resulting in more money available to spend on education, health, housing and social services ◆ Improving working conditions including longer holidays ◆ Rising incomes provide increasing personal mobility ◆ More Chinese travelling abroad for tourism	◆ The world's largest ever rural-urban migration ◆ Ageing populations in peripheral rural areas ◆ Increasing regional inequality ◆ Rapidly expanding cities with social/cultural issues ◆ Concerns over working conditions in some regions/companies ◆ The erosion of traditional values
Political	◆ China's influence at the UN and around the world generally is increasing ◆ Rising international profile helped to secure the 2008 Olympic Games ◆ Economic growth has allowed higher military/defence spending	◆ With rising affluence and awareness of freedoms in democratic countries there could be increasing pressure on the Chinese government for democratic reform ◆ Neighbouring countries concerned about Chinese military strength ◆ China's large trade surplus is straining relationships with the USA and the EU
Environmental	◆ As national income rises, China will have more money to invest in environmental improvement, as MEDCs have done in the past	◆ Air, water and land pollution from the rapidly increasing number of factories and power stations ◆ The large and increasing demand for raw materials is damaging the environment in China and many other countries ◆ There are concerns over the impact of major projects such as the Three Gorges Dam

This has made life more difficult for those in China who have not benefited from economic growth.

As China's demand for raw materials continues to rise there is huge concern over the impact this will have on the price of oil, metals, timber and other natural resources. China's share of total world oil consumption increased from 3.8 per cent in 1992 to 6.1 per cent in 2000 and 8.6 per cent in 2006, and there are still only 20 cars per 1000 people in China compared with 950 per 1000 in the USA. As China closes the gap the resource and environmental impacts will be considerable.

Figure 5.19 A large Chinese factory in Beijing

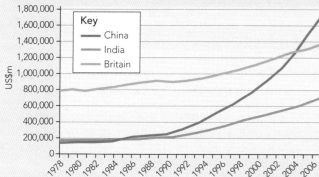

Figure 5.17 Total GDP (in constant US$): China, India and Britain

Source: Marcouse, I. 'China vs India: The world's biggest two-horse race?' *Business Review*, Sept 2007, Phillip Allan

As countries industrialise and impact increasingly on their environments, public concern grows and governments are forced to take action. Figure 5.18 shows what has happened in the past to MEDCs. Pollution in China is still increasing, but hopefully concerted action will cause pollution levels to fall or stabilise in the future. The scale of pollution in China has a global impact. China is already the world's second biggest producer of greenhouse gases, due largely to the burning of fossil fuels in industry and energy production (Figure 5.19).

After suffering a series of environmental problems due to deforestation, the Chinese government has acted to protect its own forests. However, one major side effect of this welcome policy has been to increase demand for imported forest products (see Figure 5.12). The conservation group WWF estimates that China imports more than 100 million cubic metres of wood a year with over a quarter of this total illegally felled in eastern Russia, Brazil, south-east Asia and Africa.

Activities

1. Describe the trends shown in Figure 5.17.
2. Why might some foreign TNCs decide to move investment away from China to other countries?
3. Explain the changing relationship between GDP and pollution shown in Figure 5.18.

Discussion point

Globalisation has brought about major economic change in China. Will significant political change also occur because of globalisation?

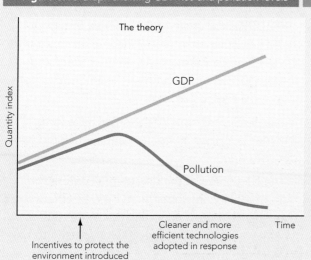

Figure 5.18 Graph showing GDP rise and pollution levels

Globalisation and the development gap

Global integration is spatially selective: some countries benefit while others, it seems, do not. A few developing countries have increased their trade and per capita incomes substantially, and overall, developing countries have increased their share of global output (Figure 5.20). However, on the other side of the coin are the 2 billion people who live in countries that have become less rather than more globalised as trade has fallen in relation to national income. This group includes most African countries. In these 'non-globalising' countries, income per person fell by an average of 1 per cent a year during the 1990s.

Figure 5.21 shows Human Development Index trends. The OECD (Organisation for Economic Cooperation and Development) is the grouping of the developed countries. The graph shows that according to this measure, most other world regions have narrowed the development gap. However, sub-Saharan Africa has made hardly any impact on the gap with the OECD over the 30-year period.

In contrast, Figure 5.22 shows the development gap in terms of GDP per capita. Under this measure Africa is again the least developed world region. Of the other regions, some have narrowed the gap in the post-1950 period but other regions have fallen further behind.

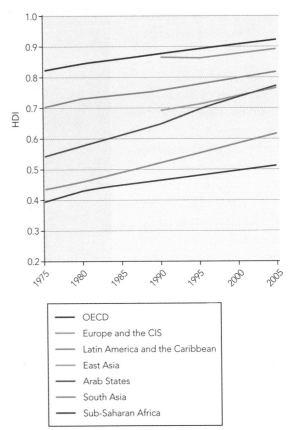

Source: Human Development Report, 2007–8

Figure 5.21 Human Development Index trends

The conditions of the world's poorest people are a serious cause for concern. For example:

- one in five of the world's population live on less than a dollar a day, and almost half on less than two dollars a day
- more than 850 million people in poor countries cannot read or write
- nearly a billion people do not have access to clean water and 2.4 billion to basic sanitation
- 11 million children under five years old die from preventable diseases each year
- in the 1990s the share of the poorest fifth of the world's population in global income dropped from 2.3 per cent to 1.4 per cent; on the other hand the proportion taken by the richest fifth rose from 70 per cent to 85 per cent
- in sub-Saharan Africa, 20 countries have lower incomes per head in real terms than they did two decades ago
- at the beginning of the nineteenth century, the ratio of real incomes per head between the world's richest and poorest countries was three to one; by 1900 it was ten to one; by 2000 it had risen to sixty to one.

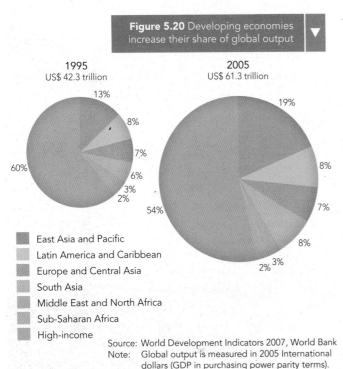

Figure 5.20 Developing economies increase their share of global output

Source: World Development Indicators 2007, World Bank
Note: Global output is measured in 2005 International dollars (GDP in purchasing power parity terms).

	GDP per capita				Ratio of GDP per capita to that of the developed world			
	1950	1973	1980	2001	1950	1973	1980	2001
Developed world	6298	13376	15257	22825				
Eastern Europe	2111	4988	5786	6027	0.34	0.37	0.38	0.26
Former USSR	2841	6059	6426	4626	0.45	0.45	0.42	0.20
Latin America	2506	4504	5412	5811	0.40	0.34	0.35	0.25
Asia	918	2049	2486	3998	0.15	0.15	0.16	0.18
China	439	839	1067	3583	0.07	0.06	0.07	0.16
India	619	853	938	1957	0.10	0.06	0.06	0.09
Japan	1921	11434	13428	20683	0.30	0.85	0.88	0.91
Africa	894	1410	1536	1489	0.14	0.11	0.10	0.07

Source: UN World Economic and Social Survey, 2006

Figure 5.22 Developing versus developed countries, 1950–2001

Least developed countries

While the NICs are an important part of the successful side of globalisation, the Least Developed Countries have largely been by-passed by the processes of wealth creation.

Key term

Least Developed Countries: the poorest and weakest economies in the developing world. LDCs are a subset of the LEDCs.

The NGO (Non-Governmental Organisations) Forum at the Brussels Conference in 2001 stated that 'Globalisation according to the free market model is making the rich richer and the poor poorer.' The Forum identified the following causal factors of this increasing gap between the world's wealthiest and poorest nations:

- the World Trade Organization has undermined the interests of LDCs

Figure 5.23 Extreme poverty in a LDC

- global official development assistance (ODA) has never reached the level of the UN Commitments of 0.7 per cent of GNP and 0.15 per cent to LDCs
- initiatives to cancel debt have advanced too slowly with too little effect: many LDCs spend 40 per cent of their GDP on debt servicing
- in many countries development has been held back or put into reverse by the impact of HIV/AIDS and conflicts.

As the gap between the richest and poorest countries in the world widens, LDCs are being increasingly marginalised in the world economy. Their share of world trade is declining and in many LDCs national debt now equals or exceeds GDP. Such a situation puts a stranglehold on all attempts to halt socio-economic decline.

Activities

1. Describe the trends shown in Figure 5.21.
2. Compare the information shown on the proportional circles in Figure 5.20.
3. Present a brief analysis of Figure 5.22.

5.3 What are transnational corporations and what is their contribution to the countries in which they operate?

Global expansion

Transnational corporations (TNCs) are the driving force behind economic globalisation. They are capitalist enterprises that engage in foreign direct investment and organise the production of goods and services in more than one country. As the rules regulating the movement of goods and investment have been relaxed, TNCs have extended their global reach. As the growth of foreign direct investment has expanded the sources and destinations of that investment have become more and more diverse. There are now few parts of the world where the direct or indirect influence of TNCs is not important. In some countries and regions their influence on the economy is huge. Apart from their direct ownership of productive activities, many TNCs are involved in a web of collaborative relationships with other companies across the globe. Such relationships have become more and more important as competition has become increasingly global in its extent.

TNCs have a substantial influence on the global economy in general and in the countries in which they choose to locate in particular. They play a major role in world trade in terms of what and where they buy and sell. A fair proportion of world trade is intra-firm, taking place within TNCs. The organisation of the car giants exemplifies intra-firm trade with engines, gearboxes and other key components produced in one country and exported for assembly elsewhere. The world's five largest TNCs by revenue in 2007 were Wal-Mart Stores, Exxon Mobil, Royal Dutch Shell, BP and General Motors, each with a revenue exceeding US$ 200 000 million.

The development of TNCs over time

Figure 5.24 shows the main stages in the historical evolution of TNCs. Although the first companies to produce outside their home nation did not emerge until the latter half of the nineteenth century, by 1914 US, British and mainland European firms were involved in substantial overseas manufacturing production. Prior to the First World War the UK was the major source of overseas investment, the pattern of which was firmly based on its empire. Between the wars, TNC manufacturing investment, particularly US investment, increased substantially. By 1939, the USA had become the main source of foreign investment in manufacturing. The USA was to become even more powerful in the global economy after the Second World War, for it was the only industrial power to emerge from the conflict stronger rather than weaker.

However, the USA does not dominate the global economy today in the way it did in the immediate post-war period. The reconstruction of the Japanese and German economies resulted in both countries playing a significant transnational role by the 1970s

Period	Type	Characteristics
1500–1800	Mercantile capitalism and colonialism	• Government-backed chartered companies
1800–75	Entrepreneurial and financial capitalism	• Early development of supplier and consumer markets • Infrastructure investment by finance houses
1875–1945	International capitalism	• Rapid growth of market seeking and resource-based investments
1945–60	Transnational capitalism	• FDI (foreign direct investment) dominated by USA • TNCs expand in size
1960–present	Globalising capitalism	• Expansion of European and Japanese FDI • Growth of inter-firm alliances, joint ventures and outsourcing

Figure 5.24 Stages in the evolution of TNCs

and this was to expand considerably in the following decades. In fact, the large Japanese TNCs were to become models for their international competitors as they revolutionised business organisation. Other developed countries such as Britain, France, Italy, the Netherlands, Switzerland, Sweden and Canada also played significant roles in the geographical spread of foreign direct investment. More recently, NICs such as South Korea and Taiwan have expanded their corporate reach, not just to lower wage economies but also into MEDCs.

Foreign direct investment

The share of developing countries in the global stock of inward direct investment is 32 per cent. However, this is very unevenly distributed – 60 per cent is in Asia, with 38 per cent in Latin American and just 2.3 per cent in Africa.

The phenomenal growth in foreign direct investment is the most obvious sign of the increasing integration of the world's economies and much of this investment is by transnational corporations. Although most FDI is in the developed world, investment in developing economies has also risen substantially in the 1980s and 1990s. Of the emerging economies, the big five in order of importance, all with an FDI stock of over US$ 50 billion, are China, Brazil, Mexico, Singapore and Indonesia. The UN expects the growth in FDI to continue as more governments are liberalising their investment rules to attract FDI in the quest for capital and growth.

Activities

1. What is FDI?
2. Use Figure 5.24 to explain how TNCs have developed over time.

Theory into practice

Produce a fact file on one of the nearest outlets of a TNC to your school.

Contrasting spatial and organisational structures

TNCs vary widely in their overall size and international scope. Variations include:

- the number of countries in which they operate
- the number of subsidiary companies they own
- the share of production accounted for by foreign activities
- the degree to which ownership and management are internationalised
- the division of research activities and routine tasks by country
- the balance of advantages and disadvantages to the countries in which they operate.

Large TNCs often exhibit three organisational levels: headquarters; research and development; and branch plants. The headquarters of a TNC will generally be in the developed world city where the company was established. Research and development will most likely be located here too, or in other areas within the same country. It is the branch plants that are the first to be located overseas. However, some of the largest and most successful TNCs have divided their industrial empires into world regions, each with research and development facilities and a high level of independent decision making. Figure 5.25 shows the locational changes that tend to occur as TNCs develop over time.

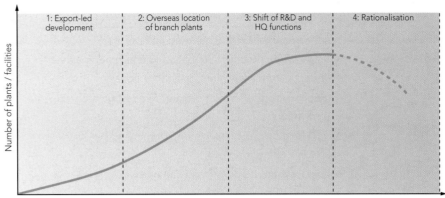

Figure 5.26 shows four simplified models which illustrate major ways of organising the geography of TNC production units. Toyota and the other global car manufacturers are closest to model 'C', a system often referred to as a horizontal organisational structure. In contrast, Nike is a good example of model 'D', illustrating a vertical organisational structure. However, Nike is not integrated in the traditional sense, in that it does not own the various stages of production because it subcontracts the manufacturing stages of its product range.

Figure 5.25 The development of TNCs: locational changes

Figure 5.26 Five ways of organising the geography of TNC production units

Another distinction can be made between producer-driven and buyer-driven production chains. In a producer-driven chain, TNCs play a central role in controlling the production system; in a buyer-driven chain, the pivotal role is played by large retailers, brand-named merchandisers and trading companies. Toyota can be classed as an example of a producer-driven production chain and Nike as a buyer-driven production chain.

a) Globally concentrated production

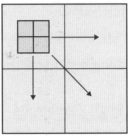

All production at a single location. Products are exported to world markets.

b) Host-market production

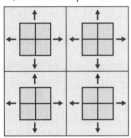

Each production unit produces a range of products and serves the national market in which it is located. No sales across national boundaries. Individual plant size limited by the size of the national market.

c) Product specialisation for a global or regional market

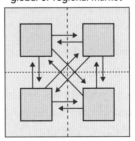

Each production unit produces only one product for sale throughout a regional market of several countries. Individual plant size very large because of scale economies offered by the large regional market.

d) Transnational vertical integration

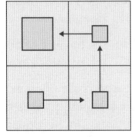

Each production unit performs a separate part of a production sequence. Units are linked across national boundaries in a 'chain-like' sequence – the output of one plant is the input of the next plant.

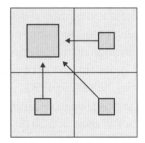

Each production unit performs a separate operation in a production process and ships its output to a final assembly plant in another country.

Source: Dicken, P. *Global Shift*, PCP (3rd edn, 2001)

Case study | Nike

Nike is the world's leading supplier of sports footwear, apparel and equipment and one of the best-known global brands. It was founded in 1972 and the company went public in 1980. The company is an example of a vertical organisational structure across international boundaries, characterised by a high level of subcontracting activity. Nike does not make any shoes or clothes itself, but contracts out production to South Korean and Taiwanese companies. Nike employs 650 000 contract workers in 700 factories worldwide. The company list includes 124 plants in China, 73 in Thailand, 35 in South Korea and 34 in Vietnam. More than 75 per cent of the workforce is based in Asia. The majority of workers are women under the age of 25.

The subcontracted companies operate not only in their home countries but also in lower wage Asian economies such as Vietnam, the Philippines and Indonesia. The company has a reputation for searching out cheap pools of labour. Nike's expertise is in design and marketing. Figure 5.27 shows Nike's 'commodity circuit'. It is a good example of the new international division of labour [NIDL].

Nike illustrates both 'Fordist' and 'flexible' characteristics. An example of its Fordist nature is the Air Max Penney basketball shoes consisting of 52 component parts from five different countries. The shoes passed through 120 people during production, on a clearly demarcated global production chain.

However, Nike also exhibits flexible characteristics. The company aims to produce new shoes on a regular basis to cater for niche markets.

To achieve this objective it utilises a just-in-time innovation structure, buying in necessary expertise at short notice. This involves short-term subcontracts, often allocated to firms based near to Nike's research and development headquarters near Beaverton in the state of Oregon, USA (Figure 5.28).

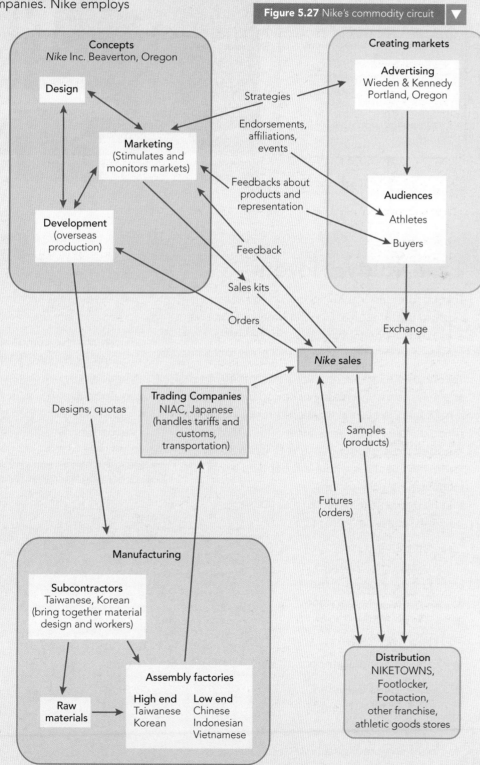

Figure 5.27 Nike's commodity circuit

Source: page 8 of *Nike Culture* by Robert Goldman and Stephen Papson, 1998, SAGE Publications Ltd

Key terms

Fordist: traditional manufacturing systems based on the techniques of mass production introduced by Henry Ford in the early twentieth century.

Flexible: modern manufacturing systems, originating in Japan in the 1960s and using new manufacturing management techniques including just-in-time technology. Flexible manufacturing makes a wider range of specialised products than traditional (Fordist) industries.

Figure 5.28 Nike headquarters in Oregon, USA

Case study | Toyota

Toyota, founded in 1937, is the world's second largest car manufacturer after General Motors. However, it is gaining fast on GM and plans to be the first car producer to sell more than 10 million vehicles in 2008.

Toyota is an example of a horizontal organisational structure across international boundaries because it has plants in several locations producing similar products, often with all or most of its suppliers grouped closely around it. Toyota has a policy of 'producing vehicles where the demand exists'. Its manufacturing philosophy is that its overseas operations become largely self-reliant.

'Lean' (flexible) manufacturing techniques were first developed in the 1950s by a Toyota manager called Taiichi Ohno. Lean production involves:

- carrying minimal stocks
- having parts delivered directly to the assembly line 'just in time'
- 'right first time' quality management
- seeing the factory as part of a supply chain with its suppliers upstream and its customers downstream.

Currently the company has 52 bases in 27 different countries (Figures 5.29 and 5.30). The world headquarters is located in Japan. The first Toyota built outside Japan was in April 1963 at Port Melbourne, Australia. A key element in the company's spatial expansion was the establishment of a growing

Figure 5.29 Toyota: location of manufacturing and assembly plants

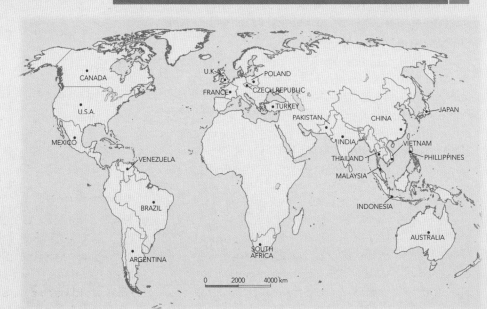

Figure 5.30 The Toyota factory at Burnaston, Derby, in the UK

presence in North America in the 1970s. In 1984, Toyota entered into a joint venture with General Motors called NUMMI, the 'New United Motor Manufacturing, Inc'. Toyota began to establish new brands at the end of the 1980s with the launch of Lexus in 1989. In the 1990s, the company continued to widen its range of products, particularly into the SUV (Sports Utility Vehicle), pickup and sports car markets. In 1997 it began production of the world's bestselling hybrid car, the Toyota Prius.

Toyota has shareholdings in Daihatsu Motors, Isuzu Motors and Yahama Motors. The Toyota Group also makes automatic looms for the textile industry and electric sewing machines.

The advantages and disadvantages of TNC operations

The operations of Nike and other companies who operate in, or outsource to, cheap labour economies have come under considerable criticism. Such criticism has not just been about the impact on poorer economies but also the effect of such TNC organisations on the headquarters' nation. However, there can be little doubt that the operations of TNCs also bring considerable benefits to both types of country. If this were not so then governments would be likely to place greater restrictions on the actions of TNCs. This would be possible despite the considerable influence TNCs have. It is also important to note that the operation of TNCs is at times very controversial and opinions may vary widely. What one expert may construe as on balance an advantage, another expert may view as a disadvantage. Figure 5.31 attempts to summarise the potential advantages and disadvantages of Nike to the USA, its headquarters'

location, and to Vietnam where 34 subcontracted Nike factories operate.

In response to years of 'sweatshop allegations', Nike produced a 108-page report in 2005 which gave the most comprehensive picture to date of the 700 factories producing its footwear and clothing. The report published the addresses of all contract factories and detailed admissions of abuses including:

- restricting access to toilets and drinking water
- denying workers at least one day off in seven
- forced overtime
- wages below the legal minimum
- verbal harassment.

The pressure group Human Rights First described the report as an important step forward in improving conditions for workers in poor countries. Phil Knight, the Chairman of the company, said he hoped Nike could become a global leader in corporate responsibility.

Country	Possible advantages	Possible disadvantages
USA: headquarters	• Positive employment impact and stimulus to the development of high-level skills in design, marketing and development in Beaverton, Oregon • Direct and indirect contribution to local and national tax base	• Another US firm that does not manufacture in home country – indirect loss of jobs and the negative impact on balance of payments as footwear is imported • Trade unions complain of an uneven playing field because of the big contrast in working conditions between LEDCs and MEDCs
Vietnam: outsourcing	• Creates substantial employment in Vietnam • Pays higher wages than local companies • Improves the skills base of the local population • The success of a global brand may attract other TNCs to Vietnam, setting off the process of cumulative causation • Exports are a positive contribution to the balance of payments • Sets new standards for indigenous companies • Contribution to local tax base helps pay for improvements to infrastructure	• Concerns over the exploitation of cheap labour and poor working conditions • Allegations of the use of child labour • Company image and advertising may help to undermine national culture • Concerns about the political influence of large TNCs • The knowledge that investment could be transferred quickly to lower cost locations

Figure 5.31 The possible advantages and disadvantages of Nike as a TNC

Activities

1. Produce a table to compare the spatial and organisational structures of Nike and Toyota.
2. Evaluate the advantages and disadvantages of Toyota to both the US and the LEDCs where its products are manufactured.

5.4 How far do international trade and aid influence global patterns of production?

Trade and aid both support and hinder the broader balance of the world's patterns of production.

The origin and continuing basis of global interdependence is trade. It is the most vital element in the operation of the global economy. World trade now accounts for 25 per cent of GDP, double its share in 1970.

The distribution of world trade

Figure 5.32 shows the spatial distribution of world trade in manufactured goods for the top 20 countries. Germany was the largest exporter of merchandise in 2005 with the USA in second place. However, the USA dominates imports by a huge margin, taking over one-fifth of the world total. The USA's trade deficit is something that worries many economists as it is a situation that cannot be maintained indefinitely.

In recent decades, trade in commercial services has increased considerably. However, in terms of total value it is still only about a quarter of that of merchandise trade. The USA is the largest importer and exporter. The UK ranks second in the export of services, ahead of Germany, France and Japan. However, Germany imports a greater value of services than the UK.

Figure 5.34 shows the share of world merchandise trade by world region for 2005. This shows the leading role of Europe as a whole for both imports and exports. In contrast, Africa has a very small share in merchandise trade.

Rank	Exporters	US$ billion	Share (%)	Annual percentage change	Rank	Importers	US$ billion	Share (%)	Annual percentage change
1	Germany	969.9	9.3	7	1	USA	1732.4	16.1	14
2	USA	904.4	8.7	10	2	Germany	773.8	7.2	8
3	China	762.0	7.3	28	3	China	660.0	6.1	18
4	Japan	594.9	5.7	5	4	Japan	514.9	4.8	13
5	France	460.2	4.4	2	5	United Kingdom	510.2	4.7	8
6	Netherlands	402.4	3.9	13	6	France	497.9	4.6	6
7	UK	382.8	3.7	10	7	Italy	379.8	3.5	7
8	Italy	367.2	3.5	4	8	Netherlands	359.1	3.3	12
9	Canada	359.4	3.4	14	9	Canada	319.7	3.0	15
10	Belgium	334.3	3.2	9	10	Belgium	318.7	3.0	12
11	Hong Kong, China	292.1	2.8	10	11	Hong Kong, China	300.2	2.8	10
	Domestic exports	20.1	0.2	0		Retained imports (a)	28.1	0.3	3
	Re-exports	272.1	2.6	11					
12	Korea, Republic of	284.4	2.7	12	12	Spain	278.8	2.6	8
13	Russian Federation	243.6	2.3	33	13	Korea, Republic of	261.2	2.4	16
14	Singapore	229.6	2.2	14	14	Mexico	231.7	2.1	12
	Domestic exports	124.5	1.2						
	Re-exports	105.1	1.0						
15	Mexico	213.7	2.0	14	15	Singapore	200.0	1.9	16
						Retained imports (a)	94.9	0.9	
16	Taipei, Chinese	197.8	1.9	8	16	Taipei, Chinese	182.6	1.7	8
17	Spain	187.2	1.8	2	17	India	134.8	1.3	39
18	Saudi Arabia (c)	181.4	1.7	44	18	Switzerland	126.5	1.2	9
19	Malaysia	140.9	1.4	11	19	Austria	126.2	1.2	5
20	Switzerland	130.9	1.3	7	20	Russian Federation (b)	125.3	1.2	29

(a) Retained imports are defined as imports less re-exports.
(b) Imports are valued f.o.b.
(c) Secretariat estimates
Source: World Trade Organization

Figure 5.32 Leading exporters and importers in world merchandise trade, 2005

Figure 5.33 A major container port in Singapore

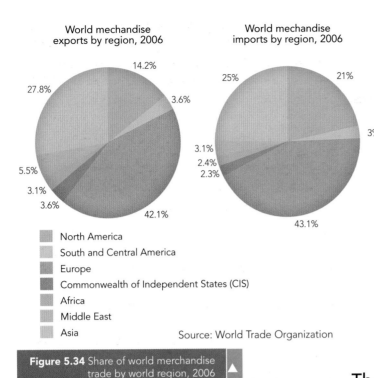

Figure 5.34 Share of world merchandise trade by world region, 2006

Trade and development

There is a strong relationship between trade and economic development. In general, countries which have a high level of trade are richer than those which have lower levels of trade. Figure 5.35 shows the value in US dollars of exports of goods and commercial services per capita for 2003 and provides good evidence for this assertion. Countries which can produce goods and services in demand elsewhere in the world will benefit from strong inflows of foreign currency and from the employment its industries provide. Foreign currency allows a country to purchase goods and services abroad that it does not produce itself or produces in insufficient quantities.

An Oxfam report, published in April 2002, stated that if Africa increased its share of world trade by just 1 per cent it would earn an additional £49 billion a year – five times the amount it receives in aid. The World Bank has acknowledged that the benefits of globalisation are barely being passed on to sub-Saharan Africa and may actually have accentuated many of its problems.

The ways in which trade can hinder production and development

The ways in which world trade works can cause significant development problems for many poorer countries. Non-governmental organisations, such as Christian Aid and Oxfam, argue strongly that trade is the key to real development, being worth 20 times as much as aid. However, Africa's trading situation will only

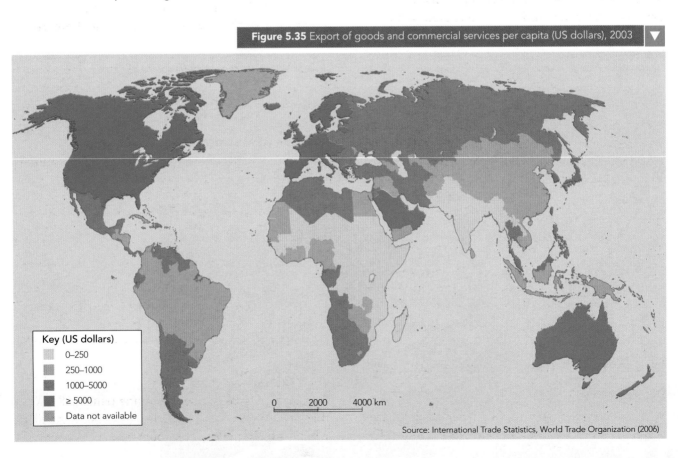

Figure 5.35 Export of goods and commercial services per capita (US dollars), 2003

improve if the trading relationship between MEDCs and LEDCs is made fairer and it can benefit from the resulting increase in trade. In fact, Africa's share of world trade has fallen in recent decades. According to Oxfam, if sub-Saharan Africa had maintained its exports at the same level as 1980, its economy would be worth an extra US$ 280 billion a year.

IMF-World Bank loans are usually conditional on African and other poor countries opening their markets. Although the situation varies across the continent, some countries such as Mali and Zambia are more open to trade than the EU and the USA. However, many countries complain that MEDCs, in particular the EU and the USA, are not implementing at home the free trade policies they expect African countries to follow. The high level of agricultural subsidies in the USA and the EU is a particular cause of concern, resulting in artificially cheap food flooding African markets.

The most vital element in the trade of any country is the terms on which it takes place. If countries rely on the export of commodities which are low in price and need to import manufactured items which are relatively high in price, they need to export in large quantities to be able to afford a relatively low volume of imports. Many poor nations are primary product dependent – that is, they rely on one or a small number of primary products to obtain foreign currency through export. The world market price of primary products in general is low compared to manufactured goods and services. Also, the price of primary products is subject to considerable variation from year to year, making economic and social planning extremely difficult. In contrast, the manufacturing and service exports of the developed nations generally rise in price at a reasonably predictable rate resulting in a more regular income and less uncertainty for the rich countries of the world. The terms of trade for many developing countries are worse now than they were a decade ago. Therefore, it is not surprising that so many nations are struggling to get out of poverty.

> **Key term**
>
> **The terms of trade:** the price of a country's exports relative to the price of its imports and the changes that take place over time.

International trade negotiations and agreements

Poorer nations are particularly concerned about the way that the more powerful trade blocs operate. For example, LEDCs complain that the EU's Common External Tariff denies them fair access to the large EU market for food and other products. The Common External Tariff is an agreed tax on imports that all EU countries charge. LEDCs also complain that excess production of food in the EU is 'dumped' in LEDCs, undermining their fragile food markets. Excess production of food in the EU has been encouraged by generous subsidies to farmers under the EU's Common Agricultural Policy. Many countries outside the EU see such subsidies as unfair competition.

In contrast, many poor countries have been put under very strong pressure to reduce their own tariffs and open up their markets to foreign competition. However, the rapid removal of tariffs can have a significant impact on a nation's domestic industry. Since India was 'forced' to accelerate the opening up of its markets, food imports have quadrupled. As a result, prices for domestically produced food have fallen sharply, as have rural incomes.

In 1947, a group of 23 nations agreed to reduce tariffs on each other's exports under the General Agreement on Tariffs and Trade (GATT). This was the first multilateral accord to lower trade barriers since Napoleonic times.

The GATT was replaced in 1995 by the World Trade Organization (WTO). Unlike the loosely organised GATT, the WTO was set up as a permanent organisation with far greater powers to arbitrate trade disputes.

Although agreements have been difficult to broker at times, the overall success of GATT/WTO is undeniable: today average tariffs are only a tenth of what they were when GATT came into force and world trade has been increasing at a much faster rate than GDP. However, in some areas protectionism is still alive and well, particularly in clothing, textiles and agriculture. In principle, every nation has an equal vote in the WTO. In practice, the rich world may shut the poor world out in key negotiations (Figure 5.36).

> **Key term**
>
> **Protectionism:** the institution of policies (tariffs, quotas, regulations) which protect a country's industries against competition from cheap imports.

The WTO exists to promote free trade. The fundamental issue is; does free trade benefit all those concerned or is it a subtle way in which the rich nations exploit their poorer counterparts? Most critics

Figure 5.36 Delegates at the World Trade Organization headquarters in Geneva, July 2008

of free trade accept that it does generate wealth but they deny that all countries benefit from it. The non-governmental organisation Oxfam is a major critic of the way the present trading system operates. The main goals of its 'Make Trade Fair' campaign are given on its website; to access this, go to www.heinemann.co.uk/hotlinks, enter the express code 7627P and click on the relevant link.

Regional trade agreements have proliferated in the last decade. In 1990 there were less than 25; by 1998 there were more than 90. The most notable of these are the European Union, NAFTA in North America, ASEAN in Asia and Mercosur in Latin America.

Activities

1. Use Figures 5.32 and 5.34 to describe the distribution of world trade.
2. How does Figure 5.35 support the view that richer countries have generally higher levels of trade than poorer countries?

The structure, direction and impact of trade

Case study | The UK: a MEDC

Figure 5.37 shows trade statistics for the UK in 2006. Although the UK is classed as a post-industrial economy, manufacturing accounts for about 60 per cent of total exports.

For merchandise trade (goods), manufacturing is by far the most important import and export sector. The UK's North Sea oil and gas fields have allowed the country to be a net exporter of fuels and mining products in the past, but this is changing as North Sea energy production continues to decline. It is not surprising that the UK is a net importer of agricultural products as its climate limits the range of crops that can be grown. In most years the UK has a trade deficit in merchandise trade.

MERCHANDISE TRADE (primary and manufactured products)	Value	Annual percentage change		
	2006	2000–2006	2005	2006
Merchandise exports (million US$)	448 291	8	11	17
Merchandise imports (million US$)	619 385	10	9	21
	2006			2006
Share in world total exports	3.71	Share in world total imports		4.99

Breakdown in economy's total exports		Breakdown in economy's total imports	
By main commodity group		*By main commodity group*	
Agricultural products	5.3	Agricultural products	8.8
Fuel and mining products	13.0	Fuel and mining products	11.7
Manufactures	77.6	Manufactures	65.3
By main destination		*By main origin*	
1. European Union	61.8	1. European Union	50.2
2. USA	13.2	2. USA	8.0
3. Switzerland	1.8	3. China	6.1
4. Japan	1.7	4. Norway	4.4
5. Canada	1.6	5. Japan	2.4

COMMERCIAL SERVICES TRADE	Value	Annual percentage change		
	2006	2000–2006	2005	2006
Commercial services exports (million US$)	227 529	11	5	11
Commercial services imports (million US$)	171 957	10	10	8
	2006			2006
Share in world total exports	8.26	Share in world total imports		6.49

Breakdown in economy's total exports		Breakdown in economy's total imports	
By principal services item		*By principal services item*	
Transportation	13.5	Transportation	21.1
Travel	14.8	Travel	36.5
Other commercial services	71.7	Other commercial services	42.4

Source: World Trade Organization

Figure 5.37 Trade statistics for the UK

The European Union is by far the most important destination for UK merchandise exports, with the USA in second place. The EU also dominates imports into the UK, but not to such a great extent. Again, the USA is in second place. However, China has narrowed the gap in recent years. The position of Norway in third place is due largely to energy imports.

Although the value of the UK's trade in services is considerably less than for goods, the gap has narrowed over time. In contrast to goods, the UK runs a trade surplus with regard to services. This helps to pay for the deficit in merchandise trade. The UK accounts for over 8 per cent of the world's services exports. The City of London plays a vital role here in terms of banking, insurance and other financial services.

Case study | China: a NIC

China's trading position (Figure 5.38) is very different from that of the UK. Merchandise is much more dominant for both total exports and imports, with services being of a much lower relative value in both cases. For merchandise exports, manufacturing accounts for over 92 per cent of the total, well above the level of merchandise imports. However, China's large and increasing demand for raw materials accounts for a relatively high import figure for fuels and mining products. The USA is China's main merchandise export market, followed closely by the EU. However, Japan is top of the import list.

China only accounts for 3.3 per cent of world exports in services, with a marginally higher figure for imports. These levels are much lower than for the UK. In terms

Figure 5.38 Trade statistics for China

MERCHANDISE TRADE (primary and manufactured products)	Value	Annual percentage change		
	2006	2000–2006	2005	2006
Merchandise exports (million US$)	968 936	25	28	27
Merchandise imports (million US$)	791 461	23	18	20
	2006			2006
Share in world total exports	8.02	Share in world total imports		6.38

Breakdown in economy's total exports		Breakdown in economy's total imports	
By main commodity group		*By main commodity group*	
Agricultural products	3.4	Agricultural products	6.5
Fuels and mining products	4.0	Fuels and mining products	20.0
Manufactures	92.4	Manufactures	73.2
By main destination		*By main origin*	
1. USA	21.0	1. Japan	14.6
2. European Union (25)	18.8	2. European Union (25)	11.4
3. Hong Kong, China	16.0	3. Korea, republic of	11.3
4. Japan	9.5	4. Taipei, Chinese	11.0
5. Korea, Republic of	4.6	5. China	9.3

COMMERCIAL SERVICES TRADE	Value	Annual percentage change		
	2006	2000–2006	2005	2006
Commercial services exports (million US$)	91 421	20	19	24
Commercial services imports (million US$)	100 327	19	16	21
	2006			2006
Share in world total exports	3.32	Share in world total imports		3.79

Breakdown in economy's total exports		Breakdown in economy's total imports	
By principal services item		*By principal services item*	
Transportation	23.0	Transportation	34.3
Travel	37.1	Travel	24.2
Other commercial services	39.9	Other commercial services	41.5

Source: World Trade Organization

of the composition of services, the transportation sector is of much greater relative importance to both exports and imports when compared with the UK.

In late 2007, China overtook Germany to become the world's top exporter. China now accounts for 8 per cent of global exports. Almost half of this total is made up of machinery, equipment and cars. For some time the EU, the USA and other countries have been calling on China to revalue its currency, the yuan. They argue that the value of the yuan is set too low, making Chinese exports so cheap that other countries struggle to compete. The EU estimated that its trade deficit with China may be more than US$ 220 billion in 2007. Some US politicians have argued for a substantial import tariff on Chinese goods entering the USA.

The key year for China's export success story was 2001, when China joined the World Trade Organization. By 2006, China's trade surplus had risen to an incredible US$ 250 billion.

Case study | Bolivia: a LEDC

Comparable data for Bolivia is shown in Figure 5.40. The poorest country in South America has a much lower level of international trade compared with more affluent countries of a similar population. Bolivia's leading merchandise exports are natural gas, tin, zinc, coffee, silver, wood, gold, jewellery and soybeans. Manufacturing plays a very minor role in exports. In contrast, manufactures dominate imports, a familiar story with virtually all LEDCs. The major imports are machinery and transportation equipment, consumer products, and construction and mining equipment.

Brazil, the dominant economy in South America, is now Bolivia's major trade partner; it used to be the USA. However, the USA features in the top three for both imports and exports of goods. About 30 per cent of Bolivia's exports to the USA qualify for preferential tariff treatment under the 2002 Andean Trade Promotion and Drug Eradication Act, which was intended to promote economic alternatives to the drug trade.

Bolivia is a member of the Andean Community, a trade bloc whose other members are Colombia, Peru and Ecuador. Bolivia is also an associate member of Mercosur (other members are Brazil, Argentina, Paraguay, Uruguay); this was an important fact in the development of its trade with Brazil in particular, as membership provides for significant trading advantages.

Figure 5.39 Pipeline taking Bolivian natural gas to Brazil

Case Study: Bolivia: a LEDC

MERCHANDISE TRADE (primary and manufactured products)	Value	Annual percentage change		
	2006	2000–2006	2005	2006
Merchandise exports (million US$)	3863	21	30	39
Merchandise imports (million US$)	2819	7	27	20

	2006		2006
Share in world total exports	0.03	Share in world total imports	0.02

Breakdown in economy's total exports		Breakdown in economy's total imports	
By main commodity group		*By main commodity group*	
Agricultural products	17.3	Agricultural products	10.6
Fuels and mining products	71.5	Fuels and mining products	10.5
Manufactures	11.2	Manufactures	78.2
By main destination		*By main origin*	
Brazil	37.7	Brazil	20.4
USA	9.8	Argentina	15.8
Argentina	9.3	USA	12.1
Japan	8.9	European Union (25)	9.3
Peru	5.9	Chile	8.3

COMMERCIAL SERVICES TRADE	Value	Annual percentage change		
	2006	2000–2006	2005	2006
Commercial services exports (million US$)	419	12	18	–11
Commercial services imports (million US$)	787	10	13	19

	2006		2006
Share in world total exports	0.02	Share in world total imports	0.03

Breakdown in economy's total exports		Breakdown in economy's total imports	
By principal services item		*By principal services item*	
Transportation	26.6	Transportation	38.0
Travel	48.0	Travel	28.7
Other commercial services	25.4	Other commercial services	33.3

Source: World Trade Organization

Figure 5.40 Trade statistics for Bolivia

Activity

Compare the trading situations of the UK, China and Bolivia in terms of (a) merchandise trade and (b) commercial services trade.

Global patterns of aid

Figure 5.41 shows the global pattern of international aid. The different types of aid by source are:

- *bilateral aid*: aid supplied directly from one country to another taking the form of money, technology, training or other benefits

- *multilateral aid*: aid given by a number of different countries together directly or through international organisations such as the World Bank and International Monetary Fund

- *non-governmental aid*: aid given by charitable organisations such as Oxfam and CAFOD – these NGOs raise money through public donations and from government grants.

International aid can also be subdivided in terms of duration:

- *Short-term emergency aid* provides immediate help to cope with the impact of natural disasters such as earthquakes and tropical storms. Such aid can come from a wide variety of sources.

- *Long-term development aid* is international aid intended to promote more equitable global development by creating long-term sustainable economic growth in developing countries.

The willingness of LEDCs to accept foreign aid is based on three deficiencies:

- the 'foreign exchange gap' whereby many developing countries lack the hard currency to pay for imports which are vital to development

- the 'savings gap' where population pressures and other drains on expenditure prevent the accumulation of sufficient capital to invest in industry and infrastructure

- the 'technical gap' caused by a shortage of skilled personnel.

But why do richer nations give aid – is it down to altruism or self-interest? Much of the evidence suggests the latter. Contrary to popular belief, most foreign aid is not in the form of a grant, nor is famine relief a major component. In addition, a significant proportion of foreign aid is 'tied' to the purchase of goods and services from the donor country and often given for use only on jointly agreed projects. Tied aid accounts for around 40 per cent of donations from OECD countries. The OECD estimates that tying aid increases a developing country's costs by an average of 15 per cent.

> **Key term**
>
> **Tied aid:** bilateral aid in which the donor country specifies conditions relating to the way the money is spent. This often involves spending the money on goods and services from the donor country.

Figure 5.41 The global pattern of international aid

	Official development assistance (ODA) received (net disbursements)			
	Total (US$ millions)	Per capita (US$)	As % of GDP	
	2005	2005	1990	2005
Developing countries	86043.0	16.5	1.4	0.9
Least developed countries	25979.5	33.9	11.8	9.3
Arab states	29612.0	94.3	2.9	3.0
East Asia and the Pacific	9541.6	4.9	0.8	0.2
Latin America and the Caribbean	6249.5	11.3	0.5	0.3
South Asia	9937.5	6.3	1.2	0.8
Sub-Saharan Africa	30167.7	41.7	5.7	5.1
Central and Eastern Europe and the CIS	5299.4	13.1	–	0.3

Source: UN Human Development Report 2007/8

Short-term emergency aid

Emergency aid is usually in the form of food, clothes, medical supplies and shelter. NGOs usually play a leading role here as their expertise ensures that aid is directed to those most in need. Emergency aid is also used to help refugees fleeing from natural or human disasters. A recent example of short-term emergency aid followed the devastating effect of a tropical storm that hit Bangladesh in November 2007 (Figure 5.42).

> **Theory into practice**
>
> Which humanitarian disasters have your school or other community organisations contributed to in recent years?

Long-term development aid

According to some left-wing economists, aid is an obstacle to development because:

- of the tied nature of much aid, benefiting the donor more than the recipient in economic terms
- of the frequently inappropriate use of aid on large capital intensive projects which may actually worsen the conditions of the poorest people
- the strengthening of political ties as a result of bilateral aid may increase dependency and hinder democracy in the recipient country
- aid may delay the introduction of reforms, for example the substitution of food aid for land reform.

DFID provides emergency aid towards Bangladesh cyclone

4 December 2007

On 15 and 16 November, southern Bangladesh was hit by Cyclone Sidr. So far, over 6 million people have been affected and 2997 people have been confirmed dead. Many more have been injured, and the death toll could reach 10 000 (the death toll following the cyclone in 1991 was 140 000). Also, around 300 000 houses have been destroyed, as have many crops and large tracts of agricultural land.

Following an initial DFID contribution of £2.5 million, which is being channelled through the UN for immediate relief efforts, a further £2.5 million was pledged on 23 November. On 28 November, an additional £2 million was committed to help survivors to rebuild their homes and livelihoods. DFID has also provided 12 lightweight boats to reach inaccessible parts of Bangladesh, and despatched over 100 000 blankets for people made homeless.

Already DFID money is helping to rebuild more than 16 000 homes, provide food to 70 000 families and clean water to 260 000 families. The UK's disaster relief aid in Bangladesh now totals almost £12 million (US$24 million) for this year, with £4.7 million having been provided in response to the severe floods that occurred in August.

Secretary of State for International Development, Douglas Alexander, said yesterday:

'Unless emergency relief supplies get to victims it is all too likely that more people will die needlessly. That is why the UK continues to provide funds to get more food, clean water, basic shelter and other emergency supplies to tens of thousands of survivors. With half a million animals killed, nearly two million acres of crops and more than a million homes destroyed, the next challenge is to help people rebuild their homes and livelihoods. UK support is meeting immediate and longer term needs as well. I continue to be admiring of the resilience and determination of the Bangladeshi people as they face these challenges.'

Source: UK Department for International Development (DFID)

Figure 5.42 An example of short-term emergency aid

Figure 5.43 International aid arriving in the wake of Cyclone Sidr

Arguments put forward by the political right against aid are:

- aid encourages the growth of a larger than necessary public sector
- the private sector is 'crowded out' by aid funds
- aid distorts the structure of prices and incentives
- aid is often wasted on grandiose projects to raise the profile of political regimes which seem to have little concern for the poorest in their societies
- the West did not need aid to develop.

However, in its best form there can be little doubt that aid can combat poverty, although all too often it fails to reach the very poorest people and when it does the benefits are frequently short-lived. NGOs such as Oxfam and CAFOD have often been much better at directing aid towards sustainable development. The selective nature of such aid has targeted the poorest communities using appropriate technology and involving local people in decision making. Many development economists argue that there are two issues more important to LEDC development than aid: changing the terms of trade so that developing nations get a fairer share of the benefits and writing off Third World debt.

An example of long-term development aid is from the UK to Bangladesh; Figure 5.44 sums up the objectives of the UK's approach. The four largest donors to Bangladesh – Japan, the Asian Development Bank, the World Bank and the UK, have set up a joint strategy approach to coordinate their efforts.

Activities

1. Explain the difference between (a) bilateral and multilateral aid (b) short-term emergency aid and long-term development aid.
2. Discuss the UK's strategy in helping Bangladesh in the long term.

The UK's long-term goal is for Bangladesh to be a stable, prosperous and moderate Muslim majority democracy, playing a positive role in the global community. Our objectives are to embed democratic values; enable prosperity for all; and engender stability.

The UK's development programme is a significant part of the UK's relationship with Bangladesh. Over the past three years we have spent over £350 million and helped to:

- lift more than half a million people out of extreme poverty
- raise more than 20 000 flood-prone homesteads on Char islands above 1988 flood levels
- construct 14 000 new classrooms and recruit 12 000 new teachers
- provide basic education to 4.5 million children through a non-government programme
- ensure 14 million urban dwellers have access to basic health services
- enable more than 100 000 farmers gain improved access to markets.

The UK remains fully committed to working with the government and people of Bangladesh to support their economic, social and political reform ambitions. We will do this by helping to:

- build better governance
- reduce extreme poverty and vulnerability to climate change, and eliminate seasonal hunger
- increase jobs and incomes through private sector development
- improve the availability and quality of basic social services for the poor.

Source: UK Department for International Development (DFID)

Figure 5.44 Department for International Development – Policy on Bangladesh

Discussion point

Analyse the reasons why MEDCs give international aid to LEDCs.

5.5 How can governments evaluate and manage the impact of globalisation?

Redesigning the global economy

Critics of the way globalisation is proceeding argue for a number of significant changes to the global system, including:

- the establishment of a global central bank
- a revamping of the International Monetary Fund to make it more democratic
- a tax on international financial transactions to reduce speculation
- the establishment of a Global Environmental Organisation to monitor and reduce the impact of economic activity
- the control of capital for the public good.

The major objective is that the economically-powerful nation states and TNCs become more accountable and that all the impacts of economic activity are taken into account in the decision-making process. The goal must be to spread the benefits of globalisation more widely so that all peoples feel included in the global improvement in the quality of life.

Not all FDI is beneficial

Any serious analysis of FDI has to look at its targets. It is the quality of FDI rather than its quantity that brings net benefits to the receiving country. To bring such benefits FDI needs to be channelled into productive rather than speculative activities. The power of governments to influence the quality of investment has been steadily declining. It is a fact that a significant proportion of FDI is made up of companies:

- buying out state firms
- purchasing equity in local companies
- financing mergers or acquisitions.

Over the last decade or so a relatively new phenomenon has gathered pace – alliances of capital. This involves a great variety of negotiated arrangements: cross-licensing of technology among corporations from different countries; joint ventures; secondary sourcing; off-shore production of components; and cross-cutting equity ownership. All these actions can result in substantial benefits for the companies concerned, but may be of little benefit to the regions and countries where the investment occurs.

Sustainable development

Present levels of consumption are creating an unsustainable demand for many resources. As the world globalises, the effects of excess demand crosses borders and affects societies economically, socially and environmentally. There are two aspects to the problem:

- the impact of developed world consumption on the environment of developing countries
- the impact on the environment of developing countries acquiring developed world consumption habits.

Getting the global community to agree on a common course of action will become even more important in the future with the scale of 'spillovers' generated by a more interconnected world and global economy.

Case study | Bolivia – managing the impact of globalisation

Bolivia has recently introduced a resource nationalisation policy. Along with Cuba and Venezuela it forms the so-called 'radical block' of nations in Latin America which are concerned about US economic power in the region and the exploitative action of TNCs in general.

- **Key term**

  Resource nationalisation: when a country decides to place part or all of one or a number of natural resources under state ownership.

Bolivia (Figure 5.45) is South America's poorest country. In the 1980s and 1990s the Bolivian government introduced free market reforms. Such reforms were required by the World Bank if Bolivia was to continue to be granted aid. Privatisation was at the heart of this agenda. Investors, usually foreign, were allowed to acquire 50 per cent ownership and management control of public sectors such as electricity, telecommunications and the state oil corporation in return for an agreed level of capital investment. However, the Bolivian government also wanted to link growth with equity so that poorer people would gain more benefit from Bolivia's participation in the global economy. The measures to achieve this included:

- a type of decentralisation called Popular Participation
- education reform to improve access to opportunities for the poor.

Figure 5.45 Map of Bolivia

These two strategies were mainly targeted at improving the lives of the indigenous and mestizo (mixed race) populations. However, very limited progress with these objectives led to frequent changes in government due to public disquiet. In December 2005 Evo Morales became the country's first indigenous president. He was elected on a pledge to challenge the free market reforms that most people felt the country had been pressurised into adopting. There was widespread concern that these policies benefited large TNCs and the rich in Bolivia to the detriment of the poor and the country's environment.

In May 2006, President Morales nationalised the country's gas and oil industry. Bolivia has the second largest natural gas reserves in Latin America, but produces only a small amount of oil for domestic use. The foreign energy companies were given six months to sign new operating contracts or leave the country. All did so, which should result in higher revenues for the government. Currently, all foreign energy companies have to deliver all their production to the state-run YPFB for distribution and processing. Bolivia has taken control of 82 per cent of the oil and gas in the country, leaving the remainder to foreign companies.

Bolivia is adopting a socialist model of regional commerce and cooperation rather than what it sees as 'US-backed free trade'. It views the concept of the Free Trade Area of the Americas as an attempt by the US to 'annex' Latin America, and is concerned that the USA has too much economic and political influence in Latin America. The government is trying to attract foreign investment while also giving the state a larger role in managing the economy.

The privatisation of water has been a major issue. The resulting large increases in water bills provoked huge demonstrations, such as in Cochabamba, Bolivia's

Year	Overall ranking	Economic ranking	Social ranking	Political ranking
2004	64	111	108	51
2003	69	115	104	55
2002	65	112	102	57
2001	67	101	93	67
2000	62	100	89	64
1999	59	95	86	55
1998	85	95	87	102
1997	85	91	83	102
1996	83	93	74	104
1995	80	91	92	100
1994	74	86	85	97
1993	70	78	84	95
1992	67	74	79	88
1991	68	76	73	86
1990	61	71	71	85
1989	53	67	55	90
1988	51	69	55	89
1987		64		87
1986		61		88
1985		50		86
1984		33		82
1983		29		87
1982		32		88

Source: The University of Warwick

Figure 5.46 Rankings for Bolivia, 1982–2004

third largest city. The Bolivian government withdrew its water contract with Bechtel and its operating partner Abengoa. As a result, the companies sued the Bolivian government for US$ 50 million. However, in 2006 the companies agreed to abandon their legal action in return for a token payment.

The USA's drug war in South America has caused problems for Bolivia. The USA wants to end the production of coca and so reduce cocaine production to zero. Although Bolivia opposes the illegal drugs trade, it wants to preserve the legal market for coca leaves and promote the export of legal coca products. The reduction in the coca crop has hit the incomes of many poorer people.

The Bolivian government is concerned by the number of people moving abroad to find work. Limited employment in Bolivia is a major problem. In many cases children are left behind with no one to care for them. In 2007, the government announced it was to create 360 000 jobs by 2010 by attracting more FDI and increasing public sector employment.

The USA's actions in 'pressurising' Colombia and Peru into free trade agreements have damaged Bolivian exports to these countries. Bolivia's main farm export is soya beans and 60 per cent of this goes to Colombia. Bolivia is concerned that cheap, subsidised US food will undercut much of Bolivia's market in Colombia.

In April 2006, Bolivia signed the people's trade agreement with Venezuela and Cuba. The latter have agreed to take all of Bolivia's soya production as well as other farm products at market prices or better. Venezuela will also supply oil to Bolivia to meet domestic shortages in production. Cuba has agreed to supply doctors to improve healthcare in the country.

Like many countries which rely heavily on the export of raw materials to earn foreign currency, Bolivia has a sizeable foreign debt. At the end of 2006, Bolivia owed US$ 3.2 billion to foreign creditors.

In the latest year for which CSGR data is available (2004), Bolivia ranked 64 in the world in terms of overall globalisation. Between 1988 and 2004, Bolivia's ranking has varied between 51 and 85. Its economic and social rankings have slipped considerably in recent years. In contrast its political ranking has risen significantly.

Activities

1. a What is resource nationalisation?
   b Why has Bolivia pursued a policy of resource nationalisation?
   c How in the long term could this policy be (i) of benefit to Bolivia (ii) a disadvantage to Bolivia?
2. Why have Bolivia's recent policies to cope with globalisation created tension with the USA?

There can be little doubt that the process of globalisation has some way to go. When it will be complete, if ever, is the subject of much debate, as are the likely outcomes. Some see the process as offering enormous opportunities, while others are fearful of its extension.

Knowledge check

1. What is globalisation?
2. Discuss the dimensions of globalisation.
3. Explain the factors responsible for economic globalisation.
4. For a named MEDC, discuss the impact of globalisation.
5. For a named NIC, discuss the impact of globalisation.
6. What has happened to the development gap as globalisation has progressed?
7. What are TNCs and how have they developed over time?
8. With reference to at least one example, discuss the advantages and disadvantages of TNC operation.
9. Compare the structure, direction and impact of trade for a MEDC, NIC and LEDC.
10. Describe the different types of aid. How can aid enhance the development process in a LEDC?
11. How can governments manage the impact of globalisation?

Exam Café
Relax, refresh, result!

Relax and prepare

What I wish I had known at the start of the year…

Student tips

Koko

"Remember that globalisation is not a single process but an interlinked set of processes – economic, social and political. Many reinforce each other such as the spread/use of the Internet and the spread of the English language, but they may conflict or even operate in different directions or with differing impacts on differing areas."

Joe

"Remember that international trade of a country isn't just about the volume and type of imports and exports and the resulting trade balance. It is also important to be aware of the direction of trade (and major trading partners), terms of trade (relative value placed on exports) and the level and characteristics of the invisible trade such as banking and shipping."

Common mistakes – Viv

"I get very confused over trade. I understand direction of trade but I do confuse visible and invisible imports/exports. Its invisible that is really confusing but it is vital especially for MEDCs as it is the aspect that better balances their trade between exports and imports. Invisibles are broadly services, such that nothing tangible exchanges hands. In the UK, for example, foreign students paying to study in its universities or foreign tourists buying holiday accommodation in London count as invisible exports. Similarly, British students and visitors abroad are an invisible import. Other invisible trade items include banking and financial services, airlines, shipping and increasingly telecommunications including the Internet, teleworking and media. It is invisible trade that has grown the fastest in the last fifty years."

Refresh your memory

5.1 Factors causing globalisation

Physical	Differences in natural resources, e.g. minerals, climate, etc.
Economic	Transport – cheap sea and air travel, bulk carriers
	Trade – cheaper to import
	Financial markets – linked so easier to raise funds
	Comparative cost advantage – areas specialise
	Economies of scale – concentration in the hands of a few large-scale producers, usually TNCs
	Ease of transferring capital – international currencies, e.g. dollar
Social	Education and cultural – increased interest in other cultures
	Media and Internet – satellites mean instant communication anywhere
	Language – increased spread of certain languages (especially English, e.g. Internet) so easier communication
	Migration – ease of migration, professional mobility
Political	International bodies, e.g. World Bank, IMF
	Spread of democracy and capitalism – common links

5.2 Globalisation brings a variety of impacts

Benefits	Economic: fall in global poverty, greater income equality, cheaper goods, greater choice of goods, less child labour, more jobs
	Social: rise in life expectancy, reduction in hunger, increased literacy/education, cultural exchange, rise of feminism
	Political: more democracy, less centralisation
Problems	Economic: poorer countries exploited, fall in price of primary produce, low wages, local industries undercut, rise of part-time workers, financial 'meltdowns' more common
	Social: exploitation of workers, weaker trade unions, unskilled become unemployed, rise of underclass, culture swamped, westernisation/Sinoisation, increased illegal migration, brain drain
	Political: controlled by unelected corporate bodies, increased centralisation on main port/capital (core versus periphery)
	Environmental – increased pollution, global warming, destruction of environments, e.g. rainforest

5.3 Transnational companies bring a variety of impacts

Impact	Could be on home area, on the 'exploited' area, or on itself	
	Advantages	Disadvantages
Environmental	Conservation	Pollution, land clearance, loss of habitats
Economic	Jobs, wages, communications, infrastructure, multiplier, larger tax base, trade, cheaper goods	Exploits labour, low pay, destroys local industry, takes workers from primary, raises prices
Social	Education, training, health, pensions, company housing	Long hours, poor health, encourages migration to city
Political	Larger tax base, more international influence	Too much political influence, regional inequality

Refresh your memory

5.4 Trade influences patterns of production

Types	Visible imports, exports: Invisible – banking, tourism, shipping, etc.
Aspects	Trade balance, terms of trade
Trade influences	Reflects value, ease of transport, bulk, value added; enables areas to specialise
Global pattern	Heavy polluting industries to LEDC and NICs. Assembly needing cheap labour to NICs. Research and development in MEDCs. Invisibles in MEDCs but increasing in Anglophile NICs, e.g. India. Raw materials processed in MEDCs or at source depending on bulk
National pattern	Concentration at communication hubs, e.g. ports, airports. Location in major cities, capital – skilled labour, political factors. Accessible areas more favoured than remote – regional inequality

5.5 Aid both supports and hinders

Timing	Emergency versus long term; loans versus grants
Source	Multilateral versus bilateral; charities; companies; UN, etc.
Constituents	Financial, capital goods, military, experts, information, food, etc.
Relationship	Former colony, military pact, part of trade bloc, etc.
Support	'Seed corn' for multiplier effect; saves lives; slows regional and rural to urban migration; healthier and better skilled workforce; transfers knowledge without the need to discover/invent it (time); helps bring into world markets, trade, etc.; reduces expensive imports; creates jobs, raises wages, etc.
Hinders	Undercuts local industry/farming; high interest rates – drain on economy, debt burden; economic colonialism; hidden costs, e.g. spares, fuel, etc.; undue political influence; corruption – aid is diverted; increases regional inequalities

5.6 How can governments manage the impact of globalisation?

Measure	Governments must first measure its impact and on which areas/groups
Evaluate	Governments must then evaluate the size, type and duration of impact. Cost/benefit analysis. Set clear criteria or goals for judging impacts
Encourage	For example, UK: remove trade barriers; encourage international links – air, sea, Internet; reduce bureaucracy for foreign investors, open stock exchange, invest abroad; free market economy with no state intervention; tax harmonisation; open borders
Cherry pick	For example, Australia, Nigeria: selective trade barriers, e.g. on products that are made within the country; selective controls on migrants, e.g. by wealth, type of job, etc.; subsidies for selected industries/agriculture; minimum percentage local ownership, indigenous managers, etc.; nationalisation of key foreign investments
Discourage	For example, North Korea: erect trade barriers; state control/ownership, nationalisation; control media including Internet; control movement of people, close frontiers; withdraw from international treaties, trade blocs, etc.; subsidies for local industries – import substitution, become self-sufficient

Exam Café

Get the result!

Example question

Study Figure 1, which shows the location of the main activities of a TNC. Suggest the possible issues indicated and appropriate strategies that could be used to manage their impact. [10 marks]

Figure 1

Activity	Countries located	Number of employees
Research and development	UK, Germany, Netherlands	34 000
Assembly	Myanmar, Indonesia, Pakistan	121 000
Manufacturing	South Africa, Argentina, Poland, Hungary	56 000
Raw material extraction	Brazil, Mauritania, Australia	67 000

Student answer

This TNC exhibits both vertical integration and horizontal. This produces expensive duplication but also there is an over-reliance (nearly 50 per cent of the labour force) on politically unstable countries such as Myanmar for the assembly stage. This makes the whole production cycle vulnerable to disruption. Extra costs are incurred by assembling in countries different to, and far from, the manufacturing countries, such as Argentina. Rationalisation of assembly into fewer plants located nearer to manufacturing, for example South Africa which has abundant cheap labour suitable for assembly work, would help reduce transport costs and gain more security. By so concentrating, extra economies of scale can be gained as well as shortening the supply chain.

Examiner says
Good use of technical terms and clear reference to the data. Identification of a number of appropriate issues. Good to see appreciation of wider issues.

Examiner says
Range of appropriate strategies well justified. Good use of economic terms.

Examiner's tips

Globalisation can be viewed as creating winners and losers. This may be between nations but equally well within a country where larger centres become global hubs. In such cases it is useful to be aware of the core-periphery model and Friedman and Myrdal's model of cumulative causation. The latter suggests areas with some initial advantage – port, resources, capital, etc., grow faster than remoter, less advantaged areas. As these centres grow the multiplier effect operates that pulls in resources, investment and people from other areas (backwash) which further disadvantages non-core areas. Hence inequalities worsen – the core wins and the periphery loses. Eventually the model suggests development spreads out from the increasingly expensive core to the cheaper periphery. In reality this often needs a lot of political will to encourage such spread. This model can be applied within many countries, regions and even on a global scale.

»» Exam Café

Chapter 6

Development and inequalities

Accelerated development of the world is being driven by the need to support a rapidly growing population. It is also fuelled by the human wish for a higher standard of living. While development is recognised as bringing benefits, it also has its costs. The modern technology and enterprise driving development are doing so at the expense of the environment and to the detriment of the global stock of resources. Development is also creating a very unequal world of the 'haves' and the 'have-nots', 'winners' and 'losers', 'rich' and 'poor'. The symptoms of this 'two-class' world are all too evident in such things as employment and housing, as well as in the quality of life and the environment. So development confronts us with a major challenge: can we reduce the economic, social and environmental inequalities and move nearer to a situation of fair shares for all? Or is development inherently a divisive process that we are powerless to change?

Questions for investigation

- In what ways do countries vary in their levels of economic development and quality of life?
- Why do levels of economic development vary and how can they lead to inequalities?
- To what extent is the 'development gap' increasing or decreasing?
- In what ways do economic inequalities influence social and environmental issues?
- To what extent can social and economic inequalities be reduced?

Consider this

Why is it that as the world becomes richer through economic development, it also becomes more unequal in the distribution of wealth? There are winners and losers, but are there also cheats?

6.1 In what ways do countries vary in their levels of economic development and quality of life?

Defining terms

From the outset, it is important to be clear about the meanings of four terms – development, economic development, inequality and quality of life – that occur throughout this chapter. The first two, development and economic development, may look similar and indeed are often taken to be one and the same thing, but they are not. Economic development is only one part of development, but the former drives the latter. To make this clearer it is important to understand what development is.

Development is a process of change within countries and their societies. As with the development of human beings, the process is one of becoming more mature and better organised. Development has been likened to an electric cable (Figure 6.1). At the core of the cable lies economic development, which is the growth of a country's economy. This growth is usually derived from the exploitation of resources and by the application and interaction of capital, technology and enterprise. The outer casing of the development cable is made up of many different strands. Each strand represents a particular aspect of the development process as a whole.

It is important to understand that:

- development is a multi-strand process
- development is much more than just economic development
- it is economic development that provides the 'power' driving progress in all the outer strands of the cable.

Defining inequality is more straightforward. Sometimes the term disparity is used instead of inequality, but the two words have exactly the same meaning. In geographical terms, the inequalities that come into focus first are spatial: the inequalities or unevenness between places. These exist at a range of scales: there are inequalities between the global regions, between countries, between regions within countries, and between and within urban and rural areas.

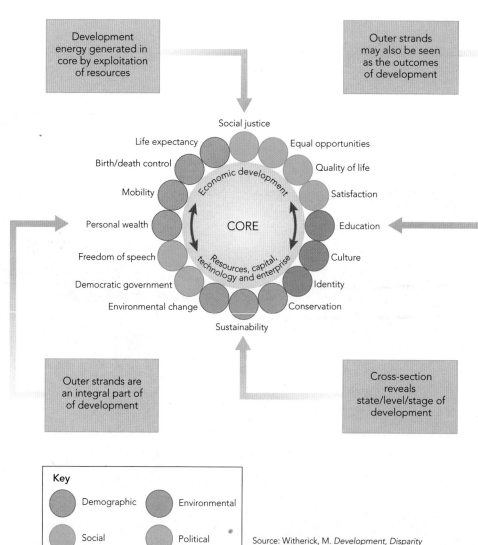

Source: Witherick, M. Development, Disparity and Dependence (1998)

Figure 6.1 The development cable

Inequality is something that can also apply to all or any one of those strands involved in the development cable and any of those quality-of-life components discussed on pages 212–16. Inequalities also exist at a personal level and these cannot be omitted from the discussions.

— **Key terms** —

Development: a process of change within countries and their societies. It is seen as involving progress in four main directions: economic prosperity, technology, well-being and social organisation.

Economic development: the growth of a country's economy leading to a rise in the general level of prosperity.

Inequality (or disparity): unevenness or difference. For example, the uneven distribution of wealth results in a basic inequality or disparity between rich and poor places and between rich and poor people.

Activity

In your own words, explain and illustrate the distinction between development and economic development.

Classifying countries

The land and the sea represent two very different physical worlds. Their 'human' equivalents might be:

- the First World – made up of the more economically developed countries (MEDCs) with their capitalist or free-market economies
- the Second World – comprising the socialist or communist countries, but now greatly reduced with the break up of the Soviet Union and its bloc of East European states
- the Third World – consisting of the less economically developed countries (LEDCs).

Alternatively, there is the widely-used distinction between the Developed World and the Developing World. These two regions, following the Brandt Report (1980), are sometimes referred to as the North and the South and are seen as being separated by the North–South Divide (Figure 6.2). Underlying this subdivision is one of the most obvious inequalities of today's world, namely the unevenness between countries in their levels of economic development. These levels might be imagined as a flight of steps rising from the least developed to the most developed countries (Figure 6.3). As they develop, countries climb up that stairway but at very different speeds. At any moment in

Figure 6.2 The North–South Divide

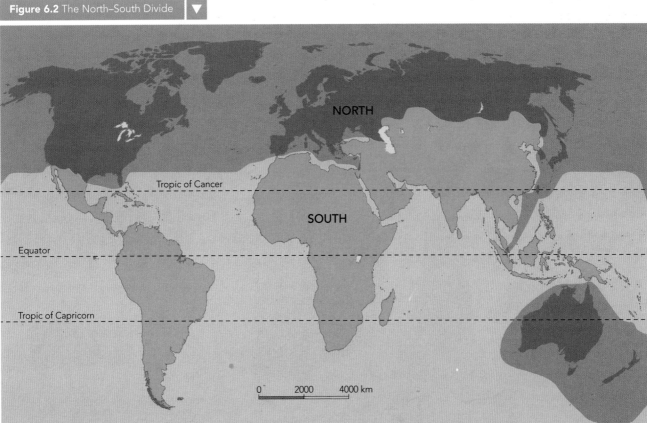

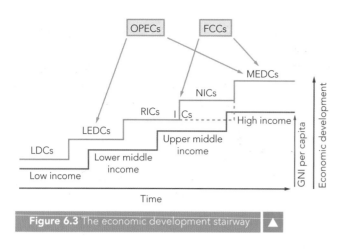

Figure 6.3 The economic development stairway

time, there are clusters of countries sharing the same step. The names given to the rising flight of steps are:

- LDCs – least developed countries (e.g. Ethiopia, Haiti, North Korea)
- LEDCs – less economically developed countries (e.g. Cuba, Morocco)
- RICs – recently industrialising countries (e.g. China, India, Mexico)
- NICs – newly industrialised countries (e.g. South Korea, Malaysia, Taiwan)
- MEDCs – more economically developed countries (e.g. UK, USA, Australia).

The distinction between NIC and RIC is vague, so these can be lumped together into one category of industrialising countries (ICs), namely countries making the transition from LEDC to MEDC. Also there are two other commonly recognised country groupings:

- OPEC members – Organization of the Petroleum Exporting Countries – oil-producing and exporting countries (e.g. as in the Middle East)
- FCCs – former communist countries (e.g. Russia and other members of the former Soviet Union).

The difficulty with these particular groups is where they should be placed on the flight of steps. The OPEC members include countries such as Algeria, Angola, Indonesia, Iran, Iraq and Nigeria which are also seen as LEDCs. At the same time, the wealth enjoyed by OPEC members such as Kuwait, Saudi Arabia and the United Arab Emirates must justify their recognition as MEDCs (Figure 6.4). As a grouping, the FCCs show similar contrasts in their levels of economic development. Some, like Russia, the Czech Republic and Hungary, might be placed among the MEDCs, while others, such as Romania and Ukraine, probably lie closer to the ICs or transitional countries.

There are two widely accepted measures of economic development, namely per capita GDP (gross domestic

Figure 6.4 City life in Saudi Arabia. Are the oil-rich Middle Eastern states MEDCs?

product) and per capita GNI (gross national income). Of those measures, the latter is now more frequently used. One of the advantages of using this measure is that it helps to solve the problem of where to place the OPEC and FCC countries on the stairway. The World Bank is responsible for calculating GNI per capita for all the countries of the world. On the basis of current values (2006), it recognises a stairway of four different income groupings:

- Low-income – US$ 905 or less
- Lower middle-income – US$ 906 to US$ 3595
- Upper middle-income – US$ 3596 to US$ 11 115
- High-income – US$ 11 116 or more.

Key terms

GDP (gross domestic product): the total value of goods and services produced by a country over a specified period, normally one year. For the purposes of international comparison, GDP is expressed in per capita (head of population) terms.

GNI (gross national income): the GDP of a country plus all the income earned from investments abroad, but less all the income earned in the country by foreigners. For the purposes of international comparison, GNI is expressed in per capita (head of population) terms.

So there are two ways of classifying the countries of the world on the basis of their economic development. A crude attempt has been made in Figure 6.3 to show a possible relationship between the two versions of the stairway. Which format is to be preferred? The problem with the first is that it is highly subjective. As already mentioned there is a problem categorising OPECs and FCCs, as well as the vague distinction between RICs and NICs. In addition, what distinguishes an LDC from an LEDC? Instinctively, we know that the former are poorer and even less developed than the latter countries. But what is the critical threshold that separates them? Yet another concern is the classification of some countries, such as Argentina, Brazil, Chile, South Africa and Turkey: are they all really MEDCs?

The second classification, as devised by the World Bank and based on per capita GNI, seems altogether more objective. Furthermore, the values used to distinguish the four income groups are updated each year. But the scheme does throw up anomalies. Looking at the high-income group, there are some surprises in the list. For example, it includes a number of Caribbean countries, such as Antigua, Barbados, Puerto Rico and Trinidad and Tobago. Few would claim that these are MEDCs. All are in the high-

Figure 6.5 The global distribution of four income groupings (2003)

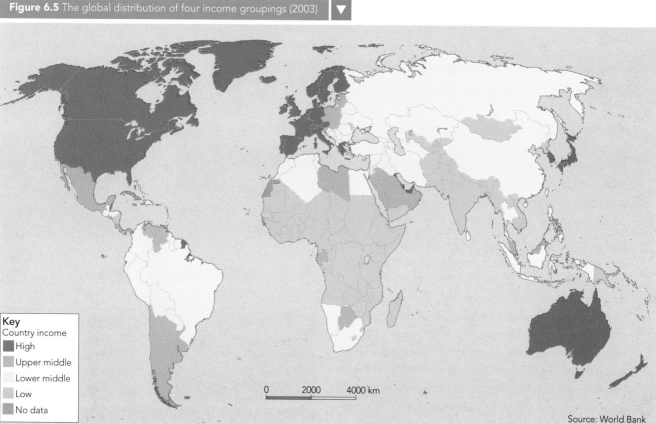

Key
Country income
- High
- Upper middle
- Lower middle
- Low
- No data

Source: World Bank

income group because they are popular tourist destinations and second-home locations. Trinidad has an alternative source of revenue in the form of valuable oil resources. All four countries have attracted much foreign investment, but relatively few native residents share in the wealth that appears in the World Bank's records.

At the other income extreme, it is surprising to find India languishing in the low-income group. Yet India is being hailed as one of the fastest industrialising countries in the world today. Clearly, India's huge population dilutes the per capita GNI figure. The mean figure conceals a major inequality: only a relatively small percentage of India's population is benefiting from the current economic boom. The appearance of Zimbabwe in this group serves as a warning. Less than 50 years ago, this was one of the most prosperous countries in Africa; it is now bankrupt (see pages 226–28).

Activity

Summarise the main features of the distribution pattern shown in Figure 6.5.

Discussion point

Which of the two versions of the economic development stairway (Figure 6.3) do you prefer? Give your reasons.

There are also two more measures to show the global pattern of economic development:

- One of the features of economic development is that the tertiary or service sector of the economy becomes increasingly important and eventually dominant. So measuring this sector, in terms of either its percentage of the working population or its percentage contribution to a country's GDP, could be helpful (see Figure 6.13 on page 218).

- The average annual rate of economic growth can give a crude indication of the rate of economic development and can be used to compare countries. To be really useful, however, calculation of the average needs to take into account the rate of inflation. A high rate of inflation can easily make the annual rate of growth in per capita GNI seem much more impressive than it really is. When inflation is taken into account, it is appropriate to refer to real economic growth. Between 1995 and 2005, one of the highest rates of real economic growth, 10.2 per cent, was recorded by China. In contrast, Zimbabwe experienced an average annual decline of 6.5 per cent.

Key term

Real economic growth: the economic growth rate adjusted for inflation.

Quality of life and economic development

Quality of life is a fairly slippery term. We may all have some sense of what it means, but it is difficult to define precisely. This is partly because it has much to do with individual perceptions and preferences. It has a psychological side in that it is about feelings, such as satisfaction, happiness, fulfilment and security (Figure 6.6). It also has a physical dimension that embraces such matters as diet, housing and mobility. In addition to these psychological and physical aspects, there are socioeconomic ones, such as employment prospects, access to services, education and leisure.

Quality of life is best thought of as the outcome or product of economic development. In order to measure it, we need to focus on the outer strands of the development cable (Figure 6.1, page 208). Figure 6.7 sets out some demographic, social and environmental indicators of three key aspects of quality of life – health, education and housing. Each of these indicators is readily measurable. Furthermore, for many countries, data is available, to show regional and local variations within countries.

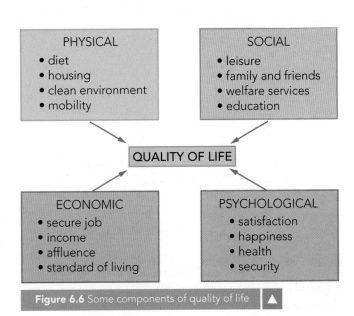

Figure 6.6 Some components of quality of life

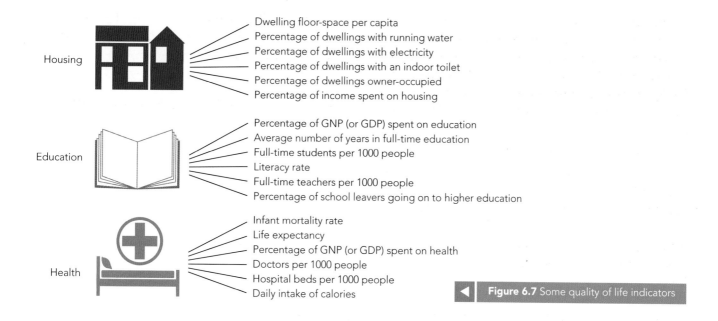

Figure 6.7 Some quality of life indicators

Given that quality of life has many different facets, it follows that there must be merit in using multivariate measures to assess it. To date, the most widely accepted of such measure is the Human Development Index (HDI).

Key terms

Human Development Index (HDI): this was devised in 1990 and takes into account three variables which are given equal weighting: income per capita, adult literacy and life expectancy. The HDI takes the highest and lowest values recorded for each variable. The interval between them is given a value of 1 and then the value for each country is scored on a scale of 0 to 1 (from worst to best). The HDI is the average score of the three variables and so too is expressed as a value between 0 and 1. The wealthy MEDCs have an index approaching 0.999 and the poorer countries range down to less than 0.300.

Quality of life: the social and psychological condition of people arising from their everyday life. It is sometimes measured by the Human Development Index but this is based on only three variables.

Activity

Compare the global distribution of HDI (Figure 6.8) with that of per capita GNI (Figure 6.5).

Theory into practice

How do you rate your own quality of life on a scale of 1 (bad) to 10 (excellent)? What aspects have carried most weight in your evaluation?

To find out whether quality of life changes with different levels of economic development, it is best to focus on a sample of countries. But how should the countries be chosen? Which of the classification schemes should be followed? In this instance, the sample has been chosen by steering a middle course between them (Figure 6.9).

One country has been taken from each of the World Bank's income groupings, except the upper-middle income one. Each country is also thought to be representative of important steps up the development stairway. So Bangladesh has been chosen as representing the Third World and the low-income, LEDC groupings. China has been selected as a Second World, lower middle-income country, a transitional economy in the early phases of industrialisation. Finally, Japan represents the First World as well as the high-income, MEDC groupings. All three countries have the added merit of being drawn from the world's most economically dynamic region – Asia.

A look at the economic data in Figure 6.9 confirms that the sample countries show some strong contrasts. In terms of per capita GNI, Bangladesh ranks 180th in the world out of 209 countries; China and Japan are ranked 129th and 19th respectively. The last figure is perhaps a little surprising, since in absolute terms Japan is the second largest economy in the world after the USA, and China's economy is now the third largest in the world. This is more a reflection of China's geographic and demographic size than its present level of economic development.

A word is needed about the importance of the tertiary sector, which is widely regarded as a good

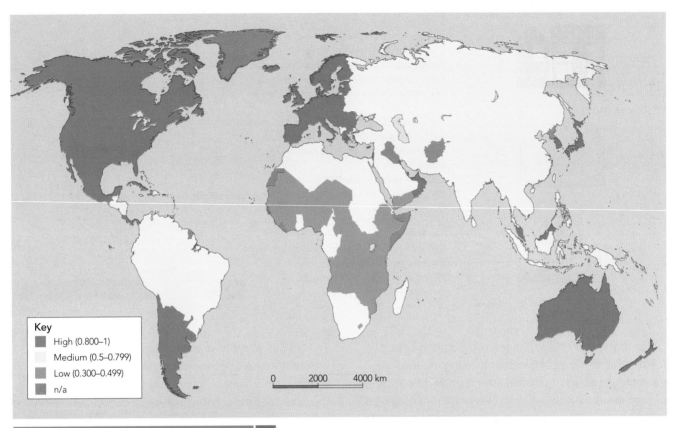

Figure 6.8 The global distribution of HDI (2006)

measure of economic development. In Bangladesh, the significance of this sector is inflated by the fact that:

- the primary sector is largely concerned with subsistence farming and therefore contributes little to GDP
- the manufacturing sector is only just beginning to expand
- many people earn a living in what is known as the informal sector – that is, providing a range of services for other people.

In contrast, note the high value for the secondary sector of 'newly industrialising' China.

> **Key term**
>
> **Informal sector:** this is made up of activities that are not officially recognised but are undertaken by poor people in order to survive. They include selling on the street, providing low-cost transport and jobs in workshops.

It is possible to pick up a number of more general benefits or spin-offs from economic development. These include:

- increased access to education and higher rates of adult literacy
- the wider provision of health services, safe water and sewage disposal
- more food security
- greater life expectancy
- lower infant mortality.

These are to be seen as basic improvements in the quality of life. Sadly, apart from access to safe water and sewage disposal, there are no other available measures that would inform us about a key aspect of the quality of life, namely housing conditions.

When looking at the quality of life measures, it is important to remember income inequality. This is measured by the Gini coefficient; the higher the value, the greater the inequality. It is significantly greatest in the transitional economy of China. The downside of this inequality is the incidence of poverty and malnutrition, even in a scenario of economic growth. China has successfully pursued rigorous campaigns to reduce both poverty and malnutrition over the last decade, but Bangladesh has a long way to go to achieve this. Even in prosperous Japan, problems

Indicator	Bangladesh	China	Japan
Population (millions), 2005	144.3	1306.3	127.4
Economic			
GNI per capita (US$), 2004	1980	5530	30040
Average annual growth in GDP per capita (%), 1990–2002	3.1	8.6	1.0
Agriculture (% of GDP), 2004	21.2	13.8	1.3
Industry (% of GDP), 2004	27.1	52.9	24.7
Services (% of GDP), 2004	51.7	33.3	74.1
Demographic			
Population density (persons per km^2), 2005	1002	136	337
Average annual population change (%), 2005	2.1	0.6	0.1
Urban population (% of total), 2005	25	41	66
Life expectancy (years), 2005	62	72	81
Health and welfare			
Food intake (calories per capita per day), 2003	2205.0	2951.0	2760.9
Population per doctor, 2003	4348	610	495
Illiteracy rate (% adults), 2003	57	14	1
Social			
Motor vehicles per thousand people, 2002	1.5	10.2	566.8
Internet usage per thousand people, 2003	2	63	483
Environment			
Energy produced (tonnes of oil equiv. per capita), 2002	0.07	0.81	0.84
Energy consumed (tonnes of oil equiv. per capita), 2002	0.10	0.83	4.31
CO_2 emissions per capita (metric tonnes), 2003	0.24	2.72	2.98
Composite indices			
Human Development Index, 2002	0.509	0.745	0.938
Gender Development Index, 2002	0.499	0.741	0.932

Source: Adapted from Philip's *Modern School Atlas*, 95th edition, 2006

Figure 6.9 Bangladesh, China and Japan: some measure of economic development and quality of life

of poverty and malnutrition linger on. Ironically, the increased consumer spending that is the outcome of economic development has its downside in the form of raised levels of obesity and alcohol consumption. It also appears that the greater the inequality of income (i.e. the wider the gap between rich and poor), the higher the incidence of crime and the larger the prison population.

Figure 6.10 The greatest global challenge – the eradication of poverty

- **Key term**

    Gini coefficient: a statistical measure of the degree of correspondence between two sets of percentage frequencies. It is frequently used as a means of assessing the degree to which a given distribution of data differs from a uniform one. The coefficient can range in value from 0 to 100; the higher the value, the more uneven the distribution.

Figure 6.9 illustrates the symptoms of three different worlds – three different steps on the development stairway (developing, transitional and developed). But it is vital to remember that the statistics used are national averages and therefore they often hide significant contrasts or inequalities within countries. This sample of three countries will be retained in the next section when the factors influencing the rate and level of economic development are explored.

Take it further activity 6.1 on CD-ROM

6.2 Why do levels of economic development vary and how can they lead to inequalities?

What factors energise economic development?

Figure 6.3 (page 210) illustrates the idea of countries climbing their way up an economic development stairway. But how is it that some countries have managed to climb further than others? The short answer lies in those factors which energise or drive economic development. In some countries, those factors are, or have been, much stronger than in others.

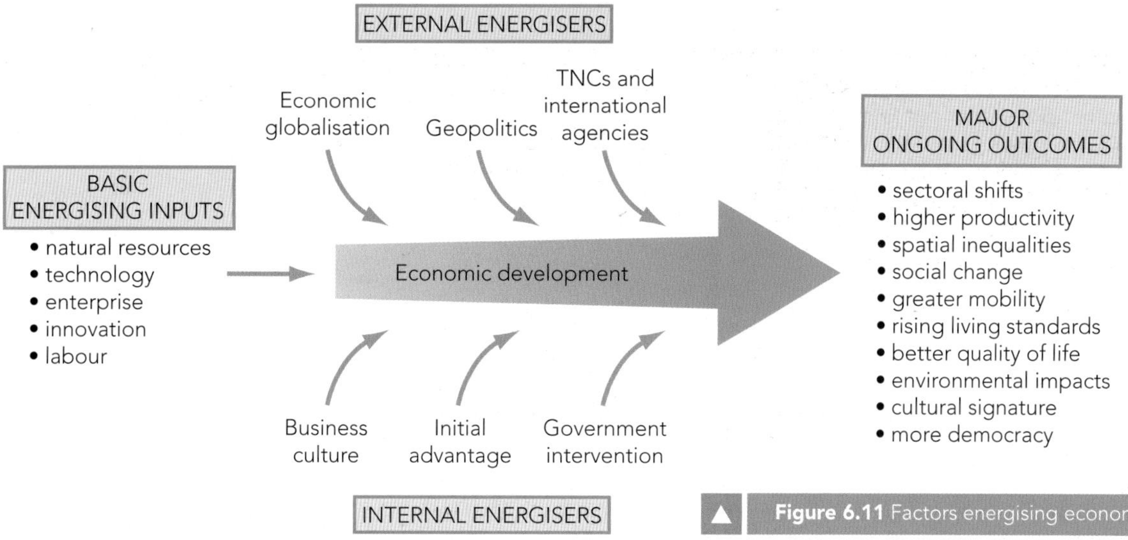

Figure 6.11 Factors energising economic development

EXTERNAL ENERGISERS: Economic globalisation, Geopolitics, TNCs and international agencies

BASIC ENERGISING INPUTS:
- natural resources
- technology
- enterprise
- innovation
- labour

→ Economic development →

MAJOR ONGOING OUTCOMES:
- sectoral shifts
- higher productivity
- spatial inequalities
- social change
- greater mobility
- rising living standards
- better quality of life
- environmental impacts
- cultural signature
- more democracy

INTERNAL ENERGISERS: Business culture, Initial advantage, Government intervention

Figure 6.11 shows three sets of energisers:

- *Basic energising inputs* – the availability and exploitability of a mix of physical and human resources.

- *External energisers* – external factors which together create a sort of scenario that has the potential to encourage economic development. The most obvious factors today are economic globalisation, trade, aid and the growing markets for a wide range of goods and services. Geopolitics and strategic considerations can also stimulate economic growth, as occurs when, for example, MEDCs look to secure future supplies of oil and other important natural resources. Undoubtedly, the major players in this external scenario are the transnational corporations (TNCs) and, to a lesser extent, the various international agencies, such as the United Nations, the World Trade Organization and the World Bank.

- *Internal energisers* – these come from within the country; particularly significant are government intervention, business culture (the practices, ambitions and effectiveness of 'native' businesses) and initial advantage. The last of these involves generating a momentum which is often boosted by the process of cumulative causation.

Thus economic development is energised by:

- meeting a range of internal prerequisites that are mainly to do with resources, politics and business culture
- taking advantage of external opportunities, such as those currently created by the expanding global economy.

The varying degrees to which these two conditions are fulfilled, and their relative timing, help explain why countries climb the stairway at different speeds. Those countries that meet them most fully, and do so from an early date, form the ranks of today's advanced countries.

Key terms

Cumulative causation: the process which perpetuates the exploitation of some initial advantage. It is a matter of capitalising on being first in the field. Success breeds more success and prosperity, and in so doing the draining of investment and labour from other areas. (See also Figure 6.24, page 230.)

Initial advantage: the advantage that accrues to a firm, city, region or country as a result of it being first – be it in the introduction of a new product, exploitation of a new resource, the commercial application of a new technology or in the opening up of a new market. Once the initial advantage has been seized, then the process of cumulative causation sets in reinforcing the benefits to be gained.

Activity

Explain how each of the 11 energisers shown in Figure 6.11 contributes to economic development.

Three case studies

The aim of the following case studies is to analyse the factors that help explain the relative levels of economic development achieved by the three countries: Japan, China and Bangladesh. The order that the countries will be discussed in will help the overall explanation.

Case study | Japan – the economic miracle

Japan is the world's second most powerful economy after the USA. That it should emerge as a front runner in terms of economic development is quite remarkable, bearing in mind three facts:

- it possesses few mineral or energy resources
- it did not start to industrialise until the second half of the nineteenth century – nearly 100 years later than the UK; before then it was an isolated and largely agrarian country
- its physical and industrial infrastructure lay in ruins at the end of the Second World War (1945).

Figure 6.12 shows the growth in the Japanese economy since 1950; two phases may be recognised. During the high-growth phase, annual growth rates exceeded 10 per cent on a number of occasions. In the current low-growth phase, there have been years when the economy has actually contracted. But it is important to note that real economic growth, as shown in Figure 6.12, is cumulative. Each year of percentage growth adds to the overall mass of the economy. So it is not surprising that over a 55-year period there should be a gradual lowering of the percentage growth rate.

Figure 6.13 shows important shifts in the sectoral balance of the economy. The secondary sector – that is, manufacturing – is widely regarded as having been the powerhouse of Japan's economic development. Interestingly, of the three sectors, it is the one that has shown least change. Its percentage share of employment increased only from 22 to 34 per cent. It has now slipped back to 25 per cent. It is the tertiary sector that has shown the greatest positive change, particularly since 1995. It now accounts for almost three-quarters of Japanese GDP. The percentage importance of this sector is widely recognised as a good barometer of economic development.

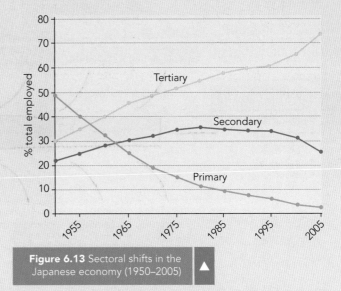

Figure 6.13 Sectoral shifts in the Japanese economy (1950–2005)

So the question to be asked is: what is the secret of Japan's economic miracle? In particular, what factors enabled Japan to achieve a spectacular recovery from destruction and debt following the Second World War within a mere 25 years? Taking a cue from Figure 6.11, answers can be found both inside and outside Japan.

Internal factors

Human resources

With a population well past the 150 million mark, Japan has plenty of labour. Not just that, but the labour supply has always shown particular strengths. For example, there is a strong work ethic and a healthy attitude to wealth. There is a willingness to embrace new work practices and new technology. There is a strong loyalty to the employer, and as a result there has been no real development of trade unions. The only problem concerning labour is that as Japan has prospered, so the labour force has expected higher wages and salaries. The rising cost of Japanese labour explains in part why so much of Japanese manufacturing has now moved abroad.

Government intervention

Japan is often cited as the supreme example of the free-market economy. In fact, this is far from being the truth. Since the Second World War, Japan's economy has been carefully managed by its government in a variety of ways. For example, it has maintained a strict control over the banks and made sure that the Japanese currency (the yen) has been slightly undervalued. This has been to the benefit of exports, though not so good for imports.

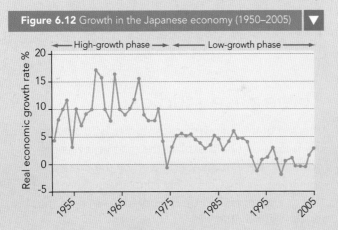

Figure 6.12 Growth in the Japanese economy (1950–2005)

Business culture

Much of Japan's economic power and wealth is in the hands of a small number of huge trading companies with considerable overseas business interests. The names of Japan-based TNCs like Sumitomo, Fuji, Matsushita, Mitsubishi and Mitsui are now known throughout the world. The TNCs have certainly played a vital role in Japan's economic development. They have done much by way of pioneering new production techniques (such as the 'just in time' system), seeing the commercial spin-offs from new technology and undertaking the most thorough market research, both at home and overseas.

> **Key terms**
>
> **Free-market economy:** an economy in which there is no government intervention and market forces prevail.
>
> **'Just in time' system:** a manufacturing system designed to minimise the costly holding of large stocks of raw materials and components. It requires an efficient ordering and reliable delivery of inputs. The technique was first introduced in the UK by Japanese car manufacturers such as Nissan and Honda.

External factors

One factor has dominated all others: the Japanese economy has grown by gradually spreading its influence. It has become global in operation, now reaching to almost all parts of the world. The life blood of the economy is now largely drawn from overseas (Figure 6.14). Japan is one of the first countries to realise that in this day and age of economic globalisation, no country can afford to be an island.

Trade

Japan is one of the world's leading trading nations. It has no other option because it lacks mineral and energy resources and has had to obtain these and other industrial raw materials from overseas. Equally, it needs to seek out overseas markets for its huge output of manufactured goods, particularly those aimed at the consumer.

Overseas direct investment (ODI)

This is a highly significant link. It is the main way that businesses from one country become involved in the economic life of another. This includes setting up branch factories and branch offices, as Honda, Nissan and Sony have, in various parts of the world. The aims are either to take advantage of cheap labour and raw materials or to gain better access to foreign markets.

Aid

It may seem a little odd to include this as the third of the tentacles binding the Japanese economy to the outside world, but Japan, like other nations, has always followed a rather pragmatic approach to aid – that is, aid with strings attached. It has never hidden its wish to turn the act of 'giving' to its own advantage. So, while Japan appears as a generous donor of technical and capital aid, it is also promoting its own interests. These interests included ensuring future supplies of energy and industrial raw materials, as well as opening up new markets for its manufactures.

It would be wrong to think that Japan's progress up the economic development staircase has been an easy one. Besides lacking minerals and energy resources, Japan has been handicapped by its physical geography. Only one-quarter of the land area is suitable for settlement. This habitable land is in the form of small lowland areas, mainly coastal in location, each separated from its neighbours by steep uplands that are difficult to traverse. The creation of national transport networks has been a real challenge, made even more so by the island character of Japan. Linking the four main islands – Hokkaido, Honshu, Shikoku and Kyushu – by bridge and tunnel has involved major civil engineering projects. Above all else, though, usable land remains a very scarce resource in Japan.

Figure 6.14 Globalisation of the Japanese economy

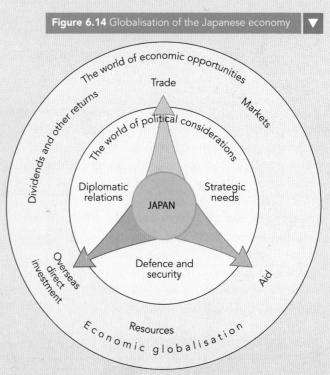

Source: Carr, M. and Witherick, M. *The Changing Face of Japan* (1993)

Case study | China – the awakening giant

From a country with too little land to a country perhaps with too much – China is a giant of a country. It is nearly 40 times the size of the UK and accounts for one-fifth of the world's population. Not surprisingly, China covers a diversity of physical environments, ranging from very high mountains to extensive coastal and alluvial lowlands, from semi-arid steppe to humid subtropical forest. A varied agriculture and food supply are two obvious outcomes of such diversity. China is also quite well endowed with energy resources (coal, oil, natural gas and abundant HEP potential). Its mineral reserves include iron ore, bauxite, tungsten, tin and lead.

Despite its wealth of resources, China is still a long way short of the level of economic development reached by Japan and other MEDCs. But that is beginning to change.

The modern state of China came into being in 1949 as a result of a communist insurrection led by Mao Zedong. Since then, the country has gone through various types of communist rule, from the bizarre and repressive Cultural Revolution of the 1960s to the more liberal and market-minded government of today. The state still retains complete control over the economy and the people. Despite the collapse of the Soviet Union and of communist rule in Eastern Europe, China seems set on remaining a communist state. Does this make China a dinosaur or the pioneer of some new mode of economic development?

What is clear is that China is fast becoming an economic superpower and shaking off its LEDC image. It is now the world's third largest economy, currently growing at annual rates around the 10 per cent mark. Much of that growth is driven by the secondary sector, which accounts for 48 per cent of GDP. It rivals India as an industrialising country.

In addition, China now plays a major role in international trade. Overall, it is the second largest exporter in the world and the third largest importer. The exports are mainly manufactured goods produced by the abundant supply of cheap labour. On the other hand, imports reflect China's voracious appetite for energy, industrial raw materials and concrete. Domestic production can no longer keep pace with demand and some reserves are running out. And here is the worry: in order to ensure these vital supplies, China has become something of a global 'wheeler dealer'. It is now extracting these vital supplies from weak and much poorer countries, particularly in Africa, in exchange for 'aid' and military equipment. In this respect, it is now rivalling Japan as an exploiter of overseas raw material sources.

At present, the service sector accounts for 40 per cent of GDP, but this is likely to rise as the economy prospers and consumer spending takes off (Figure 6.15).

Some Chinese cities, such as Beijing, Hong Kong and Shanghai, have gained the status of world cities. Shanghai has made the greatest strides since 1992, when the government allowed it to introduce a range of economic reforms. The glistening waterfront of ultra-modern, high-rise buildings symbolises the city's recent success (Figure 6.16). It is now the world's busiest port, one of its major financial centres and a shopper's paradise.

> **Key term**
>
> **World city:** one of the world's leading cities – a major node in the complex networks being produced by economic globalisation. The influence of world cities is linked primarily to their provision of high-order financial and producer services.

For all this impressive economic development, China faces a number of problems. These include:

- a rapidly ageing population – the inevitable backlash of the 'one-child' policy
- an increasing rural–urban income gap
- gaining access to foreign sources of supply
- keeping wages down so that its labour force remains cheap and competitive in the world marketplace
- serious environmental pollution.

Finally, it should be remembered that China's economic growth is being driven not just by its own huge domestic demand. It is also prospering from the ever-expanding global demand for manufactured goods elsewhere.

Figure 6.15 The links between growth in the primary and secondary sectors and the tertiary sector

Figure 6.16 Shanghai: one of China's world cities

Case study | Bangladesh – waiting its turn

Bangladesh became an independent state in 1971. It is one of two mainly Muslim countries on the Indian subcontinent, the other being Pakistan. It is a low-lying country occupying the huge delta area formed by the confluence of the Ganges, Brahmaputra and Meghna rivers before they enter the Bay of Bengal (Figure 6.17). Because of the fertile soils and tropical monsoon climate, it is possible to reap three harvests of rice in a year. Despite its huge and rapidly growing population, Bangladesh is now largely self-sufficient in rice production. Until recently, jute was the major cash crop: Bangladesh was responsible for 90 per cent of the world's production. Cotton has been another long-established cash crop.

In 1971, Bangladesh was one of world's poorest and most densely-populated countries. Since then, with the help of foreign aid, it has begun to develop an industrial base and exploit its limited energy resources. Economic development has, however, been retarded by a number of serious handicaps. These include frequent and devastating floods, a very high rate of population growth, too much inefficient state ownership, corruption and limited capital. Initially, Bangladesh relied heavily on the jute-processing industry making such things as canvas, hessian, sacking, laminated cloth and carpets. But the jute industry has since declined for a number of reasons including competition from cheaper synthetic substitutes (such as polypropylene).

Economic globalisation has provided Bangladesh with an opportunity to capitalise on what is perhaps its number one resource – an abundant supply of very cheap labour. One industry in particular seems to have benefited, the making of ready-made garments (RMG). This is partly the outgrowth from the much older cotton industry which still produces yarn and cloth. The RMG industry took off in the late 1970s. Today, it employs around 1.5 million workers, roughly 80 per cent of whom are women. As the industry has grown, so it has encouraged the expansion of linked activities which supply such things as fabrics, yarns, accessories (buttons, zips, etc.) and packaging materials. In addition, it has created a large demand for services like transportation, banking, shipping and insurance. This has led to many more jobs. Bangladesh sells most of its clothing to the European Union and the USA. RMG products (mainly shirts, jackets and trousers) account for nearly a quarter of all Bangladesh's exports.

Case Study: Bangladesh – waiting its turn

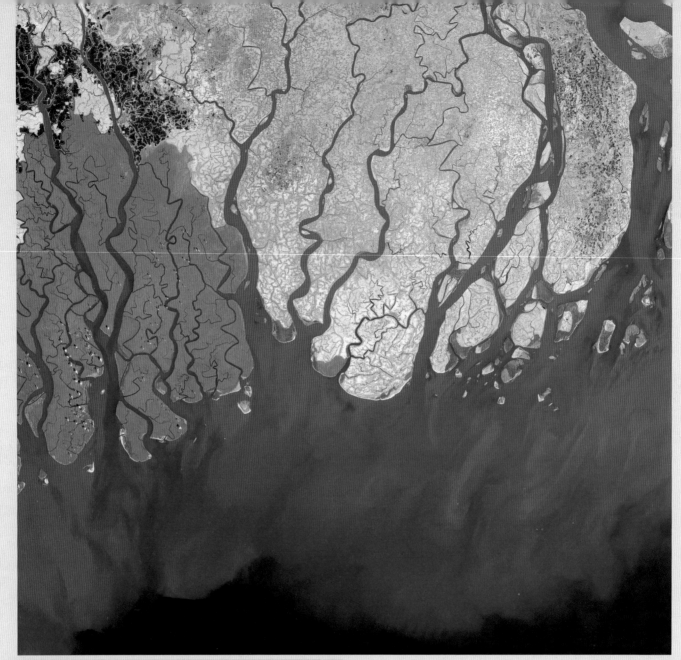

Figure 6.17 Satellite image of the delta region of Bangladesh

Key term

Ready-made garments (RMG): clothing that is ready for sale to consumers.

Despite using local raw materials, the RMG is highly dependent on imported supplies. For example, about 90 per cent of the woven fabrics and 60 per cent of the knitted fabrics used to make garments for export are imported. Clearly, it would be advantageous to Bangladesh if it was able to produce more of the raw materials it needs, but this is hindered by problems of power supply, transport and communication. In addition, the efficiency of Bangladeshi labour is much less than in competitor producers like China and Thailand.

The economy of Bangladesh remains predominantly agricultural and this depends on an erratic monsoon cycle, with periodic flooding and drought. Despite increased agricultural output, it is estimated that between 10 and 15 per cent of the population is still at risk from malnutrition. Underemployment remains a problem, explaining why the informal sector is booming. The challenge for Bangladesh is to become more integrated into the global economy. The first move must be to raise the skills levels of its abundant labour supply and to improve the physical infrastructure, particularly the energy supply.

Figure 6.18 A garment factory in Dhaka

What emerges from these three case studies is that an abundance of natural resources is not as critical to economic development as might first be thought. Rather, it is the quality of human resources – the work ethic, skills, enterprise and inventiveness – that appears to count more. But people also have a downside as there can be too many of them. Both Bangladesh and China have to devote much of their economic effort to the basic task of feeding their populations. Excessively large populations can deflect and hamper economic progress. The case studies also show that through their vision and ambition, governments can do much to promote a country's economic development. This is well shown by Japan and China. Perhaps this is a priority for Bangladesh, which has the advantages of a stable government with a sound view of how the country might best become more actively involved in the global economy.

Two other observations might be drawn from the three case studies:

- Japan may refute the concept of initial advantage – it industrialised much later than a number of European countries and yet has overtaken them on the economic development stairway.
- Both China and Bangladesh have only recently become nation states – could that be a reason for their late economic development?

These three case studies and the comparative statistics given in Figure 6.9 (page 215) demonstrate that the basic inequality between countries is largely due to where exactly countries are located relative to each other on the economic development stairway (Figure 6.3, page 210). As we shall see in the next section, inequality can increase or decrease depending on the relative speeds at which countries are moving up the stairway. The factors controlling this speed are shown in Figure 6.11 (page 217). Any inequality in terms of economic development, in its turn, is inevitably reflected in a whole range of disparities that are largely to do with the quality of life (see pages 212–16).

Discussion point

What do you see as the single greatest hurdle to further economic development in each of the three case study countries?

Activity

Read once again the case studies of Bangladesh, China and Japan and assess the physical, economic, social and political factors that have encouraged their economic development in each case.

Inequalities within countries

When it comes to economic development at the global level, we have seen that there are countries that lead and others that follow, and that these change through time. The same is the case with economic development at a national level. It is ignited in certain favoured areas of a country and then gradually spreads to other parts (see page 229). For this reason, internal spatial inequalities exist from the outset and are inclined to persist.

Case study | China – plenty of room for internal differences

Figure 6.19 provides a snapshot of spatial inequality in China. Revealed by the map is the strong contrast between certain favoured coastal areas – the five Special Economic Zones and the 17 Open Cities set up during the 1980s (Figure 6.20) – and the interior provinces. Another dimension to this contrast is the inequality between urban and rural areas, as it is the cities that are spearheading China's economic development. The rural areas are lagging behind and can only play a subordinate role as suppliers of food and labour. The poverty that prevails in rural areas is driving a massive rural-urban movement of people.

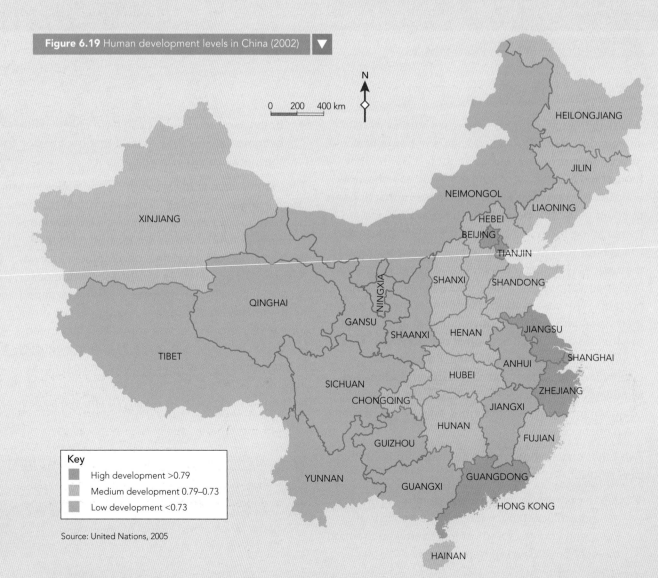

Figure 6.19 Human development levels in China (2002)

Key
- High development >0.79
- Medium development 0.79–0.73
- Low development <0.73

Source: United Nations, 2005

This migration is being reinforced by the better quality of life promised in the cities: greater personal freedom; better social services including housing; higher wages and modern lifestyles. Here is an example of the core-periphery situation and the so-called backwash effect (see pages 229–30).

One other basic inequality in modern China should be mentioned: the non-spatial disparity between males and females. Traditionally, Chinese women have been treated as second-class citizens, particularly in the contexts of education, political life, health and promotion at work. With the one-child policy, there has been a certain amount of sex-selective abortion, because Chinese fathers still wish to produce a son and heir. However, the situation is beginning to change, thanks to some government reforms that have come on the back of economic development. Even so, it will take a generation or two before equal opportunities exist between the genders.

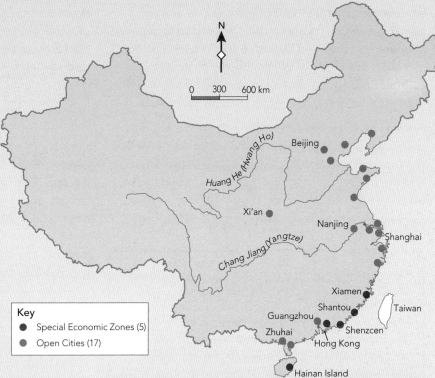

Figure 6.20 Encouragement of foreign investment in China

From the discussion and examples in this section, it would seem that two different types of inequality are the by-products of economic development:

- spatial inequalities between countries and between regions or areas within countries (see also page 230 on Japan, the Case studies of the UK on page 238 and on the CD-ROM)
- inequalities between people – between rich and poor, between men and women, and, in some countries, between ethnic groups.

 Take it further activity 6.2 on CD-ROM

Theory into practice

Make a list of what you think are significant spatial inequalities in the UK.

6.3 To what extent is the development gap increasing or decreasing?

What is the development gap?

In this section, attention turns to what has been happening over time to the inequalities that exist between countries, particularly between the richest and poorest. This is usually referred to as the development gap, or sometimes as the North–South Divide (Figure 6.2, page 209). Per capita GNI is the most obvious way of demonstrating that the gap is first and foremost an economic one. However, the inescapable link between economic development and quality of life, with the former controlling the latter, means that the development gap is also a humanitarian one. However, the idea of two groupings of countries separated by such a gap is complicated, as indicated (on page 209) by those industrialising countries (ICs) which are in the process of crossing the gap; they are in the gap rather than on one side of it.

Another possible definition of the development gap is simply the difference, say in terms of per capita GNI or the Human Development Index, between the richest and the poorest country. In other words, this definition takes into account the total length of the economic development stairway (Figure 6.3, page 210).

> **Key term**
>
> **Development gap:** 1) the general difference between LEDCs (the South) and MEDCs (the North) as measured in terms of economic attributes and quality of life; 2) the specific difference between the richest and poorest country as measured in terms of economic attributes and quality of life. The former definition is the more widely accepted.

Discussion point

Which of the two definitions of the development gap do you think is better? Give reasons.

Is what is happening to the development gap certain?

The key to determining whether or not the development gap is widening lies in the relative speeds of countries climbing the economic development stairway (Figure 6.3, page 210). If a country moves more quickly than its immediate 'neighbours' on the stairway, then two things will happen: the gap between itself and neighbours further up the stairway will decrease; at the same time, it will draw further away from those neighbours further down the stairway. The same principle applies to any groupings on the stairway. The gaps between groups can increase or decrease.

Many claims are made in the media that the development gap is widening. The poorer nations are believed to be falling further behind the rich ones. Unfortunately, it is extremely difficult to prove or disprove this popular perception as there is no widely accepted way of measuring change in the 'width' of the gap. The fact that some industrialising countries are managing to work their way across the gap does not tell us anything about possible changes in its width; all that it tells us is that some LEDCs are proving that the gap is still crossable. Equally, it is possible for countries to stagnate or even decline in terms of economic development. For example, there are around 30 countries today that the UN calls 'complex emergencies' – countries being threatened by some sort of catastrophe. The number of such countries and states has increased considerably in recent years due to various factors. They include:

- the end of communism in Eastern Europe and the break-up of the Soviet Union (e.g. Romania, Georgia)
- the outbreak of previously inhibited ethnic hostility (e.g. former Yugoslavia, Rwanda)
- weapons proliferation (e.g. Angola, Somalia)
- the loss of sovereignty (e.g. Palestine, Lebanon)
- the stubbornness of underdevelopment (e.g. Ethiopia, Malawi)
- the very slow pace of democratisation (e.g. North Korea, Myanmar).

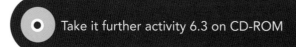

Take it further activity 6.3 on CD-ROM

Case study | Zimbabwe – economic decline and catastrophe

In the 1960s, when European colonial rule in Africa was coming to an end, Southern Rhodesia (now Zimbabwe) was one of the most prosperous parts of Africa. In 1965, the white government issued a unilateral declaration of independence from the UK, then its colonial master. In retaliation, the UK imposed trade sanctions which hit the country's economy. In 1980, the first multiracial elections were held and Robert Mugabe's ZANU PF Party won. He became prime minister (later president) and soon established a one-party, Marxist-oriented state. Trade sanctions against Zimbabwe were lifted, but the real slide in the Zimbabwean economy was only just beginning. Over the last decade of Mugabe's time in office, the economic decline has plunged into catastrophe.

- The once prosperous agricultural sector is no longer a significant exporter. Indeed, it can no longer feed the population. A particularly damaging act was the so-called land reform which dispossessed the white farmers and redistributed their land among black Zimbabweans. This was a quick way of boosting support for the regime. However, much of the land was given to supporters with little interest in farming the land or putting it to good economic use (Figure 6.21).

- As food shortages have become more acute, so rampant inflation has set in. In 2008, the annual rate of inflation was over 1000 per cent, making food and other goods about 11 times more expensive than they had been 12 months earlier.

- With few exports (some cotton, tobacco, gold, textiles and clothing) and therefore little foreign exchange, Zimbabwe is unable to buy the fuel it badly needs to generate electricity and run the transport system. The trade balance is heavily in deficit.

- In 2006, the real economic growth rate was minus 4.1 per cent (Figure 6.21) and the unemployment rate was 80 per cent. Per capita GNI had fallen to US$ 340 and there were only 15 countries with a lower figure. Since 2006, the output of the economy has continued to decline.

- The leaders of the ZANU PF Party have proved to be highly corrupt. Money originally sent to the country as aid has been either stashed away in foreign bank accounts or invested overseas. Properties once owned by foreign companies have been requisitioned for the benefit of Mugabe's immediate circle of supporters.

To add to the economic tale of woe:

- two-thirds of the population now live below the poverty line, and the number is rising rapidly each year

- people are deserting the country in large numbers and seeking refuge in neighbouring countries, particularly South Africa; the country has already lost many of its professional people who have fled abroad

- HIV/AIDS is sweeping through the country (25 per cent of children are orphans; 1.1 million have lost one parent to AIDS) and the rundown of healthcare and the lack of money to buy retroviral drugs mean that the spread of this killer disease goes unchecked; life expectancy has fallen from 65 years a decade ago to less than 40 years

- the government's urban slum demolition drive in 2005 left some 700 000 people without jobs or homes (Figure 6.22). Mugabe said it was an effort to improve law and order and for development; critics accused him of destroying the slum housing of opposition supporters.

The Mugabe regime only remains in power thanks to election rigging, the suppression of any opposition and support of the army, police and a group known as the 'veterans' (former guerrillas in the war of independence). When the present regime comes to an end, as now looks to be likely, its legacy will be a bankrupt and corrupted country inhabited by a poverty-stricken and repressed population. The recent move towards power sharing may be a small step in the right direction. However, it will take many decades for the country to recover from such blatant misrule.

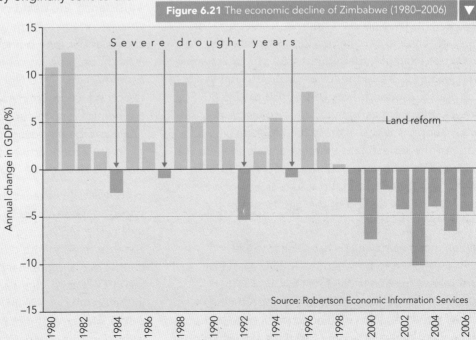

Figure 6.21 The economic decline of Zimbabwe (1980–2006)

Source: Robertson Economic Information Services

Case Study: Zimbabwe – economic decline and catastrophe

Figure 6.22 One of the slum areas of Zimbabwe bulldozed in 2005

Given that countries like Zimbabwe are slipping back in terms of economic development, it must follow that the gap between the leaders and the back-markers of the global economy is increasing. The field of runners, or 'climbers', is spreading out. In 2006, for example, the highest GNI per capita value was US$ 74 040 and the lowest US$ 710. Ten years previously, the range was US$ 40 630 to US$ 80. So, maybe the gap has widened in absolute terms. But what happens if you express the highest value as a ratio of the lowest, and thereby remove the inflation factor? The ratios were 508:1 in 1996 and 107:1 in 2006. This lowering of ratio values does not support the idea of a widening of the gap between the richest and poorest countries.

When it comes to what is happening to the development gap between the MEDC and LEDC groupings, opinion is divided. One view is that economic globalisation is creating opportunities for all those countries wishing to participate in the expanding global economy. Every nation, rich or poor, has something to offer and something to gain. Poor countries, previously isolated and preoccupied with subsistence, are being given the chance to make some economic progress. It might be supplying cheap labour for a branch factory set up by a TNC. It might be growing some cash crops for sale in the supermarkets of the North. It might be exploiting and exporting its own mineral and energy resources with the help of foreign capital and technical assistance. There can be no question that these sorts of activity generate economic wealth and may do something to close the development gap.

In contrast, however, is the view that the terms of trade, as between North and South, are distinctly unfair. The MEDCs are in the driving seat – they are able to negotiate terms that are favourable to themselves and are in a position to exploit whatever the LEDC has to offer to their own advantage. Even in the context of aid, it is pointed out that the so-called 'giving' of donor nations most often comes with strings attached. Crudely put, it is claimed that while economic globalisation is helping many LEDCs to develop and even some of them to become ICs, it is also helping the MEDCs more. If this is the case, then the inevitable outcome must be a widening of the economic development gap.

What statistics are there to support or deny these two fundamentally different views? As a general indicator, the aggregate income in 1960 of the countries with the richest fifth of the world's population was 30 times greater than the aggregate income of the countries with the poorest fifth. By 1997 that figure had risen to 74 times. However, statistics recently published by the World Bank show that developing countries grew faster between 1995 and 2005 than during the previous 20 years and faster than high-income countries (see Figure 5.20, page 178). The share of developing economies in global output increased from 39 to 46 per cent. The developing economies in the East Asia and Pacific region grew the most, doubling their output and increasing their share of global output from 13 to 19 per cent. The following changes contributed to this improved performance of the developing countries:

- the rapid industrialisation of the two 'giants' – China and India
- rising global demand for many commodities
- further integration of these countries into global markets
- stronger internal markets
- more financial stability and higher investment confidence at a global level.

Are the two sets of statistics set out in the previous paragraph really contradictory? Or might they be suggesting that the development gap was widening up to around 1995 and that since then it has begun to narrow? It is impossible to be certain. However, when looking at the statistics about shares of global output, it is important to remember that the high-income countries' 54 per cent share in 2005 was for the benefit of a mere 16 per cent of the world's population. If nothing else, this underlines the gross inequality either side of the North–South Divide.

Activity

Are you persuaded that the development gap between LEDCs and MEDCs is increasing or decreasing? Identify the evidence that leads you to your conclusion.

Theories of economic development and the development gap

Having explored what may have been happening to the development gap between the North and the South, we need to ask – why did this inequality spring up in the first place? Was it simply due to initial advantage and cumulative causation (page 217)? Do the theories of economic development lead to any answers? There are three models or theories:

- the core-periphery model
- the stages of growth model
- dependency theory.

The core-periphery model

In the late 1950s and early 1960s, development economists such as Myrdal and Friedmann argued that spatial inequalities are an integral part of economic development. Economic development is born in favoured locations (cores) which happen to have some initial advantage, such as resources, labour or capital. As a core prospers, so it creates backwash effects which draw in more resources, capital and labour from surrounding areas (peripheries). In this way, a core's growth soon becomes at the expense of its periphery (Figure 6.23). That growth is reinforced by the process of cumulative causation (Figure 6.24). The dependence of the periphery on the core intensifies. Thus, core-periphery inequalities are established at the very foot of the economic development stairway (Figure 6.3, page 210).

The core-periphery idea can be applied at a range of spatial scales. Cities may be viewed as cores and

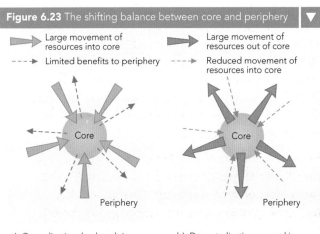

Figure 6.23 The shifting balance between core and periphery

a) Centralisation: backwash is greater than spread effects

b) Decentralisation: spread is greater than backwash effects

the remoter rural parts of their spheres of influence as peripheries. Within most countries, it is possible to detect a core region and a periphery. The Japanese core is an elongated region extending from Tokyo along the coastlands of southern Honshu to the northern part of Kyushu (Figure 6.25). The remainder of the country constitutes the Japanese periphery. The Dhaka region is Bangladesh's core; China, because of its size, has several. At present, the North is the global core and the South its periphery.

Figure 6.24 The process of cumulative causation

hope for the periphery, whether it is the South or that rural area remote from some city. It is just a matter of waiting and being patient. It has to be said, however, that there are those who consider this argument to be flawed; differences in levels of economic development are not gradually eliminated.

Key terms

Backwash effect: the concentration of resources, people and wealth in the core and at the expense of the periphery.

Core: a favoured location, usually possessing some form of initial advantage, in which economic and demographic growth become concentrated.

Dependence: the condition in which a country, region or group of people is only able to survive and progress by reliance on the support provided by another. This support can take various forms including trade, aid, political protection and defence.

Periphery: a relatively retarded area in terms of economic development and one that supplies core areas with resources and labour.

The core-periphery model argues that eventually growth and development in the core will trickle down to the periphery by what are called spread effects (Figure 6.23). This is encouraged by the congestion and the high costs of space and labour in the core, making it easier and cheaper for businesses and people to move out. Thus the early success of the core leads eventually to a downside. This part of the core-periphery model clearly holds out

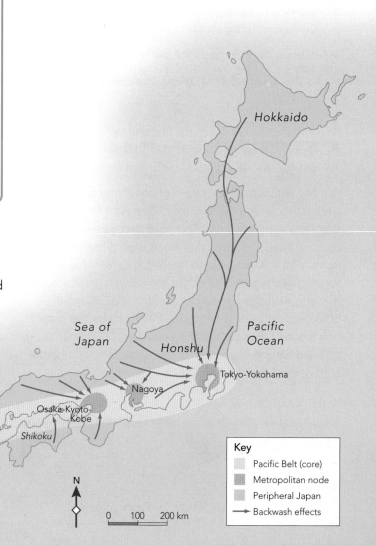

Figure 6.25 The Japanese core and periphery

> **Key term**
>
> **Spread effect:** the spatial spread of economic development and population growth from core to periphery. It is encouraged by: 1) the growth created by meeting the needs of the core; 2) the voluntary decentralisation of businesses and people; and 3) government intervention aimed at evening out the inequality between core and periphery.

The five stages of growth model

The stages of growth model (Figure 6.26) was first put forward in the early 1970s. It is a sequence of five stages of economic and social development through which all countries might pass in the fullness of time. Rostow's view was that transition from Stage 1 to Stage 5 was driven by economic development. He believed that the nations of the world would follow the slow but inevitable path to MEDC prosperity. However, the model really does not offer an explanation for why some countries have moved through the transformation and others have yet to do so. The development stairway idea discussed earlier (pages 209–10) bears some similarities to Rostow's model, but it is less prescriptive. The stairway model merely assumes that over time most countries will undergo some form of economic development that will ultimately raise per capita GNI. How that growth is achieved, and whether it is sufficient to lift to another step on the stairway, will vary from country to country.

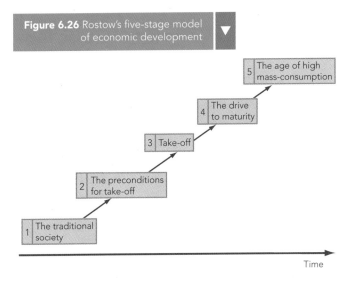

Figure 6.26 Rostow's five-stage model of economic development

Dependency theory

Dependency theory, mostly based on Latin America, was first put forward about the same time as the Rostow model. It sees the relationship between the North and South in a particular way, namely that the links between them are the cause of worsening conditions in the South. The underdevelopment of the South is seen as the consequence of the way in which the global economy works to the advantage of the North and makes the South increasingly dependent on the North. So dependency theory puts the blame for the development gap on colonialism, but it does not explain how and why European countries were able to become colonial powers in the first place and command the power to exploit other countries.

The outstanding feature of today's global economy is that it increases the dependence of all participating countries, whether strong or weak, as they become inextricably involved in a highly complex network. Even the world's most powerful nations now have to accept some degree of dependence. Perhaps it would be more appropriate to talk of interdependence rather than dependence.

> **Theory into practice**
>
> Relate the groupings of countries along the economic development stairway (Figure 6.3, page 210) to the five stages shown in Figure 6.26.

> **Activity**
>
> Identify the ways in which the USA is dependent.

Summing up

Despite living in an increasingly interdependent world, it is not clear if the development gap is increasing or decreasing; it partly depends on how one defines the gap. If it is the margin of separation between the LEDC and MEDC groupings, the gap may well be stable, perhaps even narrowing. This is evidenced by the number of industrialising countries that appear to be crossing the gap today. On the other hand, if it is the overall margin that separates the world's wealthiest country from the poorest, the evidence such as we have is somewhat contradictory.

When it comes to finding explanations for what might be happening to the development gap, the key factor would seem to be the development opportunities offered by economic globalisation. The rich nations may be becoming richer, but economic globalisation is also encouraging even the poorest countries to participate. By beginning to supply the global market, whether it is with agricultural products, minerals, energy or cheap labour, they are igniting or accelerating their own economic development process.

6.4 In what ways do economic inequalities influence social and environmental issues?

The point was made in Section 6.3 that the economic inequalities created by economic development are of two types: spatial and personal. In both instances, the inequalities relate to scales that range from growth (rising prosperity) to decline (increasing poverty); from advantaged to disadvantaged; and from rich to poor. In this section, the spotlight falls on the social and environmental issues arising from such economic inequalities. Figure 6.27 outlines some of the issues that arise in the geographical context of core and periphery regions, as well as at a personal level between rich and poor people.

Activity

Take one issue from each of the eight categories in Figure 6.27 and, in your own words, explain the nature of the issue and its economic roots.

Social issues

Many of the issues shown in Figure 6.27 relate broadly to what is often referred to as the quality of life. As discussed on page 212, this is strongly 'social' in character but it also has an important 'environmental' dimension. The latter is largely to do with living conditions, such as housing, air and water quality, health and safety. Reference has already been made to the Human Development Index (HDI) (page 214) as a possible measure of the quality of life. Figure 6.8 (page 214) illustrates the great difference in the quality of life that separates Europe, Australasia and North America on the one hand from much of Africa on the other. Much of the middle ground in the range of HDI values is occupied by Latin America and Asia.

The Economist Intelligence Unit has developed a new and more broadly-based Quality of Life Index (QLI). It involves a unique methodology that links the results of subjective life-satisfaction surveys to the objective determinants of quality of life across countries. The index has been calculated for 111 countries for 2005. Figure 6.28 shows the nine quality of life factors and the indicators used to represent these factors. The total scores for each country are then expressed on a global scale running from 10 (best) to 0 (worst). Figure 6.29 shows the QLI scores for a sample of 30 out of the 111 countries involved in this investigation.

Figure 6.27 Social and environmental issues arising from economic inequalities

	Core region	Peripheral region
Social issues	Housing costs	Loss of community
	Pressure on services	Loss of services
	Personal safety	Unemployment
	Leisure and recreation	Sense of abandonment
	Social and ethnic segregation	Second-homers
Environmental issues	Pollution	Neglect of land
	Erosion of green space	Maintaining physical infrastructure
	Making growth more sustainable	Redundant buildings
	Outdoor leisure and recreational needs	Conserving wilderness
	Protecting important ecological areas	
	Rich	**Poor**
Social issues	Protection of property	Housing
	Personal security	Personal security
	Social stability	Healthcare
	Upward mobility	Access to services
	Maintaining exclusivity	Mobility
Environmental issues	Protecting fashionable places	Minimising environmental hazard risks
	Minimising environmental hazard risks	Unsanitary living conditions
	Unsightliness of poor areas	Need for living space

Figure 6.28 Quality of life indicators

1. **Material well-being**
   GDP per person (US$)
2. **Health**
   Life expectancy at birth (years)
3. **Political stability and security**
   Political stability and security ratings produced by the Economist Intelligence Unit
4. **Family life**
   Divorce rate (per 1000 population), converted into an index of 1 (lowest divorce rates) to 5 (highest)
5. **Community life**
   Dummy variable taking value 1 if country has either a high rate of church attendance or trade-union membership; zero otherwise
6. **Climate and geography**
   Latitude, to distinguish between warmer and colder climes
7. **Job security**
   Unemployment rate (%)
8. **Political freedom**
   Average of indices of political and civil liberties on a scale of 1 (completely free) to 7 (unfree)
9. **Gender equality**
   Ratio of average male and female earnings, latest available data

Source: Economic Intelligence Unit

The interesting feature of QLI results is that the ranking of countries does not accord at all closely with the ranking based on income (per capita GDP) (Figure 6.29). In fact, the QLI values appear to make nonsense of the development groupings identified in Figure 6.3 (page 210). The lack of correspondence between the two sets of rankings is largely explained by one critical factor – the distribution of wealth within individual countries. The more uneven that distribution (the greater the polarisation of rich and poor and the greater the number of poor), the lower will be the overall or average quality of life.

Activity

Calculate the rank correlation coefficient of the two rankings. What does the result tell you?

Unequal social conditions

The social inequalities at an international level shown by Figure 6.29 are the outcomes of unevenness in the distribution of economic wealth both between countries and between the citizens of the same country. There is also a third unevenness – that which occurs spatially within countries, as between regions, between rural and urban areas and even between different parts of a town or city. These spatial inequalities can be well illustrated within the UK (see case studies of multiple deprivation in the UK and wealth and poverty in London on the CD-ROM).

Figure 6.29 QLI and GDP per capita rankings of a sample of 30 countries (2005)

Rank QLI	Country	Rank per cap GDP	QLI	Rank QLI	Country	Rank per cap GDP	QLI	Rank QLI	Country	Rank per cap GDP	QLI
(1)	Ireland	(4)	8.333	(32)	Mexico	(54)	6.766	(92)	South Africa	(50)	5.245
(3)	Norway	(3)	8.051	(40)	Argentina	(42)	6.469	(93)	Pakistan	(101)	5.229
(6)	Australia	(14)	7.925	(48)	Poland	(43)	6.216	(95)	Ghana	(100)	5.174
(13)	USA	(2)	7.615	(50)	Turkey	(61)	6.171	(101)	Uganda	(108)	4.879
(15)	New Zealand	(25)	7.436	(53)	Peru	(77)	6.162	(104)	Botswana	(52)	4.810
(17)	Japan	(16)	7.392	(57)	Bulgaria	(59)	6.083	(105)	Russia	(55)	4.796
(19)	Portugal	(31)	7.307	(60)	China	(74)	6.022	(108)	Nigeria	(110)	4.505
(22)	Greece	(27)	7.136	(69)	UAE	(33)	5.899	(109)	Tanzania	(111)	4.495
(26)	Germany	(21)	7.048	(73)	India	(96)	5.759	(110)	Haiti	(107)	4.090
(29)	UK	(13)	6.917	(77)	Bangladesh	(105)	5.646	(111)	Zimbabwe	(106)	3.892

Source: Economic Intelligence Unit

 Please refer to the CD-ROM for a case study of multiple deprivation in the UK.

 Please refer to the CD-ROM for a case study of wealth and poverty in London.

Environmental issues

First, take another look at Figure 6.27 (page 232) and compare the environmental issues of core and peripheral regions. Ironically, the environmental issues generated by economic success are considerably greater than those created by economic stagnation or decline. Why should this be? The simple answer is that economic success encourages the agglomeration of people and activities at favoured locations. This agglomeration, in turn, triggers a whole series of environmental consequences, most of them of a highly undesirable nature. They include the pollution of air and water, noise and light pollution, as well as the pollution that comes from the disposal of solid waste, being it by burning, dumping at sea or using landfill sites. It is probably true to say that such environmental pollution is worse in ICs than MEDCs. In the former, the prevailing attitude tends to be growth at all costs, no matter the environmental consequences. In the latter, while the same attitude applied in previous decades, there is now some realisation of the need to minimise the environmental impacts of economic growth. This is borne out by Figure 6.30, which compares Mexico (a recently industrialising country) and the UK in terms of ten environmental indicators used by the World Bank. The main conclusions to be drawn are:

- the biodiversity of the UK is slightly less threatened and more of the land area is protected
- energy consumption is greater in the UK and, as a result of its higher CO_2 emissions, the UK is a greater contributor to global warming
- Mexico still has some way to go before all of its people have access to proper sanitation; although the situation with respect to safe drinking water is better.

While there is a fair amount of data available on the subject of environmental pollution at a national level, there is very little that paints a picture of spatial variations within individual countries.

> Please refer to the CD-ROM for a case study of spatial variations in pollution within the UK.

Figure 6.30 World Bank environmental indicators for Mexico and the UK

Indicator	Mexico	UK
Annual deforestation (% change 1990–2005)	0.5	−0.6
Mammal species, threatened	72	10
Bird species, threatened	57	10
Nationally protected areas (% of total land area)	10.2	20.8
Energy use per capita (kg oil equiv)	1564	3893
Electric power consumption per capita (kWh)	1801	6209
CO_2 emissions per capita (metric tons)	3.8	9.2
Access to improved water source (% of total population)	91	100
Access to improved sanitation (% of total population)	77	100

Source: *The Little Green Data Book 2006*, Development Economics Data Group and the Environment Department of the World Bank

Case study | Mexico City – a global pollution hotspot

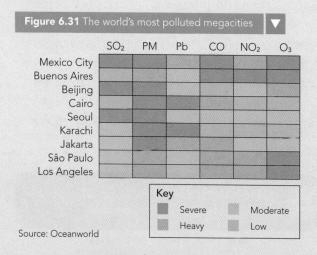

Figure 6.31 The world's most polluted megacities

Source: Oceanworld

Mexico City has the distinction of being not only one of the world's largest cities but also one of the most polluted (Figure 6.31). This is the outcome of a number of factors:

- the sheer size of the city – a population of around 20 million generates a substantial amount of pollution of all types, including light, noise and smell
- a heavy reliance on motor vehicles – more than 3.5 million vehicles now ply the city streets, 30 per cent are more than 20 years old and the most widely used form of public transport is the motor bus

- economic success – the concentration of many polluting industries unhindered by lax anti-pollution laws
- its site – located in the crater of an extinct volcano, Mexico City is about 2240 metres above sea level. The lower atmospheric oxygen levels at this altitude cause incomplete fuel combustion in engines and higher emissions of carbon monoxide, hydrocarbons and volatile organic compounds. Intense sunlight turns these noxious gases into higher than normal smog levels. In turn, the smog prevents the sun from heating the atmosphere enough to penetrate the inversion layer that blankets the city.

Although pollutant emissions have been reduced in Mexico City, approximately 4 million tonnes are still emitted each year. The main source of most pollutants is the internal combustion engine (75 per cent), followed by natural sources (12 per cent), services (10 per cent) and industries (3 per cent). Sulphur dioxide is related to industrial activity, while carbon monoxide, nitrogen dioxide and hydrocarbons arise mainly from motor vehicle emissions. For much of the year, the city lies shrouded in a dense and deadly smog caused by cold air sinking down into the city trapping pollution beneath it.

The two most serious pollutants in Mexico City are ozone (produced when nitrogen oxides and volatile organic compounds react in sunlight) and particulates (minute specks of dust that come from various sources, including road and building construction, smoke-belching diesel trucks and buses, forest fires and burning refuse in the open air). Both pollutants irritate eyes, cause or aggravate a range of respiratory and cardiovascular ailments, and lead to premature death. As a consequence of its high levels of pollution, Mexico City has the unenviable reputation as being 'the most dangerous city in the world for children'. The residents of Mexico City have become the victims of the city's economic success.

But Mexico City's tale of woe involves more than air pollution. Over 2 million of the inhabitants have no running water in their homes. More than 3 million residents have no sewage facilities: tonnes of waste are left in gutters or vacant lots to become part of the city's water and dust. Mexico City produces about 14 000 tonnes of garbage every day but processes only 8000. Of the rest, about half gets dumped in landfill, and half is left to rot in the open attracting legions of rats.

 Take it further activity 6.4 on CD-ROM

This section has illustrated some of the social and environmental consequences of economic inequalities. It is natural to think of poor areas and poor people being the principal victims. However, reality is not quite that one-sided. Economic success brings with it a whole range of social and environmental problems, particularly environmental, which afflict the whole population, whether rich or poor (Figure 6.27 on page 232). The poor endure material discomfort and a low quality of life, but the rich also have problems to bear. For example, in the social context of crime, it is the rich who are the targets of burglary and robbery. In many parts of the world, their personal security is threatened by kidnappers and ransom seekers.

6.5 To what extent can social and economic inequalities be reduced?

Reducing global inequalities

Inequalities are the differences that exist between countries, particularly between the rich and the poor, the developed and the developing nations. Currently, there are a number of opportunities that might lead to a more equitable development of the world. Thanks to economic globalisation, there are few parts of the world today that remain untouched by the influence of the ever-expanding global economy. Even the most remote and least developed areas can offer and gain something from the global marketplace. It might be cheap labour, minerals and other raw materials, food for the supermarkets of the North or tourist attractions such as sun, sea, sand and wilderness. Without doubt, being able to offer any of these 'resources' promises a payoff in the form of economic development. There are three main ways in which those opportunities are exploited: direct investment, aid and trade (Figure 6.32).

> **Key term**
>
> **Global economy:** the worldwide exploitation of resources, and the worldwide production and marketing of goods and services. Growth of the global economy is making countries increasingly interdependent.

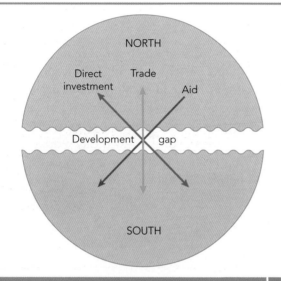

Figure 6.32 Three ways of reducing global inequalities?

Direct investment

Common targets include:

- setting up factories to take advantage of lower costs, such as cheaper labour, energy and land
- opening plants that process basic raw materials before they are shipped to the North
- improving transport infrastructure, particularly ports, airports and roads
- building hotels.

However, in most cases the economic development spin-off is not as great as it should be. The root of the problem is that direct investment is very much in the hands of transnational companies (TNCs). As a consequence, there is a substantial leakage of profits from the countries in which the investments are made. Money that might be ploughed back in support of more economic development is 'lost' either to support a new investment in some other country or to be paid out as dividends to TNC shareholders. It might be argued, however, that this profit-leakage situation is preferable to no TNC investment whatsoever.

Trade

Responsibility for ensuring fair international trade rests with the World Trade Organization (WTO). This was set up in 1995 and now has 151 member countries. The activities of the WTO are guided by a number of principles:

- to trade without discrimination
- to treat foreign and local goods equally
- to promote a freer trade by lowering trade barriers such as tariffs and taxes
- to support transparency and fair competition
- to encourage development.

Although these principles sound very laudable, in practice they favour the stronger, more developed nations. Many might argue that if developing countries are to be helped, there needs to be some concessions – but concessions run against the principle of trade without discrimination. So far, the various agreements reached by the WTO have achieved disappointingly little. As with economic globalisation, the potential development spin-off in developing countries from increased global trade has yet to be realised.

Another international organisation concerned with the promotion of trade is the Fairtrade Foundation (set up in 1992 by CAFOD). This non-governmental organisation seeks to obtain a fair price for a wide variety of goods exported from developing to developed countries, including: handicrafts, coffee, cocoa, sugar, tea, bananas, honey, wine and fresh fruit. The aim is to work with marginal small-scale producers and help make them more economically secure. While the WTO appears as rather moribund, the Fairtrade movement has gone from strength to strength. In 2006, Fairtrade's certified sales amounted to approximately US$ 2.3 billion worldwide, a 41 per cent increase on the previous year's figure. While this represents less than one-hundredth of a percentage point of world trade in physical merchandise, Fairtrade products generally account for 0.5 to 5 per cent of all sales in their product categories in Europe and North America. In October 2006, over 1.5 million disadvantaged producers worldwide were directly benefiting from Fairtrade, while an additional 5 million benefited from infrastructure and community development projects funded by Fairtrade.

Appropriate aid

Investment and trade both yield profits for the North – the benefits are two-way (Figure 6.32). In theory, the benefits associated with aid should be one-way. For decades now, many of the world's least developed countries have been receiving aid from both individual MEDCs (bilateral aid) and from international organisations (multilateral aid), such as the various agencies of the United Nations, the World Bank and the International Monetary Fund (Figure 6.33). During this time, a number of serious issues have arisen:

- Too much aid has been in the form of loans rather than technical assistance. Loans have to be repaid and loans attract interest charges. As a consequence, many countries have been plunged into ever deeper debt.

- Much bilateral aid has come with strings attached, such as agreeing to sell raw materials at below-market prices or to buy military equipment in return for aid.

- Aid has been used as a backdoor to obtaining the resources of poorer countries. For example, China is using the resources of Africa to support its large and increasingly affluent population. In previous centuries, the colonial powers, such as France, the Netherlands and the UK, were doing the same thing.

- Very little aid has reached the most needy. Much of it has been skimmed off and found its way into the bank accounts of corrupt governments and their officials.

It is now widely recognised that aid in the form of loans is inappropriate and that something should be done to ease the accumulated debt burdens of the poorest countries. There is now talk of debt relief, but as yet little has been granted and there is considerable opposition to the idea. Opponents of debt relief argue that it is like offering a blank cheque to governments. They fear that the benefits will not reach the poor in countries plagued by corruption. Others argue that countries will go out and contract further debts, under the belief that these debts will also be forgiven at some future date. They further argue that it would be unfair to developing countries that manage their credit successfully and do not get into debt. Others argue against the conditions attached to debt relief. These often widen the gap between the rich and the poor, as well as increase economic dependence on the North. Another concern is that the money will be used to enhance the wealth and spending ability of the rich, many of whom will spend or invest the money in rich countries rather than creating a trickle-down effect in their own country. Opponents argue that the money would be far better spent in specific aid projects which actually help the poor.

Key terms

Bilateral aid: aid arranged directly between two countries, a donor and a recipient.

Debt relief: the partial or total forgiveness of national debt, or the slowing or stopping of debt growth.

Multilateral aid: aid that countries provide through some intermediary or international organisation, such as the UN or OECD.

Trickle-down effect: the gradual spread of growth or wealth, as from core to periphery or from rich to poor.

Figure 6.33 Aid pathways

Discussion point

Suggest reasons why it might be better to promote multilateral rather than bilateral aid.

Take it further activity 6.5 on CD-ROM

Reducing inequalities within countries

Spatial, social and economic inequalities

It is a basic fact of geography that few phenomena are actually evenly spread. This, as we have seen throughout this chapter, is particular true of economic development. Governments of all political shades have found it necessary to apply policies aimed at achieving a more equitable distribution of growth and prosperity. The UK government has been no exception. Planning, backed by regional subsidies and taxation, has done much to reduce spatial inequalities. However, reducing those social and economic inequalities that exist more at a personal level has proved altogether more challenging. The following methods have all been tried in many MEDCs:

- the law – a concession, such as legal aid, should help to protect the civil rights of the poor, but in most legal matters it is money that ultimately buys the best defence.
- education – has the power to make people more aware of society's inequalities and, in a situation of free education for all, it should help people break out of the cycle of poverty. Sadly, there is little evidence that the latter happens to any significant degree.
- subsidies – the provision of such things as social housing, income support and free school meals may help to improve the living conditions of the poor and reduce deprivation, but they do not eliminate poverty.
- taxation – taxing the rich to pay the poor has done little to narrow the basic income gap. However, it can raise revenue to finance subsidies and concessions that benefit the poor, such as those illustrated immediately above.

Finally, pause a moment and just reflect on the situation in the developing world. Few LEDCs command the resources necessary to begin to fight their domestic inequalities, be they spatial, social or economic.

Case study | Making the UK more uniform

Although the UK's economy is essentially a capitalist or market one, since 1945 there has been a considerable amount of government intervention. Much of this has been aimed at reducing three main expressions of spatial inequality:

1. regional growth versus regional decline (the UK has its own North–South Divide) (Figure 6.34)
2. urban versus rural areas
3. inner city versus outer city.

Important contributors to these inequalities during the first half of the twentieth century included:

- the heavy concentration of population and economic growth in London and the south-east
- the decline of heavy industry based on the coalfields
- the depopulation of rural areas
- the decay and obsolescence of inner-city areas
- the proliferation of urban sprawl.

Action to deal with the first has been of a stick-and-carrot variety. The growth of London was curbed by physically corseting it inside a green belt and imposing restrictions on the creation of new jobs in offices and service industry. At the same time, mechanisms were put in place to transfer this surplus growth (or overspill, as it was termed) to the old coalfield areas (South Wales, north-east England, Lancashire, etc.) in the hope that somehow this overspill would bring them back to life and reduce the high levels of unemployment consequent on the shutdown of traditional industries. All sorts of incentives were offered to tempt businesses and people to move to these ailing regions. Although many of the coalfield areas have now adjusted to the post-industrial scenario, there are still parts of the UK receiving assistance, but nowadays this comes from the European Union.

A start was made on tackling the problems of obsolescence and dereliction in inner-city areas. Comprehensive redevelopment became the buzz word, but the process was also known as 'slum

Figure 6.34 The UK's North–South Divide

Manchester, Liverpool, Sheffield and Newcastle is clear evidence of this. These and other cities have managed to survive devastating deindustrialisation by finding new means of livelihood.

The tables have been turned in the suburbs versus inner-city scenario. As the appeal of the suburbs begins to wane and as transport becomes increasingly expensive, large inner-city areas are once again undergoing redevelopment. A thorough makeover of housing, services and the built environment has produced an image and a location that are proving highly attractive to younger homeowners.

The process of counterurbanisation has helped breathe new life into rural areas. In an age of leisure and recreation, rural areas find that they have much to offer urbanites in the context of tourism. The countryside is no longer just about farming.

clearance' or 'the bulldozer approach'. Existing residents were re-housed in one of three different ways – in high-rise apartment blocks built on slum rubble; in large estates of social housing mainly built towards the edge of the built-up area; and in distant, freestanding new towns and expanded towns.

Throughout the second half of the twentieth century, urbanisation continued to leave its mark on remoter rural areas. Such areas were being drained of people, particularly those in the economically-active age range (16 to 65 years). Much government funding was directed to the improvement of services (now concentrated in key settlements), to maintaining the farming industry and to ways of improving the accessibility of such areas.

Although government intervention and legislation played some part in reducing these inequalities, in the event, market forces have really come to the rescue. The high costs associated with living and working in London gradually persuaded many people and businesses to look to other cities and even beyond. The current prosperity of cities like Birmingham,

— **Key terms** —

Counterurbanisation: the movement of people and employment from major cities to smaller settlements and rural areas located just beyond the city, or to more distant smaller cities and towns.

Overspill: the relocation of surplus population and employment away from areas where they can no longer be effectively accommodated.

Personal inequalities

When it comes to reducing economic and social inequalities at a personal level, there is a range of actions that can be taken (Figure 6.35). No single action is capable of providing a miracle cure, but concerted action of a stick-and-carrot nature in both arenas could well begin to reduce the affluence gap. There will be some who argue that nothing will happen unless there is some form of political revolution. However, the communist experience perhaps serves as a salutary reminder that even the most egalitarian societies have their rich and poor. On the other hand, it has to be admitted that little will be achieved without government support and appropriate legislation.

Figure 6.35 Some possible ways of reducing personal poverty in the UK	
Economic	◆ Setting a minimum wage that bears a sensible relationship to the true costs of living ◆ Reducing income differentials by graded income tax bands ◆ Promoting equal opportunities ◆ Ensuring job security ◆ Encouraging self-reliance and enterprise
Social	◆ Making social and welfare services, particularly education and healthcare, accessible to all ◆ Ensuring a secure and safe living environment ◆ Subsidies to help the needy, not the lazy ◆ Promoting marriage and the extended family ◆ Protecting civil liberties and equal rights ◆ Encouraging care in the community

Activity

Go to www.heinemann.co.uk/hotlinks, enter the express code 7627P and click on the relevant link to access information on currently assisted areas in the UK. Identify the particular challenges of the parts of the UK receiving assistance.

A postscript

Throughout this chapter, the view of economic development that has been taken is the one that prevails in the North, the more developed part of the world. However, alternative views of development are beginning to appear from the South. One of these argues that there is more than one development pathway or stairway and that development in general should be more sensitive to the specific conditions that exist within a particular country. So rather than thinking that all countries should jump through the same set of hoops, better that countries be left to follow the course that best suits their circumstances. In such cases, it would be inappropriate to judge the success or otherwise of the adopted development pathway using standard economic measures, such as per capita GNI.

Until now, development has been seen as providing one overarching solution to the problems of people living in the South. It is seen by some as transforming traditional societies into modern Westernised countries. But might this view be a rather arrogant one? Might it be that the people of traditional societies are perfectly content with their lot? Might they have rather different and less material aspirations? Many in the South are beginning to realise that development,

defined in certain ways, might have a downside. They see development as perpetuating poverty and poor living conditions; as creating dependence and spatial inequalities; as eroding local cultures and traditions; and much of it as being environmentally unsustainable. In short, there is a growing anti-development groundswell (Figure 6.36). This leaves three intriguing questions:

◆ Are these emerging anti-development attitudes ever likely to be converted into real policies?

◆ Are such policies likely to reduce or increase the economic and social inequalities that characterise today's world?

◆ Can there be development without economic growth?

Over the last 50 years, the North has enjoyed unparalleled economic growth. We have better homes, cars, holidays, jobs, education and above all health. According to economic theory, this should have made us happier, but surveys show otherwise. When Britons and North Americans are asked how happy they are, they report no improvement over the last 50 years. Research shows that once basic material needs are met, extra income becomes much less important than

Figure 6.36 Opposing views of development

Pro-development	Anti-development
Development leads to economic growth.	Development is a dependent and subordinating process; it is something handed down by the powerful to the less powerful.
Development brings overall national progress.	Development creates and widens spatial inequalities.
Development leads to modernisation along Western lines.	Development undermines local cultures and ways of life.
Development improves the provision of basic needs.	Development perpetuates poverty and poor working and living conditions.
Development can lead to sustainable growth.	Development is often environmentally unsustainable.
Development leads to the improved management of people and the environment.	Development infringes human rights and undermines democracy.

Source: Morgan, J., (EPICS) *Development, Globalisation and Sustainability*, Nelson Thornes (2001)

our relationships with each other: with family, with friends and in the community. So the danger is that too keen a pursuit of economic development and higher income may threaten other important aspects of development in its fullest sense.

Discussion point

Debate the motion that 'development is not possible without economic growth'.

Knowledge check

1. Explain what is meant by the economic development stairway?
2. To what extent do you agree with the view that quality of life is the best indicator of development?
3. 'Economic growth is the powerhouse of development.' Discuss with the aid of examples.
4. Critically examine the view that economic development relies as much on human resources as it does on physical resources.
5. Referring to examples, describe ways in which governments can encourage economic development.
6. Explain why we cannot be certain about what is happening to the development gap.
7. 'The trouble with economic development is that it creates winners and losers.' Discuss.
8. Explain and illustrate what is meant by dependence.
9. Assess the relative merits of direct investment, trade and aid as ways of reducing the North–South Divide.
10. To what extent do you agree that spatial inequalities within countries are easier to reduce than those between countries?

Exam Café
Relax, refresh, result!

Relax and prepare

Student tips

What I wish I had known at the start of the year...

Radjekr

"Remember that inequalities may be environmental, demographic, economic, social and political. Many of these are interlinked or connected, as one aspect causes or results from another aspect. At one end of the spectrum this leads to multiple deprivation and at the other end it results in affluence. Try to apply the vicious and virtuous spirals to explain these interconnections."

Pritti

"Governments seek to reduce inequalities once this becomes a priority. Often, inequalities are initially part of the 'backwash' mechanism that helps power development of growth poles – usually the core area or capital. Countries with limited resources are often best advised to concentrate these limited resources (financial, technical expertise, infrastructure, etc.) to get the greatest impact on the economy."

Daisy

"When selecting a country to exemplify regional inequalities, try to choose one where there is a distinctive difference in the physical geography, as often this is the initial cause of the inequalities which economic, social and political factors reinforce. Larger countries that extend across a number of degrees of latitude, such as Brazil and Nigeria, are ideal as they often straddle a number of climatic types or different reliefs."

Common mistakes – Leevan

"I am still not sure what the term 'development continuum' means. I understand the terms LEDC, MEDC and NIC, but often I get confused as to which categories to put particular countries into, such as China and India. My teacher says that is the very reason why we should use the idea of a continuum or spectrum of levels of development rather than distinct groups – development is so dynamic that few countries remain conveniently in one category. The continuum also implies that countries may be at different levels of development in terms of demographic, economic, social and political characteristics. Cuba is an excellent example of this with many features suggesting high demographic and social development, but also many economic and political characteristics suggesting lower levels of development. This makes using examples much more difficult as it was so much easier when there were only two distinct groups – LEDCs and MEDCs. However, it does reflect the real world more accurately."

Refresh your memory

6.1 In what ways do countries vary in their levels of development?

Quantitative	Should relate to total population
Demographic	Level of birth rate, death rate, infant mortality, life expectancy in years; daily calorie supply (per cent of needs), per cent malnourished
Economic	Income or wealth, e.g. per capita GDP. Number of possessions, e.g. cars per 100 people, per cent with TVs/computers. Employment: per cent in primary, per cent unemployed or underemployed. Level of saving/investment per head. Infrastructure, e.g. miles of road, per cent with phones. Consumption levels, e.g. power consumption
Social	Percentage of total income/wealth the richest 10 per cent have. Adult literacy (per cent); per cent of 5-year-olds in school; per cent at university. Per cent of teenage pregnancies; per cent on drugs. Services: population per doctor/dentist; number of library books taken out. Crime/violence: murders per 1000 people; number in prison
Political	Percentage voting in elections; size of police force/army; number of political prisoners; per cent of industry state-owned.
Combined	HDI – combines life expectancy, adult literacy, purchasing power
Qualitative	Freedom, justice, peace

6.2 Various factors influence the rate and level of development

Physical	Level and type of mineral resources; type of climate; steep versus flat relief; prone to natural disasters; water quality and quantity; type and fertility of soil; biotic resources; disease prone or not; accessibility (centrality)
Economic	Natural source of energy; mineral types and quality; building materials; agricultural and forestry resources; level of transport; route centre; port facilities; trade blocs
Social	Population size; type and growth rate; importance of and type of religion; level and types of education; health; attitude to risk and innovation; level of unionisation; attitude to elderly
Political	Level and quality of political control; security; fiscal policy; attitude to trade and aid; international relations; attitude to international migration
Historical	Colonial development/exploitation; role of inertia; former links

6.3 Economic development can increase or decrease inequalities

Models	Myrdal's cumulative causation; Friedman; Rostow; growth poles
Inequalities	Between areas, between groups, between sectors
Increase by:	Backwash – investment, resources and labour pulled into core from periphery so stagnates
	Multiplier effect/virtuous spiral in core – upward spiral of wealth and employment. Reverse multiplier/vicious spiral in periphery – downward spiral of poverty and unemployment
	Limited resources so concentrated where growth potential is greatest – economies of scale
	Urban life equals better education, services, etc., so birth rate falls in core but remains high in periphery
Decrease by:	Spread – core spreads its wealth, employment and demand out to remote cheaper areas
	Construction of national integrated infrastructure, e.g. roads, power
	Diseconomies of scale set in the wealthy core
	Positive political decisions to reduce inequalities possibly to prevent unrest – government schemes set up outside core

Refresh your memory

6.4 The factors increasing or decreasing the development gap	
Models	Myrdal's cumulative causation, backwash versus spread
Physical	Relief – unsuitable sites, e.g. steep slope versus flat firm land
	Drainage – water supply: shortage, pollution, floods
	Vegetation – economic uses, loss of habitats
	Pollution – air, water, land, noise, visual
Economic	Settlement – housing quality and quantity, cost
	Power – availability, reliability, cost
	Industry – type of jobs, lack of jobs or low pay, migrant labour
	Services – level of provision of schools, shops, clinics, etc.
	Transport – level of connection, cost, public transport, flexibility
Social	Wealth – relative levels, inequality and deprivation
	Cultural change – role of religion, tradition
	Age profile – dependency ratio, birth rates, social services
	Migration – emigration, characteristics, depopulation
6.5 Economic inequalities may result in social and environmental inequalities	
Economic	Fuel poverty – poor spend higher proportion on heating. Transport poverty – rely on public transport. Food insecurity – quality and quantity. Financial exclusion – banks, building societies, credit agencies limit funds; often have to borrow at high interest
Social	Digital divide – fewer with Internet access. Inverse care law – those in most need get least. More on social service support. More crime and violence – greater fire risk, vandalism. Stigmatisation – postcode discrimination. Poorer schools – social exclusion. Increasing homelessness
Environmental	Dereliction and decay of property, parks, etc.; pollution more common
Multiple	Many of these interconnect to form multiple deprivation
6.6 Management of social and economic inequalities in an attempt to reduce them	
Reasons	Moral – equal opportunities, freedom from deprivation
	Economic drain on resources, loss of potential consumers and workers
	Political – reduce alienation, cut social costs, gain votes
Methods	Top-down versus bottom-up schemes; government versus self-determination
Political	Re-development schemes, regeneration schemes. Social housing, subsidies, taxation, cheap loans, improved education, improved public services and transport, better policing
Social	Gentrification, self-help schemes, refurbishment schemes, re-branding, community development
Economic	Credit unions, cooperatives

Top tips...

Use models or theories to explain why inequalities seem to be an inevitable initial stage in any kind of development process but then reduce as levels of growth accelerate (this is the theory put forward by Myrdal and others). It is worth questioning the later stages as some areas and groups, etc., seem not to share in the 'spread' or 'trickling down' processes. Again, this is an example where models are helpful but A2 candidates are expected to use them with reservations and exceptions.

Get the result!

Example question

Study Figure 1, which shows the percentage of national wealth in the main regions of a LEDC. Suggest the possible issue(s) indicated and appropriate strategies that could be used to manage its impact. [10 marks]

Figure 1

Regions	1951	1981	2001
Southern	5%	3%	1%
Central-west	10%	15%	12%
Central-east	8%	12%	15%
North-west	18%	6%	5%
North-east	9%	14%	10%
Capital	50%	50%	57%

Student answer

Figure 1 shows that over the fifty-year period, southern and north-west regions have steadily declined in their share of the country's wealth, two regions have declined in the last twenty years and two increased — central-east has doubled its share and the capital has reinforced its share by 7 per cent. This leads to typical regional inequality with a parasitic primate capital and declining peripheral regions. To reduce this inequality the government should start resource development outside the core, for example minerals or power production, encourage spread by building communications but also limit growth in the capital. 'Trickle-down' strategies should be developed such as relocation of government offices and colleges to poorer regions.

Examiner says

Good use of the data to identify a number of issues, not just the problem of the capital region.

Examiner says

Good use of technical terms and practical strategies. The role of constructing communications to encourage economic forces of spread is a key point in making this A-grade quality.

Examiner says

Range of appropriate strategies suggested which go beyond simply developing the declining regions.

Examiner's tips

Issue questions are relatively new to A-level. They provide a range of data in a variety of forms and the candidate has to identify one or more issues from the data and then suggest appropriate solutions or remedies. There is always one overwhelmingly obvious issue but there will be others. The secondary ones are not there to fool you but rather they are there so higher-achieving candidates can demonstrate their ability to analyse and interpret — two high-level skills. If you can't see a secondary issue, usually linked or connected to the obvious one, then think about the nature of the data — its origin, its accuracy, its date and how it was collected.

Exam Café

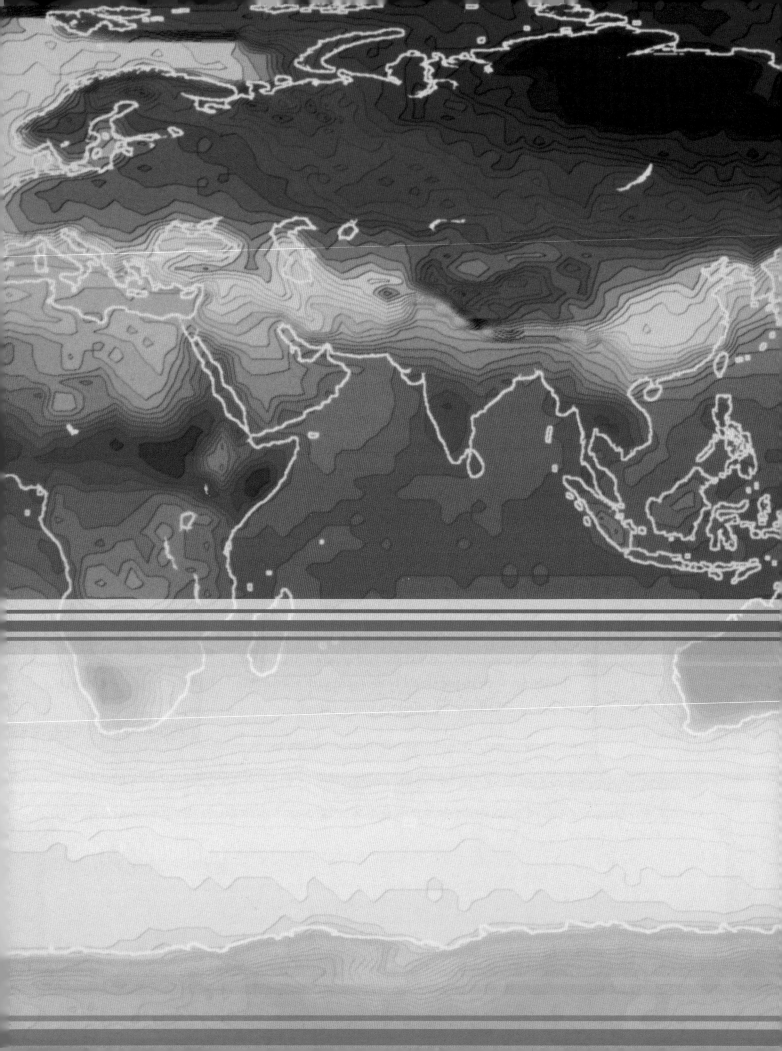

UNIT 4

Geographical skills

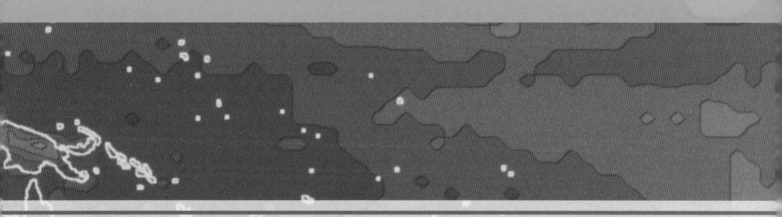

The aims of this unit are to develop:

- knowledge and an understanding of the process of geographical research, including fieldwork
- the skills necessary to complete a piece of individual geographical research
- the use of technology, e.g. GIS, remote sensing, etc., as research tools
- a knowledge and understanding of the potential of ICT and its relevance to geographical change
- the ability to select and use appropriate GIS skills and techniques to explore geographical issues including decision making and problem solving
- an awareness of the problems involved in undertaking individual geographical investigation/research
- an understanding that interpretation and evaluation of research results should reflect the links and connections between diverse elements of geography.

Chapter 7
Geographical skills

This chapter explores the six stages of geographical research. This research may be as fieldwork – practical work in a small-scale environment – or as an investigation carried out in class making use of data from a range of sources such as the Internet. It sets out the main aspects that you would need to consider when selecting particular skills or techniques for use in your research. The key aspects of any research are the accuracy, reliability and appropriateness of your chosen approach and of the techniques you use. This is examined in the separate Geographical skills paper. Section A of this paper gives you a choice of one question from three based on various aspects of the six stages of an investigation. Data is given to you and you are expected to comment on its accuracy or appropriateness. Section B consists of two compulsory questions, based on two of the six stages, which require you to refer to the geographical investigations you undertook during your AS and A2 courses.

Questions for investigation

This unit is different from the other units in that it invites you to investigate any of the questions posed in the other units, at AS or A2, or any other geographical question or topic. The important thing is that you should understand and follow the six stages of an investigation:

1. Identify a suitable geographical question or hypothesis for investigation.
2. Develop a plan and strategy for conducting the investigation.
3. Collect and record data appropriate to the geographical question or hypothesis.
4. Present the data collected in appropriate forms.
5. Analyse and interpret the data.
6. Present a summary of the findings and an evaluation of the investigation

Consider this

Geography is all about where things are located so research should seek to identify, analyse and explain spatial patterns. This module is about the skills that are used in geographical research whether this is by fieldwork or by classroom-based investigations. At all stages it is assumed that a logical approach to research is followed.

Formulate reasonable geographical question/hypothesis

↓

Plan data collection strategy

↓

Collect data

↓

Process data collected – representation

↓

Analyse the data

↓

Seek interpretations and explanation

↓

Draw conclusions

↓

Evaluate the investigation

7.1 Identify a suitable geographical question or hypothesis for investigation

'A successful geographical investigation is dependent upon identifying a clear geographical question at a scale that is practical in research terms.'

Choosing a topic area

This first part is the most crucial stage. Before you can start to think about exact questions or hypotheses to investigate, you need to find a general topic or geographical area that you are interested in and which would be suitable for investigation. Local newspapers, reading of geographical books, magazines and Internet sites can be good starting points to get some ideas. Remember that it must be something that can be investigated given the time and resources at your disposal. Often the topic area links into what you are doing in class, such as a study of a local woodland if you are studying the 'Ecosystems and environments under threat' option. The teacher may select the topic for you or you may choose to work with colleagues, in which case a group of you decide on the topic.

Turning a topic into a question or hypothesis

Once you have decided on a topic to be researched, it is vital that it is fine-tuned and a suitable question or hypothesis developed. This makes analysis much easier.

A good-quality geographical question/hypothesis should include the features described in Figure 7.1 below.

Figure 7.1 Features of a good geographical question/hypothesis

Feature	Description
A suitable scale	Small-scale studies are more accessible and more accurate than larger scale studies. For example, a land-use survey of a city such as Birmingham is far too large to be practical and unlikely to give an accurate overall picture, while a microclimate survey around the school or college is far more practical.
Readily researched	Some topics can't be researched effectively. This may be due to: • the time needed – possibly years in longitudinal studies, for example a study exploring changes in a local ecosystem following modification in the type of farming • the number of researchers needed – for example, a micro-climate study needs a lot of people to take readings at exactly the same time • the data being unavailable or of a sensitive nature – for example, data obtained during surveys of crime rates or levels of deprivation.
Clearly defined	Terms need to be carefully thought out to avoid ambiguity and to make it clear what is being researched. It is pointless having a vague topic such as 'The shopping patterns in Oxford' – this needs to be more precise, such as 'How does shopper mobility impact on shopping patterns in Oxford?'
A clear geographical nature	There should be some focus on spatial or locational patterns. For example, 'To investigate the pattern of vegetation on sand dunes' is better expressed as 'Does the nature of the vegetation on sand dunes change with distance inland from the high water mark?'
Based upon wider geographical theories, ideas, concepts or processes	Typically titles may refer to specific theories or models, for example 'Does the land use patterns of urban area X resemble those of the Hoyt sector model?'
A clear aim	There should be a single clear purpose and focus for the research. For example, 'To investigate the impact of second homes in rural areas' is given a clearer aim by 'Has the growth in second homes in rural areas had a negative effect on rural communities?' This rewording also enables a clear outcome and conclusion to be stated.

A good title is also SMART – Simple, Measurable, Achievable, Realistic, Timed.

- Simple – not a multiple set of tasks; a single question or hypothesis is most effective.
- Measurable – can it be measured? This helps suggest how the results may be analysed.
- Achievable – in the place chosen and in the time available with the resources/equipment available. Make sure it can be done!
- Realistic – can it be done? Try to avoid those 'what if' type of questions as they are difficult to prove, especially if there is not a 'control group'.
- Timed – how long will it take? It is often difficult to carry out research over a long time period or at repeated intervals.

Deciding on a question for investigation

Gemma enjoys sports so she decided to investigate the catchment area of her local sports centre in Evesham. After carrying out some Internet research and asking some questions at the local sports centre, she narrowed her area of investigation down to two questions. To decide between them she looked at each one in turn and measured them against the criteria above.

- Question 1 – 'Where do people come from to visit the centre?' The advantages of this question are that it has a definite aim, is clearly geographical, is measurable and should be achievable. The disadvantages are that it might be difficult to collect confidential data on where people live and it might vary over the time of the day or week.
- Question 2 – 'Do the type of activities undertaken at the centre reflect the age of the people attending or the time of day?' The advantages of this question are that it has a clear aim, is measurable and can be achieved by simple observation. The chief disadvantage is the lack of geographical relevancy.

Gemma decided to investigate Question 1.

Activity

All of the titles below have been used by students in the past. Discuss with a partner what makes these titles effective or poor investigations.
- Is there a relationship between salinity and vegetation type?
- Investigate the variations in land values along a road
- The CBD of Worcester
- What would be the impact of opening a new sports centre in Cambridge?
- Has the building of the M25 had an impact on Guildford?
- An investigation into the origin of the beach at Porlock Weir, Devon

Theory into practice

In the examination, you may be asked to identify possible lines of research that could be carried out in an environment shown in a photograph or on an Ordnance Survey map. Having chosen your question or hypothesis for testing, you are usually asked to justify your choice. This may reflect the nature of the landscape or environment, the scale or size of area, accessibility, time available, equipment available, the number of people in the research team, and the likelihood of conditions remaining little changed during the research. Equally, it may be important to say why alternate questions might not be so suitable.

Study the Ordnance Survey extract (Figure 7.2) of an area in North Devon.

Source: Ordnance Survey

Figure 7.2 Braunton and Croyde

1. Outline and justify an appropriate geographical investigation title that could be undertaken within the area shown on the map.
2. Why might a dune survey be unsuitable in map square 4435?
3. Why might Braunton be unsuitable to test the Burgess model of urban land use?
4. Why might the time of the year be a crucial factor when considering a human geography topic in this area?
5. What limitations does the scale of the map have on the choice of your investigation title?

7.2 Develop a plan and strategy for conducting the investigation

'A successful geographical investigation requires careful planning which balances the needs for accuracy and reliability against limitations imposed by time, resources and the environment in which the investigation is being conducted.'

This stage is about deciding how you are going to do your investigation. Through this process, you may encounter problems that persuade you to go back and change your title or topic. This stage is very much about practical issues – the strategy you are going to adopt. The time you spend on the initial planning is time well spent as it can identify and avoid many of the pitfalls that so typically face researchers once the research is underway or when you go out into the research location. It is important not to plan too rigidly as even the best plans can be disrupted by a change in the conditions or your circumstances. Salim had planned an effective stream survey but the day he went to carry it out it snowed and he couldn't even see the stream.

The issues that need to be part of the planning include:

- subsidiary questions
- data
- sampling
- location
- time
- risk assessment.

Subsidiary questions

You need to establish a limited number of subsidiary questions/hypotheses to ask within your title. You should keep these to a minimum – no more than three; any more will take too long or may conflict or create issues not linked to the main aim. Make sure your subsidiary questions support and relate to your initial question or hypothesis.

Developing hypotheses for a research project

Alison decided she was going to look at traffic congestion in the neighbourhood of her school in an urban area. Her title was: 'Does traffic flow in Erdington vary with the time of day?' To develop this further and help look for explanations she supported it with the following hypotheses:

- 'Peak congestion occurs during the morning rush hour (7.30–8.30).'
- 'The most congested roads are the main roads in the area.'

These two simple hypotheses can be tested and provide a geographical context to her title.

Data

You need to identify the type of data you will need to examine your hypothesis and where you are going to get this data from. Whatever hypothesis you have chosen, you will need a combination of primary data (which you will analyse yourself) and secondary data (which has been analysed by someone else).

Primary data

Many investigations involve some collection of primary data so the next step is to establish the strategies and methods for how you are going to collect it. It is important to consider the time that you have available and the location that you have chosen. Sampling may be appropriate (see page 254).

It is important to evaluate at this planning stage how reliable any data you collect is likely to be and how to make it more reliable. In some cases this will mean repeating your data collection at least a couple of times. Also, carrying out a pilot survey is often good practice to find out whether any amendments are needed before you start the actual survey. It is especially useful to pilot any of your questionnaires.

If you use primary data which you have not collected yourself it is important to acknowledge its source, but if you use primary data from colleagues who have processed it, such as a table of results, it is secondary data.

Secondary data

Sometimes it is not easy to find what secondary data you need and where it is easily available. Libraries, the Internet and previous investigations are good starting points, but many geographical aspects have agencies with their own database such as the Department for Environment, Food and Rural Affairs (DEFRA). However, do not just include such data for the sake of it – it should add to your investigation, perhaps as a comparison over time or space.

> ### Using secondary data
> Javan did an investigation on his local urban stream's characteristics. He obtained secondary data from the river authority on a similar stream in a rural area so he could compare the variation in flow on the same day as he measured that of the urban stream.

There are several things to bear in mind when selecting secondary data for your investigation. Even statistics can be biased so it is important to think about who has compiled and analysed the data and why they have compiled it. Be particularly careful over data which you have found on the Internet. Also remember that the date of information is important; usually you want to use the most recent data available.

All sources of data will need to be specified in the bibliography or reference section of your investigation, so it is imperative that you make a note of it.

> ### Key terms
> **Primary data:** unprocessed information, which means information collected through fieldwork investigation or material derived from other sources that has not been analysed and/or interpreted in any way. Such material might, in addition to data collected in the field, include census, telephone directories, electoral rolls, trade directories and remote sensed data.
>
> **Secondary data:** information which is derived from published documentary sources that has been analysed and/or interpreted, such as processed census data, research papers, published maps and textbooks. It also encompasses sources of specific techniques and the formulae of their calculation.

Sampling

Often an entire population cannot be surveyed so a representative sample has to be taken. The larger the sample the more representative and accurate are the results, but the availability of time and resources limit this. Some issues need deciding at this stage. The most important are those associated with how you are going to develop your sampling procedure.

How much should I measure? – sample size

Sample size is important. Generally the larger your samples the more representative of the whole it is and the more accurate are your results. Usually at least 30 samples should be taken. Some candidates only sample five or six points or areas and then try to draw valid conclusions. These would not be reliable. It is possible to calculate the size of sample needed to provide a given level of accuracy using the equation below.

$$SE\% = \sqrt{\frac{PQ}{n}}$$

P = percentage of items in a given category
Q = percentage not in this category
n = number of points in the sample
SE = standard error

> ### Activity
> If P = 20% and Q = 80%, what must *n* be to give a 5% error (or 95% accuracy level)?

Sample type

This covers two distinct aspects. First, the sample unit needs to be decided on. In most cases this is **point sampling** such as names on a list or coordinates on a map, but for some types of research **line or linear sampling** is used. This is usually along a line or transect that has some meaning, for example along a road or across a line of dunes. **Areal sampling** is also used, usually in vegetation surveys where a quadrat is used to measure the contents of a 1-metre square area or it could be a grid square on an OS map (see Figure 7.3).

Sampling method

Having decided the unit then the method of sampling needs to be chosen from those shown in Figure 7.4.

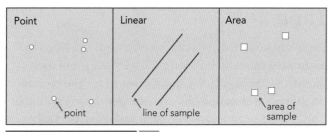

Figure 7.3 Sample unit

> **Theory into practice**
>
> Ayza was doing an investigation of the catchment area of her college. She decided to ask her friends at college where they lived, how they got to college and how long it took. Is she correct? Explain your answer. How could you have improved this sampling strategy?

Some of these methods suit particular investigations. Pragmatic sampling tends to be used in stream studies when river bank access is limited or dangerous. Systematic sampling is often used when sampling from an OS map (using grid line intersections) and wherever speed of sampling is needed. Stratified sampling is used where there are distinctive differences in something, such as in rural land use surveys where you need to ensure sample data is taken from different rock types, soils or relief in an area. Pure random sampling is quite difficult to achieve but is commonly used in sampling stone sizes in a river or glacial deposit.

In reality there is a blurring of these methods. For example, it is usual to use a random method to sample your data within the areas selected for a stratified sample, and at systematic points along a transect data is often collected randomly.

> **Key terms**
>
> **Areal sampling:** where the unit of sampling is a pre-defined area such as a quadrat.
>
> **Linear sampling:** where sampling is along a line. (Transect sampling is where sampling is along a line but it has some breadth such as 1 metre either side of the line.)
>
> **Point sampling:** where discrete or separate points or individuals are chosen as the sampling unit.
>
> **Pragmatic sampling:** method of sampling in which the samples are chosen from those that are easily and safely available.
>
> **Random sampling:** method of sampling in which all the elements of data have an equal chance of selection.
>
> **Stratified sampling:** method of sampling in which the sampling of the data from subsections of the whole is proportional to their share or size of the whole.
>
> **Systematic sampling:** method of sampling in which the sampling is at a set interval from the initial point, which is chosen at random.

Figure 7.4 Methods of sampling

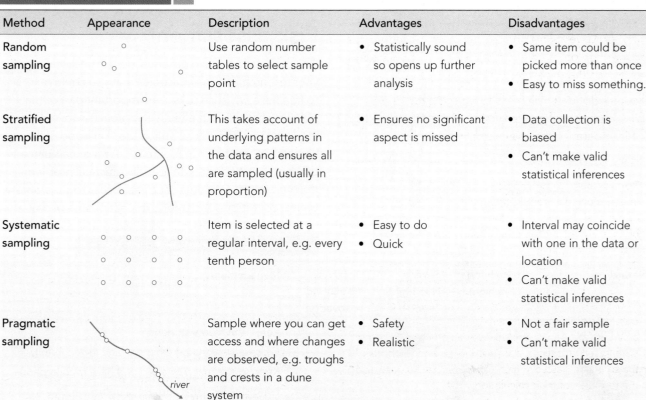

Method	Appearance	Description	Advantages	Disadvantages
Random sampling		Use random number tables to select sample point	• Statistically sound so opens up further analysis	• Same item could be picked more than once • Easy to miss something.
Stratified sampling		This takes account of underlying patterns in the data and ensures all are sampled (usually in proportion)	• Ensures no significant aspect is missed	• Data collection is biased • Can't make valid statistical inferences
Systematic sampling		Item is selected at a regular interval, e.g. every tenth person	• Easy to do • Quick	• Interval may coincide with one in the data or location • Can't make valid statistical inferences
Pragmatic sampling		Sample where you can get access and where changes are observed, e.g. troughs and crests in a dune system	• Safety • Realistic	• Not a fair sample • Can't make valid statistical inferences

Location

When you decided on a title for your investigation, you will have thought about the area you were going to use. At the planning stage you need to specify the exact boundaries of the location and ensure the location of your research is clearly identified for the reader. Be wary of merely including a map from the Internet without using it in some way. For example, include a map to show the location of planned sampling points. Figure 7.5 shows how an Internet map was adapted.

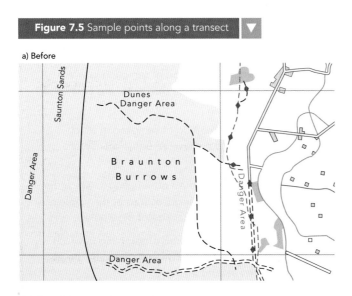

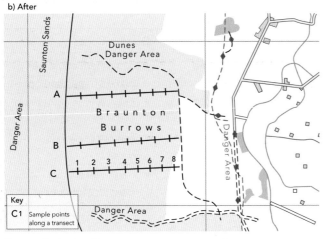

Figure 7.5 Sample points along a transect

Source: Ordnance Survey

Time

Time is an essential part of your planning. You need to accurately estimate how long different stages of the process will take you. It is good practice for you to draw up a timeline or action plan to help identify any potential bottlenecks or aspects where you have to rely on other people. Cameron's time plan is shown in Figure 7.6.

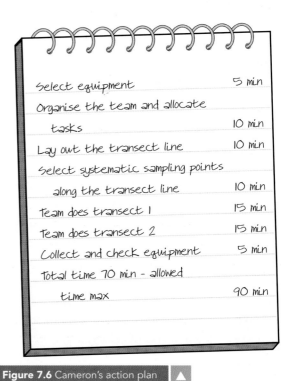

Figure 7.6 Cameron's action plan

Risk assessment

An appreciation of the potential risks in undertaking your research and the strategies that you could take to minimise any such risks is an important part of the planning process. An effective approach is to look at the severity of the risk as well as the likelihood of it happening, as shown in Figure 7.7. A reconnaissance of the area is important to identify practical problems that you might encounter, such as access to sites, safety issues, conditions under foot, impact of time of day, etc.

Figure 7.7 Risk assessment and mitigation

Hazard/risk	Likelihood (1–5 high)	Severity (1–5 high)	Total risk (likelihood × severity)	Management
Being mugged	1	5	5	Keep with fellow students
Crossing busy roads	4	3	12	Cross at designated crossings and look carefully
Getting lost	5	2	10	Keep to the mapped areas and have mobile phone
Sunburn	4	1	4	Long sleeves, caps and use sunblock

Activities

1. In the table below, identify to which type of data the sources belong.

Data source	Primary	Secondary	Either/not sure
Data collected by other teams			
Satellite image			
Types of vegetation survey			
Traffic survey			
Last week's survey results			
Ordnance Survey map			
Your photo of a location			

2. Look at Figure 7.5.
   a What aspects make version b) a more 'useful' map of the planned location?
   b What else might help improve the usefulness of the map?
   c Why do you think this location was chosen to answer the hypothesis: 'Plant species diversity does not change with distance from the HWM'?

3. Figure 7.6 shows an action plan for the dune survey shown in Figure 7.5.
   a Can you suggest any likely problems or bottlenecks in this plan?
   b Are there any aspects or stages that you would add?
   c Could this plan be shown in a more effective way? If so, what would you change?

Theory into practice

Can you spot the potential problems in the shopping survey questionnaire in Figure 7.8? Rewrite an improved version of the questionnaire.

Figure 7.8 Questionnaire on shopping habits

Name: _____ Age: _____ Sex: _____
Where do you live? _____
How often do you shop here? _____
How did you get here? Was it by walking, bus or car? _____
Why do you shop here? _____
Which shops have you visited today? _____
Is this your main shopping centre? Yes/No _____
Where else do you regularly shop? _____

The finished plan

So what could a finished plan look like? Zena drew up the plan in Figure 7.9 to help her assess the impact of the number of walkers on part of the Pennine Way footpath.

Aspects	Detail	Comment
Hypotheses	The greater the number of walkers the greater the erosion. Narrow paths are more eroded than broader ones.	Need to decide on definitions here
Data	Primary – walker count and erosion survey Secondary – local field study centre for data from previous surveys	Is it getting worse?
Location	On Malham Moor – range of locations	Need to think of sample type
Time	Start at 8.00 a.m. finish at 5.00 p.m.	Lunch break?
Risks	Do pilot survey to identify risks but mustn't be by myself	Weather is probably key

Figure 7.9 Zena's plan to assess impact of walkers on the Pennine Way

Activities

1. How effective is Zena's plan in Figure 7.9? How would you improve it?
2. Is Zena's plan more useful than Cameron's plan in Figure 7.6?
3. What other ways could you set out the plan for your investigation?

7.3 Collect and record data appropriate to the geographical question or hypothesis

'A successful geographical investigation is based upon thorough methods of data collection and recording, which consider accuracy and reliability in relation to the data being collected.'

This stage is very much about what actually happened. You planned it in Stage 2 but now you are doing it. Did it work out the way you planned, were their limitations, or were practical problems encountered? Clearly such issues are central to the accuracy of your investigation's outcomes.

Data collection

At the planning stage you should have thought about how you will collect your data.

The type of data used

The balance of primary/secondary data varies with the nature of your studies. Some will be nearly all primary data collected by yourself (for example, a dune survey) while others may be all primary data collected by someone else or secondary sources (for example, comparing population structures). This should have been decided on, together with the reasons for choosing the types of data and their sources, at the planning stage of your investigation. At this stage you might find you need to access further sources or you may discover data sources you had not considered in your planning.

How you collected it

It is important that you explain why you chose your method of data collection and answer questions like:

- How does the equipment work?
- Why does it measure the variable effectively?

You then can consider whether your method did work effectively or whether there were problems, such as the quadrat sample grid distorting or breaking when thrown a bit energetically, or thermometers not all reading the same when calibrated.

Many studies show this as a table (see Figure 7.10) or as an annotated photograph (see Figure 7.11).

Data recording

This is simply putting down the data collected in a form that can then be processed in some way. If it is raw data you collected, data recording can be greatly helped if you use prepared recording frames or tally sheets, such as the ones in Figure 7.12. This also

Data needed	Equipment used	Method	Limitations
Height and length of dunes	Clinometer tape measure	Measure foresight and backsight readings, average them and measure the distance between them.	Clinometers are not easy to read accurately and the height of the person doing the sighting may distort readings.
Vegetation cover	Quadrat	Lay quadrat and count the squares that contain vegetation.	How much of the square has to have vegetation to count?
Soil pH	pH meter	Probe is pushed into soil and meter read after 30 seconds.	It is important to be consistent in the depth of the probe.

Figure 7.10 Dune transect survey

Figure 7.11 Field equipment – 1-metre² sample grid (10 × 10)

helps when you are collecting the data as you know how it will be set out. It could even influence the way you plan your data collection, such as the design of questionnaires.

Field sketches (or photographs) and sketch maps

These can be useful at this stage as they help to remind you of the conditions and the context of your measurements. Don't just include them for the sake of it. You should carefully annotate and title any photographs or sketches that you use.

Figure 7.12 Tally sheets

Date: Time:
Location:
Questions: Questions
                                    Yes   No
1. A) ////////   ///   1   ////////
   B) ////////////
   C) ////   ////   2   ///////

2. A) ///   ///   3   ////////
   B) ///////
   C) //////////////   /////// 4   ////

3. Y) ///////////////////   ///   5   ///////
   N) ////

4.   0-10 10-20 21-30   /////// 6   //
     //// ///// /////////

Accuracy and reliability

Reliability is essential for your investigation. At all stages – both before, during and after you have collected your data – you need to be thinking about how reliable it is and how you can ensure that it is as reliable as practicable. Reliability is about how dependable the data is. Does it reflect the whole population, would it be the same if the sampling or survey was repeated? You can increase the reliability of your data collection by the use of pilot surveys and repeating the measurements or readings at different times. An important part of making your data as reliable and as accurate as possible will be the size, type and method of sampling that you use.

Methods of data collection

Kofi did a survey of the use of his local urban park. He planned to use a questionnaire to sample 50 people from when the park opened to when it closed at 7 p.m. He found that he could only ask 35 people as it rained in the afternoon and no one came to the park. His investigation was not reliable as he failed to sample in the afternoon so missed children and others who normally came to the park after school. The fact he had collected 35 of his 50 questionnaires by the end of the morning suggests he would not have sampled the afternoon users in numbers that would have enabled reliable observations or conclusions about the use of the park to be made.

Sally used a metre-square sample quadrat to measure the differences in ground cover under coniferous and deciduous trees. She took ten sample quadrats under pine trees, oak trees and beech trees in her local wood. She did this randomly by standing under each tree, closing her eyes and dropping the quadrat behind her. This is not an approach likely to produce reliable data. The sample method was not really random as not all areas under each tree had an equal chance of being sampled and the results will probably vary with distance from the tree as canopy thickness and hence light penetration to the woodland floor would vary.

Activities

1. What could Kofi have done to make his data collection more reliable?
2. What could Sally have done to make her data collection more reliable?

Accuracy is slightly different from reliability. Methods may be reliable but results can still be inaccurate. This inaccuracy can come from a number of sources. Often the equipment isn't accurate enough (or it may be faulty or damaged) or different members of your group read the measurements differently. It may be that other, unexpected factors affect your measurements such as the weather. Another source of inaccuracy is using other people's primary data when you don't know the exact conditions and circumstances under which it was collected. Sometimes the raw data itself suggests an inaccuracy. Such 'out of line' measurements are termed anomalies and the problem is in distinguishing those that result from poor data collection from those that reflect a real phenomenon.

Key terms

Accuracy: the level at which data is exact and free from error.

Anomaly: data that does not fit into the common pattern or trend.

Reliability: the extent to which sample data reflects the greater whole.

Using a tally sheet

Javan measured slope angles on three different rock types on south-facing slopes in the Pennines. This produced the data collection tally sheet in Figure 7.13.

Javan had to ask himself if the shale slope of 28 degrees was accurate or an anomaly (possibly, for example, due to mass movement or river erosion in the area). He used a clinometer to get the slope angles by sighting on a ranging pole. You can see how important it is to note any unusual situations when collecting data. Javan concluded it was an inaccurate measurement. This decision (correct though it was) caused him some major problems. As he knew one measurement was inaccurate, how many others were? This compromised the accuracy of his whole investigation.

Activities

1. Can you identify any problems with Javan's tally sheet?
2. In what ways might altering this sheet change his attitude to the anomaly?

Figure 7.13 Javan's tally sheet

Rock type/angle	0–10	11–20	20–30
Limestone	✓✓✓	✓✓✓✓	✓✓✓✓✓✓✓✓✓✓
Shale	✓✓✓✓✓✓✓✓✓✓	✓✓✓✓	✓
Gritstone		✓✓✓✓✓✓	✓✓✓✓✓

Activities

This section will tend to be examined in section B of the examination, where you are asked questions on your own geographical investigation. Typically the question might be one of the following:

- Describe your methods of data collection and assess their reliability and accuracy.
- Describe and justify your choice of primary and secondary data you used in your investigation.

In section A you are likely to be given some form of recording sheet and be asked to comment on its effectiveness. For example:

The table below was used to record data in an investigation of the nature of rural communities in an area. Suggest how the data would be obtained for each of the aspects listed in the table. Comment on any problems that are likely to arise when collecting the data.

Figure 7.14 Data collected on the nature of rural communities in an area

Village	Harbury	Southam	Ufton	Napton
Population				
% over 60				
% under 20				
% with no car				
Number of shops				
Number of services				
Number of buildings				
Average house price				

> **Theory into practice**
>
> When data collection is written up it is important to refer to sources of secondary data. This can be done by using footnotes, referencing in the body of the text or by using a list of references as an appendix (as opposed to a bibliography that lists some supportive reading – sources that were not directly used or quoted in your investigation).
>
> How would you improve this set of references?
>
> ♦ A–Z Birmingham map book
> ♦ Wild Flower Book of Birmingham by D Sutton

GIS

Increasingly, GIS is being used for geographical investigations and is an important application of ICT to research tools. GIS stands for Geographic Information Systems and covers a wide range of programs such as AEGIS, many of which are available for use by schools and colleges, with many different applications. The common feature to all of these is that data can be located exactly to any spot on the Earth's surface. It can bring together and integrate a wealth of data but all of this data is related to an exact location – see Figure 7.15.

> **Key term**
>
> **GIS:** a tool for storing, organising, displaying and analysing geographic information.

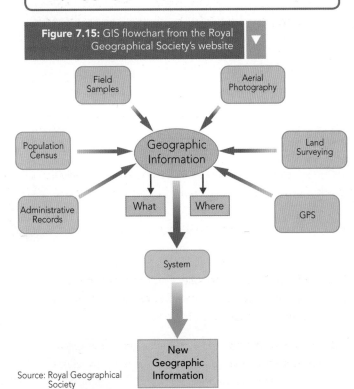

Figure 7.15: GIS flowchart from the Royal Geographical Society's website

Source: Royal Geographical Society

A geographic information system uses a computer to store, organise, display, combine and analyse multiple sets of geographic information. The advantages of GIS are that it:

- can cope with large amounts of data
- can cope with a wide variety (or types) of data
- can cover large study areas (the whole world if necessary)
- can easily change the scale of the area or select any subdivisions of the area
- is dynamic so can cope with frequent changes
- is easier to keep up to date and so findings are current
- is faster and more efficient
- can avoid political boundaries and other practical barriers to research.

The ability to investigate the ways in which two or more sets of located information interact with or relate to one another is the main aim in much of the work undertaken in geographic information systems. Combining different sets of data is a lot more powerful than separate sets of data. The place or location, is used as the common denominator (the link). It has the potential to generate new information on patterns and relationships between multiple sets of geographic information that would otherwise be missed, and to aid in answering more complex questions or decision making.

Why do patterns exist and what impact might they have?

Essentially, combining geographic information adds value to an analysis by providing new information that would not be detectable otherwise. It enables spatial data to be organised and investigated to show patterns and relationships.

What are the uses of GIS in geographical research?

GIS can be used in most of the six stages of your investigation.

Stage 3: Collecting and recording data

1. GIS can be used in the field to measure and record as information can be linked to a specific point or location. For example, palmtops can be used to enter data directly and then transferred to GIS. Weather monitors and environmental data loggers can collect information and transfer them to GIS.

2. It can be used to measure distances in a straight line (or along a feature) or areas or outlines more accurately than in the field or from a map.
3. Secondary data can be accessed from a number of sites such as census returns. Aerial photographs and satellite images can be imported as secondary data. Ground-level digital photographs can be linked to locations on a map.
4. Recording data is GIS's greatest strength since all data is located so is tied to an exact point on the Earth's surface.
5. Data is usually stored in a GIS system in two forms – **raster** data which is in the form of cells and their subsets or **vector** data which includes points, lines and outlines.

(See pages 258–64 for Stage 3.)

Key terms

Raster: representation as a regular grid of cells.

Vector: representation in the form of points, lines and polygons by coordinates.

Using GPS

Kirsty used data loggers to measure air temperature and wind speed in a micro-climate study. She used GPS to link these readings to the exact point in the school field for her study of the micro-climate around the school building.

Stage 4: Presenting the data

1. Area, line and point data can be displayed on the computer screen and then manipulated by the researcher.
2. When presenting mapping data, you can visualise features and manipulate symbols and colours to create an output map with title, scale bar, north arrow, etc. It is possible to draw and edit maps and plans in a GIS editor. This saves a lot of time and creates a high quality map.
3. Your data can be presented in a vast range of styles, colours and dimensions. For example, three dimensions can give a powerful visual impact.
4. Your data can be overlaid or integrated to enable comparisons to be seen or demonstrated.
5. Your data can be linked, layered and shown in three dimensions to make patterns and trends jump out. It is easy to change the vertical scale thereby exaggerating the differences which may make the pattern clearer.

(See pages 265–76 for Stage 4.)

Presenting data

Yusef used GIS to compare a range of styles of presentation of his data on urban land-use patterns. It enabled him to select the most effective from over 20 different versions. He would not have had the time to do this without using ICT.

Stage 5: Analysing and interpreting data

1. GIS greatly increases the ability to ask questions and allow data interrogation. For example, you can ask questions of feature attributes such as 'Where is…?', 'What's the nearest…?', 'What intersects with…?'
2. GIS enables you to select specific aspects and identify features and their attributes that meet particular criteria you have set.
3. It is possible to measure distances between any nominated features or locations thereby giving exact measurements.

4 Raster analysis: GIS may also store geographic information as a grid, with each cell (or raster) in the grid containing a subset of the geographic information There is a whole separate suite of tools for raster analysis that includes classifying cells, deriving aspect and slope, mosaicing and calculating new cell values, among others.

(See pages 277–92 for Stage 5.)

Comparing distances

Becky used GIS to compare accurately the difference in straight line distances and route distances for commuters into Salisbury from the surrounding rural areas.

GIS images are increasingly common and are used in a number of everyday activities. Figure 7.16 shows a GIS map of the quality of bathing water for part of the Thames estuary. This is taken from the Environment Agency, which has a range of data linked to specific locations. Topics include air pollution, drinking water, flood risk and waste disposal. You can check on the quality of your local environment via their website; to access this, go to www.heinemann.co.uk/hotlinks, enter the express code 7627P and click on the relevant link.

A GIS map can be keyed to a place or postcode and enables the researcher to zoom in and out on the map, thereby giving greater location detail. By clicking on coloured indicators, detailed information about that particular site can be called up and interrogated. There are a number of agencies that use GIS to key in data to specific locations. The census also stores information in this way.

Activities

1 a Visit the Environment Agency website (go to www.heinemann.co.uk/hotlinks, enter the express code 7627P and click on the relevant link) and find out about the level of air pollution around your home.

   b Can you suggest the advantages and disadvantages of using such a site for researchers?

2 a Look at Figure 7.16. What does it tell you about bathing water quality in the area?

   b What doesn't it tell you about the quality of the bathing water?

   c What could you do to improve the way the map presents information on bathing water quality?

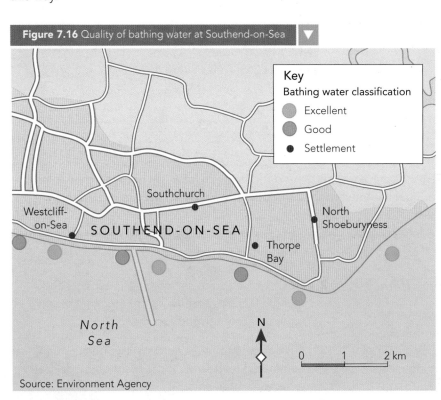

Figure 7.16 Quality of bathing water at Southend-on-Sea

Source: Environment Agency

7.4 Present the data collected in appropriate forms

'A successful geographical investigation involves the selection of techniques that are appropriate to the data collected, and their presentation to a high standard.'

Presenting data to a high standard means using appropriate and effective ways or techniques to show the data you collected in a visual form (including maps, diagrams, annotated photos and graphs). These should turn raw data (usually figures) into visual forms that hopefully show patterns and/or trends in the data. Remember, you should include ways to show spatial patterns where the study has a spatial element (and ideally it should).

So what forms of techniques to display your data are available to you? To some extent this is controlled by the nature of your data, purpose of your analysis and the nature of the area or location you choose. Data can be either qualitative or quantitative, although sometimes qualitative data can be made to look quantitative by converting opinions into scores. A typical example of this is the environmental street quality survey shown in Figure 7.17.

Key terms

Qualitative data is a subjective description in words which varies depending on who is recording it, for example a description of housing quality.

Quantitative data is statistical data which consists of numbers. It is objective and should be value-free.

Activities

1. The street in Figure 7.17 has a score of six. What do you think are the limitations of such a survey? How could you improve it?
2. In Figure 7.17 a range of five values was used. It is important to have decided what you are trying to find out before you decide on whether you have an odd number – this means people can score a middle value (and a lot of people will do this if they can't make up their mind) or an even number – where people will have to allocate it to positive or negative. This is true for all such scales including those in questionnaires. Why do you think it is not a good idea to use zero in any such scale?

There are some problems with this type of qualitative approach.

- It is subjective – different people have different views and opinions on things.
- Values are unclear, for example what does 'heavily' mean?
- Are all the measures of equal importance?
- What does the scale mean and is it appropriate?

Qualitative data can be useful to get subjective views. Descriptions can often convey more important or meaningful information than statistics.

Types of quantitative data

There are various forms of data which geographers might use. Figure 7.18 lists some of the common forms.

When selecting a technique to present your data, you should always consider the factors shown in Figure 7.19.

Figure 7.17 Environmental street quality survey

District: Gresham Ward	Score					
Positive	0	1	2	3	4	Negative
Roads and pavement well repaired		✓				Roads and pavement dangerous
No obvious litter			✓			Heavily littered
Graffiti free		✓				Lots of graffiti
Unpolluted	✓					Heavily polluted
Free of dog faeces		✓				Heavily fouled
Lots of trees and shrubs		✓				No trees and shrubs
Total score = 6	0	4	2	0	0	

Figure 7.18 Common forms of data

Discrete data	Data with distinct separate parts, e.g. crop yield.
Ordinal data	Data that has an order, e.g. ranked data.
Continuous data	Data where there are no breaks but instead something happens continuously, e.g. temperature.
Areal data	Data that applies to an area rather than a point or individual, e.g. population per km^2.
Time-series data	Data where something occurs at intervals, e.g. rainfall.
Period data	Data where a phenomenon repeats itself at intervals, e.g. census returns.

Figure 7.19 Factors to consider when selecting a technique to present data

Factors	Comment
Scale/interval	• If there is too big a scale or interval, e.g. 1–190, it will hide patterns. • If the scale is too small, e.g. 1–3, it would take too long and can produce over-complex patterns. • Is the scale interval arithmetic, e.g. 0–5, 5.1–10, or geometric, e.g. 0–5, 5.1–15.1, or a log, etc? • All figures should have a clear scale with no overlapping values such as 0–12, 12–24 (where would 12 go?). • Remember that scales always start at 0 but they can be 'broken' if there are only high values to show.
Location	Where should the figure be placed to best represent the area or feature to which it applies? Always make sure comparative diagrams are adjacent (and at the same scale).
Size of symbol	If you make symbols too large you may hide information on the map or they may overlap; if too small it may not be seen or its value judged inaccurately.
Shading	Shading should always get darker as values increase. White implies no data. Totally solid shading implies that it is full up. Try to avoid mixing stipple and lines as this confuses the eye and can't be compared accurately.
Colour	Colour is best avoided as it is difficult to compare different colours, e.g. is blue denser than green? Also, certain colours imply certain things, e.g. blue always suggests water. Colour can be useful when comparing patterns, such as red and black dots on a map, but do not use too many colours as this confuses.
Lettering	Lettering varies in alignment, size, spacing and style. It must not obscure other data and should be capable of being read easily so try to keep it horizontal. Do not use too many different sizes and styles on any one figure.
Dimension	3D can cope with higher values but the value is proportional to its volume, so this involves careful calculation and it may mislead visually by looking less than it actually is.
Key	All maps and diagrams should have a clear key. The convention is to locate it at the bottom right of maps and/or diagrams.
Legend or caption	All diagrams, maps, etc., should have a title and a figure number (conventionally written as 'Figure 1', etc.).
Scale	All maps should have a scale.

Non-spatial displays of data – techniques that show numbers

Tables

These are simple but can be an effective way to show the raw data you collected. They are a good way of showing a large amount of data in a concise way, such as census material, but they have limited visual impact.

Diagrams

This is a rather vague blanket term that covers a range of images. There is one type of diagram that uses pictures or symbols to represent the values of the data. These are called pictograms and are widely used in the media but they can be easily exaggerated or distorted (see Figure 7.20).

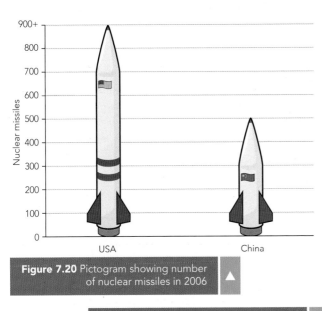

Figure 7.20 Pictogram showing number of nuclear missiles in 2006

Figure 7.21 Types of proportional symbol

Activity

Conrad drew Figure 7.20 to show the relative size of the two figures for the two countries. The height of the pictograms show the value but visually it should be the area of the symbol that should be compared. Can you suggest who might find such a distortion useful and for what reasons?

Charts

These are more precise than tables or diagrams and there are a vast variety of forms, most of which are available on your computer. These are forms of proportional symbols, so called, as their length, area or volume is proportional to the value of the data. A range of these is shown in Figure 7.21.

Form	Use	Limitations
Bar chart	Used for most types of data but especially discrete or time series data such as rainfall, pedestrian counts, number of services. They are simple, quick and give instant visual impression. Can be horizontal or vertical. Length of the bar is proportional to its value.	Width of bar can mislead over the value represented by the bar. Gaps should be left between the bars otherwise it is known as a histogram.
Pie chart	Used with percentage data to show the constituent parts of a whole, such as employment types, ethnicity, vegetation cover. Always start sectors at the top (12 o'clock) and do smallest first as it is easier to cope with errors in the larger sectors. Very visual.	Too many sectors can be silly. It is difficult to use colours as these distract from the sector size. It is difficult to label so that all the sectors can be easily read. Can only use it for percentage data – do you put per cent on the diagram? Needs calculating as 1 per cent = 3.6 degrees.
Divided bar	Used to show constituents of a whole (may be absolute per cent), such as the employment make-up of a population Must keep the divisions in the same order if you are comparing.	Too many sections can be silly. Difficult to use colours. Not always easy to compare.
Rose or star	Used to show directions, e.g. wind directions, stone orientations. Length of bar usually reflects frequency and width or some other aspect.	Time-consuming to draw and takes time to read as three aspects shown (direction, frequency and number/strength/volume).
Proportional circle, square, triangle	The area of the symbol is proportional to square root of data value. Can cope with large numbers such as town populations in an otherwise rural area. Squares are easier to draw than circles!	Time-consuming to calculate and draw. Not easy to compare accurately. Scale complex to draw.
Proportional sphere, cube, pyramid	The area of the symbol is proportional to cube root of data value. Copes with very large numbers such as large towns in an otherwise rural area. Gives good visual appearance.	Time-consuming to calculate and draw. Not easy to compare accurately. Scale even more complex to draw.

Many of these charts are available on your computer via Excel and this can add impact by creating three-dimensional images or 'exploding' the sectors in a pie chart. Remember that this can give false ideas of importance and scale. If a 3D image is used, then the data should be made proportional by taking the cube root. This is a way to obscure or reduce the appearance of large differences between values as it reduces the visual difference compared to using 2D methods, such as proportional circles.

When using computer-generated charts make sure they are fit for purpose rather than simply looking attractive. Often the computer changes or distorts scales making comparison difficult.

Activities

Joanne couldn't decide what type of pie chart she should use to show her data. She used a computer program to generate some possible images. Figure 7.22 shows her three favourite attempts – normal, exploded doughnut, and exploded 3D pie.

1. Which do you think shows Joanne's data the most effectively? Explain your answer.
2. In what way could you improve each of these charts?

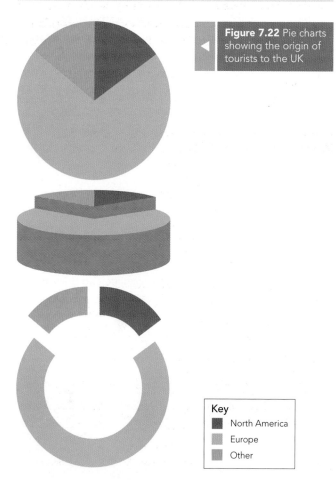

Figure 7.22 Pie charts showing the origin of tourists to the UK

Key
- North America
- Europe
- Other

Theory into practice

Figures 7.23, 7.24 and 7.25 show a selection of charts used to show data collected during fieldwork. This is how they appeared in the final report.

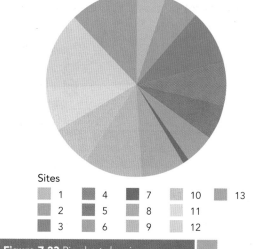

Figure 7.23 Pie chart showing average curvature of sediment at each site

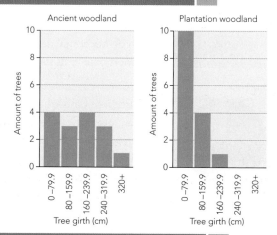

Figure 7.24 Two bar charts showing tree girth of ancient and plantation woodland

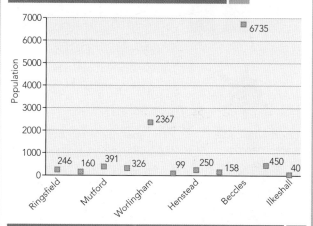

Figure 7.25 Chart showing population of local villages

1. Outline the limitations of each of these charts.
2. a How would you improve them?
   b What other techniques could show this data more effectively?

Graphs

This term normally refers to line graphs but it can apply to scatter graphs or cumulative frequency graphs as well. These are ideal for continuous data or when looking for patterns or trends between two or more variables. The x (horizontal) axis has the independent variable, e.g. rainfall, and the y (vertical) axis has the dependent variable, e.g. crop yield. The implication is x may cause or influence y. If you muddle them up then the implication can appear nonsensical, e.g. crop yield influences rainfall!

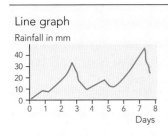
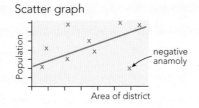
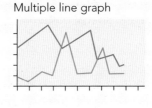
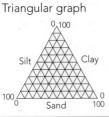
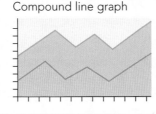
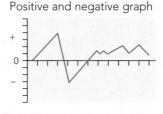

Figure 7.26 Types of graph

Choosing a graph

Elle used a scatter graph to show two sets of variables she had measured in her stream investigation. She plotted the data measured at the sample points along the stream for stream flow against stream depth. She decided that stream depth was the independent variable. She chose a scatter graph as the data was not continuous and she wanted a quick visual impression of whether there was a relationship or trend shown for these two variables.

Activities

1. There are a number of variations on the simple line or scatter graph as shown in Figure 7.26. The example above indicates when a scatter graph is best used. Suggest what type of data would be best shown using each of the other techniques.

2. Triangular graphs require careful and accurate plotting but are useful when trying to show and compare three variables, such as employment sectors – primary, secondary and tertiary. Figure 7.27 shows you one such diagram.
   a. Plot the location on the graph of country X with 10 per cent Primary, 54 per cent Secondary and 36 per cent Tertiary.
   b. Compare the data for countries Z and Y – what might this suggest about the countries?

Figure 7.27 Employment sectors for selected countries

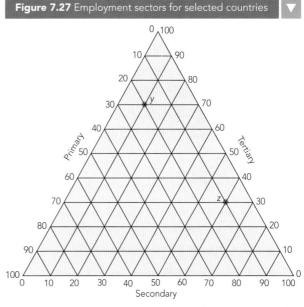

z = 10 Primary, 30 Tertiary, 60 Secondary
y = 20 Primary, 70 Tertiary, 10 Secondary

Spatial displays of data – techniques that show areal patterns

Sometimes it is important to show data in a way that particularly shows spatial patterns. Clearly these techniques are key to a subject that is all about location, place and spatial patterns.

Maps

The simplest form of displaying data over space is a map with the relevant features/data marked on.

An Ordnance Survey map is in reality a display of the spatial pattern of a vast range of data, much of which has been reduced to a symbol. A red triangle represents a youth hostel and a blue line a motorway. You can use much simpler maps to show patterns.

Don't forget that all maps need a scale, a title and north should be indicated. The choice of map scale is particularly important. A plan of a building or small woodland area needs to be at a very different scale from that of a city centre or rural area (e.g. 1:100 for a building but 1:25 000 for a rural area).

A simple map

Mary carried out an investigation of where litter accumulated in her local park. She used a simple outline map of the park and added crosses wherever she recorded litter. She also added red dots representing the litter baskets provided. This map instantly showed that there was most litter around the litter baskets.

Isopleths or isolines

This technique makes a spatial pattern out of point form data by using lines to join up points with the same value. Isopleths are lines that represent the same value at the points in an area through which it passes. An example of an isopleth is shown in Figure 7.28. The best known example is the contour but there are many others, such as isobars and isotherms. Sometimes colour or shading can be used to fill in the area between the isopleths to stress changes in value, as shown in the case of relief maps in atlases.

There are a number of issues with the drawing or use of isopleths:

- Selection of the interval value between isopleths is crucial: if you make it too big, you may place the isopleth inaccurately; if too small then the map becomes obscured by lines.

- You need to consider the shape and size of the units for which data is available – points are easiest so what do you do if the point is a small area such as a quadrat sample? The scale of the map is critical.

- Where are the locations of the plotting points? Are they merely central to the area or are they located at the point the data was measured?

- What is the method of interpolation (working out where the line should go when there is no value recorded)? Most systems assume a uniform change, so if one value is 10 and the other 20, then the 15 line would go halfway between them. That may not be true – the slope may not be uniform.

- Isopleth lines should be smooth, not angular or erratic, as we assume a smooth transition in values, but this may not in reality be accurate. Patterns may be angular instead.

Key term

Isopleth: a line joining points of equal value, for example a contour line.

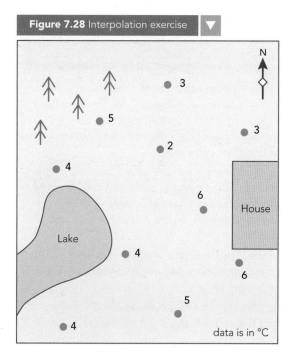

Figure 7.28 Interpolation exercise

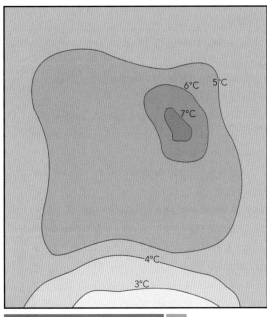

Figure 7.29 Shaded isopleths

Activities

1. Draw the isopleths on Figure 7.28.
2. Suggest reasons for the pattern your isopleth map shows.
3. In what ways might a shaded isopleth map, like Figure 7.29, help your analysis?

Choropleths

Choropleths are density maps where areas are shaded in to represent the average number per unit area. This is a very common tool as it gives a quick visual impression of patterns, but again there are a number of issues you should consider before using this technique:

- It treats the area as a whole so hides variations within the area. Figure 7.30 shows how this can mislead.
- The scale of the areal divisions is crucial. For example, enumeration districts give a more accurate picture than ward level maps for census data. The smaller the scale, the more accurate the choropleth map.
- The value interval that you use for plotting your data is crucial. If you change it, do you get a different pattern?
- Colours are difficult to use so should be avoided, as they rarely show a smooth progression in density and colours often carry unintentional meanings, for example blue represents water.
- Your choice of shading range is important. A range such as that in Figure 7.31 is visually the same, so they all appear to be the same density. Shading should get denser with lines closer together as the values being represented increase.

Key term

Choropleth: a quantitative areal map calculated on the basis of average number per unit area.

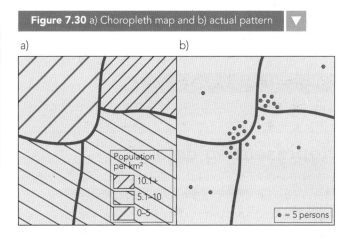

Figure 7.30 a) Choropleth map and b) actual pattern

Figure 7.31 Shading range

Located symbols

Many of the proportional symbols (see page 267) can be placed on a base map to show patterns. An effective method is to use overlays to compare patterns provided the same symbols at the same scale are used. Your choice of which proportional symbol to use may be influenced by their relative advantages and disadvantages (Figure 7.32).

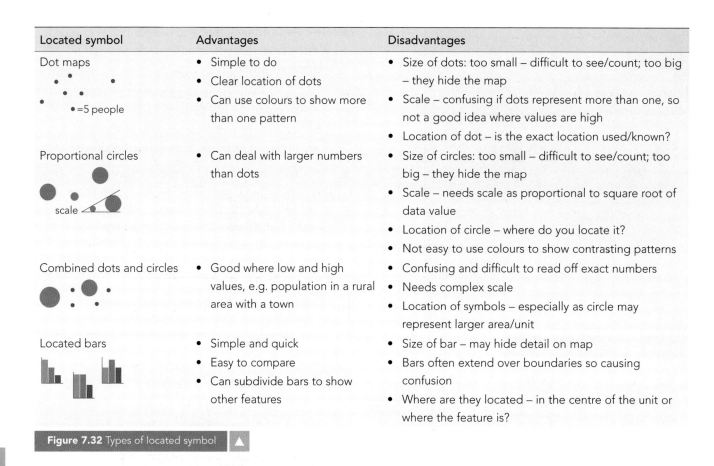

Located symbol	Advantages	Disadvantages
Dot maps	• Simple to do • Clear location of dots • Can use colours to show more than one pattern	• Size of dots: too small – difficult to see/count; too big – they hide the map • Scale – confusing if dots represent more than one, so not a good idea where values are high • Location of dot – is the exact location used/known?
Proportional circles	• Can deal with larger numbers than dots	• Size of circles: too small – difficult to see/count; too big – they hide the map • Scale – needs scale as proportional to square root of data value • Location of circle – where do you locate it? • Not easy to use colours to show contrasting patterns
Combined dots and circles	• Good where low and high values, e.g. population in a rural area with a town	• Confusing and difficult to read off exact numbers • Needs complex scale • Location of symbols – especially as circle may represent larger area/unit
Located bars	• Simple and quick • Easy to compare • Can subdivide bars to show other features	• Size of bar – may hide detail on map • Bars often extend over boundaries so causing confusion • Where are they located – in the centre of the unit or where the feature is?

Figure 7.32 Types of located symbol

Representing data

Nuria wanted to show how employment varied between wards in a major urban area. She used information from the census to classify employment into primary, secondary, tertiary and unemployed. She then drew pie charts to show the percentage in each group and located them in an appropriate place on a ward map.

Activities

1. Do you think Nuria used the most effective technique to represent her data? Explain your answer in terms of its advantages and disadvantages.
2. What alternative methods could she have used? Justify your choice.

Flow lines

Sometimes you need to show movements. Flow lines are used to show the volume, frequency and make up of movements along routes, such as roads and river channels. The flow line is a variation of the bar chart but follows the route, with its width reflecting the size of flow. Sometimes this is simplified (topological – like the London Underground map) to show direction and distance but not the precise actual route which may be too complex. Examples are shown in Figure 7.33.

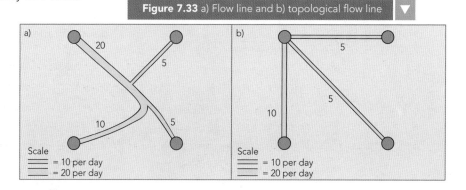

Figure 7.33 a) Flow line and b) topological flow line

Key term

Topological: a map where the points remain in the same order or relationship but other aspects such as distance are altered.

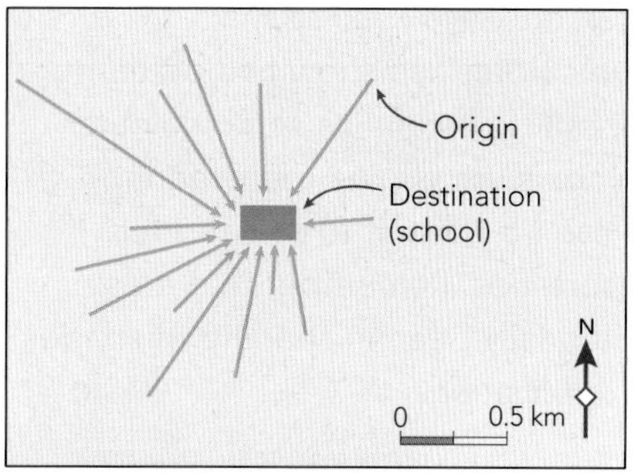

Figure 7.34 Trip lines for journey to school

Activities

1. What do you think are the advantages and disadvantages of using topological flow line maps rather than real route flow lines?
2. Alan wanted to show how much of the traffic flow was from lorries and public transport. How could he show the different constituents of the flow on such a map?

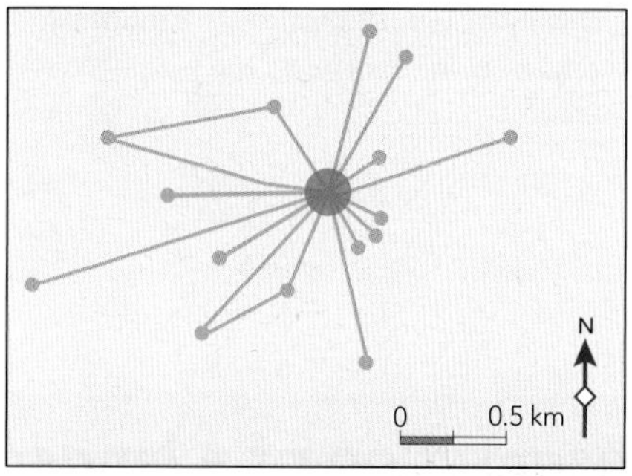

Trip lines

Trip lines are a type of flow line in that they show direction and volume but are straight line drawings from origin to destination (as shown by the use of an arrowhead in Figures 7.34 and 7.35).

Key term

Trip lines: a straight line joining the origin of a journey to its destination.

Figure 7.35 Trip line map showing daily commuting patterns for a town

Using trip lines

Zara wanted to show the sphere of influence of her local town. She carried out a shopping survey in the town centre. She then plotted where each shopper lived and drew trip lines from there to the centre. She then joined the ends of the trip lines to show the area from which people travelled into the town – its sphere of influence.

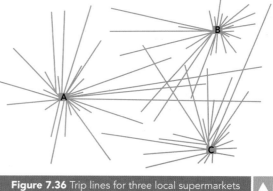

Figure 7.36 Trip lines for three local supermarkets

Activities

1. Study Figures 7.34 and 7.35.
   a. Why might Figure 7.35 be misleading?
   b. Can you suggest from Figure 7.34 why putting arrowheads on has advantages and disadvantages?
   c. What could explain the lack of trips to the school from the south-east area in Figure 7.34?
2. Figure 7.36 is an attempt to show shopping trip information for three local supermarkets.
   a. What does this map indicate about the shopping patterns?
   b. What would you do about the overlapping trip lines? What are the implications of your approach?
   c. In what ways might this map be misleading?

The use of colour can enhance such spatial representations. Two colours can show two different variables on a map, but be careful as some colours have implied meanings such as blue representing water, while yellow can be difficult to see and red can exaggerate importance. Remember that most diagrams have traditionally been in shading or shades of grey for ease of duplication.

> **Discussion point**
>
> 1 Do you think it is better to use colour in spatial representations? Justify your answer.
> 2 Under what circumstances would the use of colour be a) advantageous and b) disadvantageous?

Pictorial displays of data

This can be a useful way of showing qualitative data by locating images or photographs on a base map, but they should always have a purpose and so they should be well annotated (labelled with an arrow to the relevant aspect or feature shown) as they need to be clear to someone who doesn't know the area or site.

Sketches

Field sketches are a useful way of recording an impression of an area as well as showing its main features. They tend to be used mainly in physical geography fieldwork or rural investigations to convey some aspect of the landscape or context. Many reports contain sketches but often little use is made of them and so they detract from the investigation. Don't include a sketch unless you are going to use it effectively. Remember that the correct annotation is more important than the artistic merits of the drawing.

Visual representation of a landscape

Chani did a coastal investigation on the factors influencing the rate of coastal erosion. He took a photograph of an area of his investigation and wanted to turn it into a fieldsketch to show the main coastal features as he forgot to do one at the time of his investigation.

Activities

1 Look at Chani's photograph (Figure 7.37). Draw a fieldsketch and label the main coastal features. If you prefer, use tracing paper to get the broad outline. Remember: appropriate labels are more important than a good drawing!
2 Do you think such a sketch would help Chani's investigation? How else could he have used such a sketch in his report?

Figure 7.37 Coastal landscape

Photographs

You can use photographs of your study area or its features, taking them yourself or using ones from a range of Internet sites or via a search engine such as Google image. (To access a photographic Internet site which you may find helpful, go to www.heinemann.co.uk/hotlinks, enter the express code 7627P and click on the relevant link.) Such photos from the Internet can be satellite, aerial or oblique images which add new dimensions to your report. The use of ICT means that it is easy to select areas from a digital photograph to produce relevant images, enlarge them and to add labels, shading, etc.

Too many reports contain excessive photographs that add little and get in the way of the investigation. Don't just put them in; they must be used. They are useful to set contexts such as showing the flow and load of a river at a particular location, but are most useful in human geography investigations such as environmental quality surveys where they might even be used as a method of data collection.

Photographs capture an instant of time image so your photographs taken during your investigation are most relevant. However, they must be annotated (i.e. used) to be effective. Only include photographs that can add information to your investigation.

Figure 7.39 Potteries Shopping Centre, Hanley city centre

Figure 7.38 View of Hanley city centre

Activities

1. What advantages do you think an annotated photo has over a field sketch?
2. In what ways could you improve the annotation of Figures 7.38 and 7.39?
3. In what ways might such photos give a misleading image of the research area?

Other important aspects

It is important to organise your presented material in a logical way in relation to the analysis. You should put the presentation in a logical order that matches your original aims and analysis. You should integrate your diagrams into the main body of the text and not leave them as an appendix or kept as a distinct separate block, as this makes it difficult to connect text and image in the reader's mind. It is not good to keep referring to the appendix to find out what a particular diagram mentioned in the text looks like.

Don't forget that all your forms of presentation need numbering in order, e.g. Figure 1 (and they should be referred to in the analysis, etc.), and should be given titles, scales and appropriate labels. If you use comparative diagrams they should always have the same scales and labelling. Sometimes acetate overlays help present such comparisons more clearly.

7.5 Analyse and interpret the data

'A successful geographical investigation involves a variety of analytical approaches, ranging from the purely descriptive to detailed analysis, involving attempts to explain patterns that have been identified.'

Analysis

This is where you try to identify and investigate the patterns shown by the data you have collected. This can be done in two main ways: qualitative or quantitative.

Qualitative

This is a description of the findings shown by the diagrams drawn in the data presentation stage. At the

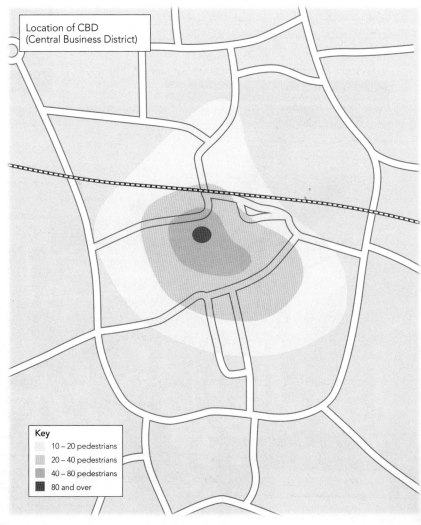

Activities

Look at Figure 7.40, which is an attempt to show the volume of pedestrians in a town centre.

1. Do you think that this is a suitable technique for showing this data? Explain your answer.
2. What do you see as the limitations of this example?

Figure 7.40 Map and overlay of a town centre

simplest, this could be an analysis of or comparison of graphs, for example. This is quick, visual and shows the obvious patterns and anomalies (odd values that don't seem to fit the overall trend or pattern).

One of the simplest approaches for qualitative analysis is to produce a scatter graph to see if two variables relate. You should be careful not to select inappropriate pairs such as temperature variation with soil pH and be wary of best fit straight lines (with the same number of values above as below the line) that ICT programs often produce. Many relationships are not straight lines (see Figure 7.41). Scatter graphs, with or without best fit lines, indicate the likely direction of a relationship.

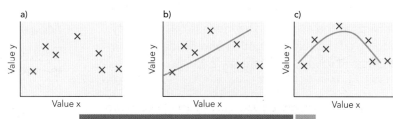

Figure 7.41 Three types of scatter graph: a) scatter, b) best fit – straight, c) best fit.

Comparing two graphs

Kirov compared two graphs that showed pedestrian flow versus property value for the city centre and a suburban shopping centre. His analysis was that both graphs showed that as pedestrian flows increased so did property value, and this pattern was more pronounced in the city centre. The pattern shown by the graphs was so clear that Kirov would have gained little in doing any calculation.

Quantitative

Where appropriate, such as when it is not very clear if there is a pattern or trend, the analysis of the data can be taken further using statistical techniques which are explored below. These can identify trends, groups, anomalies and offer predictions often from a welter of data that is too big or complex for a mere qualitative analysis. Statistical analysis is likely to be useful in most cases as it indicates if a relationship or difference really exists (and the strength of it) or whether it is just a chance coincidence. It can also simplify down large sets of data into a single result that can give a measure of the level of accuracy or reliability.

There are several things to bear in mind before you begin quantitative analysis with statistic techniques:

- Keep it Simple (KIS) – don't simply do statistical analysis for the sake of it. Often the result is obvious without the need for a mathematical calculation. Do the simpler statistical techniques first.
- Remember what the hypothesis in your investigation was and align your statistical methods to suit this (see null hypothesis).
- It is a convention that you work to a 95 per cent significance level. This means you can be confident that it is 95 per cent accurate. Anything less is unacceptable, although it may hint at a likely significance.
- Check that you have any formulae correct – always show formula and your workings (which are often best placed in an appendix). Most errors are made in minor slips in the calculation.
- If you don't understand how to do it or what the result means, then don't use that technique.

Key term

Null hypothesis: that a pattern or relationship is due to chance.

Basic description of the data

These are relatively simple but give a quick picture or summary of the data.

Measures of central tendency

This is used to calculate the average in one form or another. This can be very useful when you are trying to identify simple differences between sets of data, such as comparing average stream flow, rainfall or population density. Some give quick simple comparisons. The simplest is the mode, which is the most frequently occurring value. The median is the middle value of the data when all of the items have been placed in order of value from lowest to highest. The most useful, but one that does require a calculation, is the mean, which is what most people think of as the average. The mean is calculated by adding all the values together and dividing by the number of values. Alternatively it can be written as:

$$\bar{x} = \frac{\Sigma x}{n}$$

$\bar{x}$ = mean x = values
Σ = sum of n = number of items in the data set

Key terms

Mean: or arithmetic mean; found by dividing the total by the number of items.

Median: the middle value in a list of data that has been put in order of size.

Mode: the most frequent value in a data set.

Measures of dispersion

Data sets vary in how they are scattered around the average. Dispersion refers to various different measures which indicate the extent to which the data is grouped around the mean.

The simplest way of looking at the spread of values in a set of data is to look at the range, but this can be misleading if these values are extreme, so an improvement is the interquartile range. This takes the range of the middle half of the data range either side of the median. It is quick and easy to calculate and avoids the distortion caused by extreme values.

Comparing stone sizes

Georgina wanted to compare the mix of stone sizes she had sampled along the beach at increasing distances from the groyne. She wanted to see if the sizes of stones changed with distance. Unfortunately, six of her ten samples (each of ten stones) had the same mean. So she had to analyse further as she was convinced that her six samples were not the same overall. She then looked at the range of stone sizes in her samples. When Georgina did this she found that the range in stone sizes became greater with distance from the groyne.

A more robust method is the standard deviation. This involves a calculation but it does look at the variation of all of the data from the mean. The larger the standard deviation, the larger the variation around the mean. This is very useful when comparing two data sets with the same or similar mean, but it is more complex and requires the use of a formula and calculation.

$$\sigma = \sqrt{\frac{\Sigma x^2 - \bar{x}^2}{n}}$$

Σ = sum of n = number of items in the data set
$\bar{x}$ = mean σ = standard deviation

Key terms

Interquartile range: the range of values in the inner half of the data once the data has been put in order of size.

Range: the difference between the highest and lowest value in a data set.

Standard deviation: a statistical measure of the sum of how far from the mean is each value in a data set.

Activities

Study Figure 7.42.
1. Calculate the mean, mode and median for the two sets of data.
2. Explain why the mode and median may not be very useful in this case.
3. Calculate the range, interquartile and standard deviation for each of the data sets.
4. Which set of data is the more varied? What does this tell you about the two beaches? What would you do next?

Figure 7.42 Two sets of data – stone sizes on two beaches

Site	Beach A	Beach B
1	22	4
2	12	13
3	7	5
4	5	4
5	21	8
6	4	4
7	21	7
8	21	15
9	3	17
10	2	36
11	3	11
12	5	2

Frequencies

It is often useful to plot the distribution of the values in your data set as a frequency diagram. This gives a good visual impression of whether the two sets of data differ and the frequencies can be compared. To make such a comparison more quantitative, we use some key terms to describe the distribution, such as skew. Data is said to be skewed if the distribution is not symmetrical. If the mode is lower than the mean it is said to be negatively skewed, and if higher than the mean it is positively skewed (see Figure 7.43).

The other characteristic of a distribution is its wavelength/amplitude or kurtosis – its shape. Distributions can be peaked, flat or in between. This can be quite important as demonstrated by the shape of stream hydrographs, where the peaked ones are termed 'flashy' and show a stream prone to flash flooding.

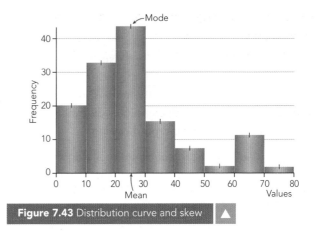

Figure 7.43 Distribution curve and skew

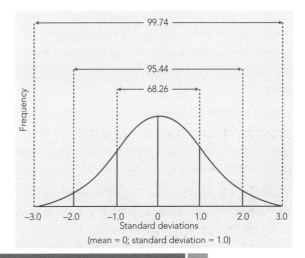

Figure 7.44 The normal distribution

> **Key terms**
>
> **Kurtosis:** the 'peakiness' of a graph of the data values. It could be very pointed or virtually flat.
>
> **Skew:** the extent to which the mode and mean differ. It can be positive or negative skew.

> **Key term**
>
> **Normal distribution:** when the data forms a symmetrical curve in a frequency graph around the mean.

Activity

You can convert the frequency diagram in Figure 7.43 into a line graph by joining the midpoints with a smooth line. If you were to do it in this case, what problems would you encounter? Is this example negatively or positively skewed?

There is one frequency pattern that is considered to be important. Figure 7.44 shows the normal distribution. This so-called normal curve or frequency is said to be a 'bell-shape' where the mean, mode and median are the same value. This is frequently used as a comparison against which a distribution can be compared, in that the normal distribution is considered to be a 'model' distribution.

The normal distribution has some statistical properties. It has the characteristic that 95.44 per cent of the values in the particular data set lie within two standard deviations of the mean. This explains why we normally work at a 95 per cent level, or two standard deviations either side of the mean, as doing this removes the extreme (probably anomalous) values. Many other more complex statistical tests are based on the assumption that the distribution of the data is normal rather than strongly skewed.

Analysing a frequency pattern

Robyn compared her frequency pattern of household incomes in an inner-city area with the normal curve. This showed clearly that her data had a negative skew with a much longer tail of low incomes. It was clearly not normal so an explanation then had to be looked for.

Testing for differences

There are a range of statistical tests that not only tell you if sets of data are different but also give you a statistical measure of how far, or the extent to which, they are different. Comparing the means, modes, ranges, etc., is the simplest technique, but this only tells you if they are different, not how far.

When you start statistical testing there are some common steps or stages that are usually followed:

1. You firstly state a null hypothesis – such as there is no significant difference between the two sets of data.
2. Then you state the alternative hypothesis or proposition – the one that must be true if the null hypothesis is not true, i.e. there is a significant difference.
3. Next you select a statistical test. This should be done on the basis of whether the data fits the assumptions made by the test, such that it has a normal distribution.
4. You must then decide on a significance level – this is the level of confidence that you must have to know that the null hypothesis has not been incorrectly rejected.
5. Next, work out the critical value for the test result beyond which the null hypothesis will be rejected.
6. Carry out the test.
7. Finally, make the decision on its significance on the basis of the result.

Remember that such tests do not tell you why a data set differs, merely that it does differ more than it would by chance or accident.

There are a number of such tests but there are two commonly used ones: the Mann-Whitney U-test and the Chi-squared test.

Mann-Whitney U-test

This is a test of difference between the medians of two sets of data and is a useful test when data is in an awkward form, such as samples of very different sizes. The test starts out by assuming that both sets of data are alike and then establishes whether it is safe to reject that assumption or not. It uses the ranks of the data so can be used for any data that can be put into a rank order. It uses the following formulae:

$$U \text{ for sample } x = n_x n_y + \frac{n_x(n_x+1)}{2} - \Sigma r_x$$

n = sample size
Σr = sum of ranks (derived from ranking all the data in both sets of observations together)

Comparing two sets of data

Samir wanted to see if the sample of stones he had taken on a northward facing beach on the Isle of Wight was different from a sample he had taken on a westward facing beach in the same area. He decided to use the Mann-Whitney U-test to show if the two sets of data were distinctly different. The values in both sets of data are put in a common rank order together, as shown in Figure 7.45.

The Mann-Whitney U-test was used to calculate whether there is a significant difference between Samir's two sets of data shown in Figure 7.45:

$$U \text{ for Sample A} = n_A n_B + \frac{n_A(n_A+1)}{2} - \Sigma r_A$$

$$= 12 \times 12 + \frac{12(12+1)}{2} - 150.5$$

$$= 144 + 78 - 150.5$$

$$U_A = 71.5$$

Figure 7.45 Stone samples from two different beaches

Site	Beach A	Rank	Beach B	Rank
1	22	2	4	18.5
2	12	9	13	8
3	7	12.5	5	15
4	5	15	4	18.5
5	21	4	8	11
6	4	18.5	4	18.5
7	21	4	7	12.5
8	21	4	15	7
9	3	21.5	17	6
10	2	23.5	36	1
11	3	21.5	11	10
12	5	15	2	23.5

> **Activities**
>
> Now repeat this for sample B using the following formula:
>
> $$U \text{ for Sample B} = n_A n_B + \frac{n_B(n_B + 1)}{2} - \Sigma r_B$$
>
> The two results are then compared to the control values at U in the table below (Figure 7.46), to see if the smallest value is larger (not significantly different) than that caused by chance. This is found by finding where n_A (12) and n_B (12) intersect – 42 in this case. If the smallest value for U_A or U_B is smaller than the critical value in the table (42), then the two sets of data can be considered to be significantly different. So are they significantly different in this case?

Figure 7.46 Significance table for Mann-Whitney U-test

n_x \ n_y	1	2	3	4	5	6	7	8	9	10	11	12	13	14	15	16	17	18	19	20
1																			0	0
2					0	0	0	1	1	1	1	2	2	2	3	3	3	4	4	4
3				1	2	2	3	3	4	5	5	6	7	7	8	9	9	10	11
4			1	2	3	4	5	6	7	8	9	10	11	12	14	15	16	17	18
5		0	1	2	4	4	6	8	9	11	12	13	15	16	18	19	20	22	23	25
6		0	2	3	5	7	8	10	12	14	16	17	19	21	23	25	26	28	30	32
7		0	2	4	6	8	11	13	15	17	19	21	24	26	28	30	33	35	37	39
8		1	3	5	8	10	13	15	18	20	23	26	28	31	33	36	39	41	44	47
9		1	3	6	9	12	15	18	21	24	27	30	33	36	39	42	45	48	51	54
10		1	4	7	11	14	17	20	24	27	31	34	37	41	44	48	51	55	58	62
11		1	5	8	12	16	19	23	27	31	34	38	42	46	50	54	57	61	65	69
12		2	5	9	13	17	21	26	30	34	38	42	47	51	55	60	64	68	72	77
13		2	6	10	15	19	24	28	33	37	42	47	51	56	61	65	70	75	80	84
14		2	7	11	16	21	26	31	36	41	46	51	56	61	66	71	77	82	87	92
15		3	7	12	18	23	28	33	39	44	50	55	61	66	72	77	83	88	94	100
16		3	8	14	19	25	30	36	42	48	54	60	65	71	77	83	89	95	101	107
17		3	9	15	20	26	33	39	45	51	57	64	70	77	83	89	96	102	109	115
18		4	9	16	22	28	35	41	48	55	61	68	75	82	88	95	102	109	116	123
19	0	4	10	17	23	30	37	44	51	58	65	72	80	87	94	101	109	116	123	130
20	0	4	11	18	25	32	39	47	54	62	69	77	84	92	100	107	115	123	130	138

The Mann-Whitney U-test is also very useful when comparing two data sets of very different sizes.

Comparing two data sets of different sizes

Chico and Paulina were comparing their traffic survey results for the ring road and two of the radial urban routes of their town, to see if the flow of the traffic was significantly different. Chico had data from 10 sample points on the ring road and Paulina had 17 sample points on the radials. They used the Mann-Whitney U-test and found their data sets were significantly different.

Chi-squared test

This test does not look at ranks but rather compares the frequencies you actually observed (O) with those expected (E). As you decide the expected, you must be clear what you expect and why. In most cases the expected value is that which would have arisen by chance.

There is more than one version of this technique (see the relationships section for a more complex use of this technique). The simplest can be used to compare an observed pattern against a pattern expected due to some reason such as chance.

It uses the following formula:

$$\chi^2 = \sum \frac{(O - E)^2}{E}$$

O = observed value
E = expected value
Σ = sum of

Observing numbers of cattle

Ian surveyed a farm with five identically-sized fields. The farmer has 100 cows which he grazes in these fields. Ian carried out a survey of the fields and the observed numbers of cattle noted (O).

Figure 7.47 Results of cattle survey

Field	Observed	Expected	O–E	(O–E)²	(O–E)²/E
1	34	20			
2	0	20			
3	0	20			
4	45	20			
5	21	20			

Now chance would suggest an even spread of cattle between the fields, so each field is expected to have 20 cattle. (Just going this far is of some value as it shows which fields have more or less cattle than expected. In this case the last field is almost what you would expect.)

Activities

Complete the Chi-squared test calculation for Ian. The higher the value the greater the difference between the two sets of data. Its significance can be found using a table or graph (see Figure 7.48).

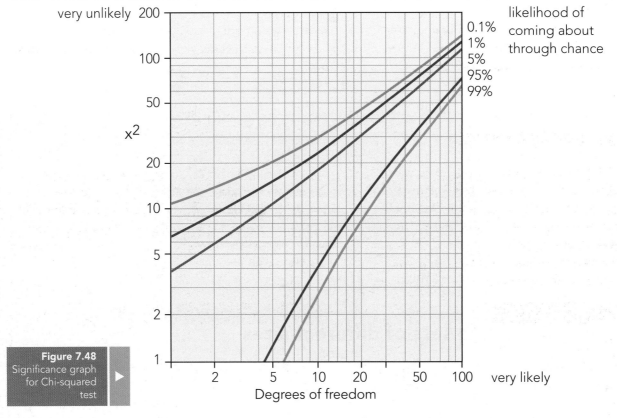

Figure 7.48 Significance graph for Chi-squared test

Measurement of patterns

A number of techniques have been devised to measure whether a pattern exists and what type it might be.

Nearest neighbour

This involves a comparison between the observed spacing of a set of points (for example, trees, farms, shops) and the spacing that would be expected if the pattern had been random. This technique relies on measuring the distance between neighbouring points, usually on a map. The observed spacing is expressed as the average distance of all points from their nearest neighbour in straight line measurement. The formula used is:

$$NNA = 2\overline{d_o}\sqrt{\frac{n}{A}}$$

$\overline{d_o}$ = observed mean distance
n = number of points in the pattern
A = area over which the points are distributed

The answer will be on a scale ranging from 2.15 (regularly spaced), where 1 indicates perfect randomness and below 1 suggests increasing clustering.

Activities

1 Figure 7.49 shows the pattern of shoe shops and banks/building societies mapped by Maggie in a small market town.

   a Calculate the nearest neighbour statistic for each of the patterns. Which is more clustered?

   b Can you think why this happens?

2 From the activity you have just done, can you identify some of the limitations of this technique?

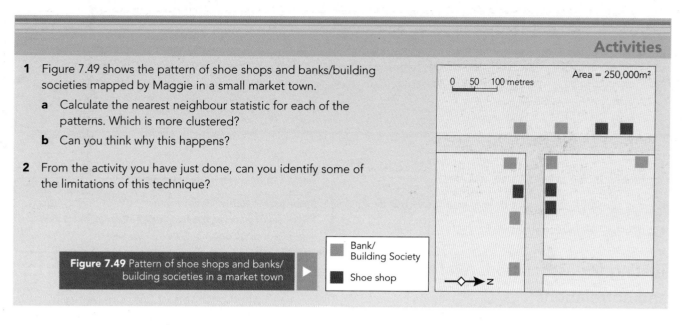

Figure 7.49 Pattern of shoe shops and banks/building societies in a market town

Lorenz curve

This technique doesn't look at the horizontal pattern of points but at pattern in a different dimension – it is a measure of the degree of concentration. These curves are used commonly in economic geography on data linked to areas and involve the comparison of percentage frequencies.

Using a Lorenz curve

Figure 7.50 shows the percentage share of the national income (GNP) in the regions of a developing country. Alberto plotted these in rank order on the graph as a cumulative percentage, with the largest plotted first. He then drew in a diagonal line as this represents a pattern where there is an even distribution of the wealth between the regions. This can be made more exact by turning it into a statistic by measuring the longest perpendicular from the diagonal line to the curve. This is a useful technique for comparing different features for the same set of areas, such as employment in different sectors or industries, as shown in Figure 7.51.

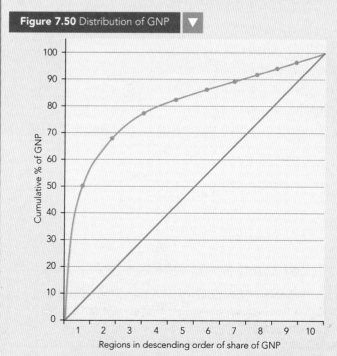

Figure 7.50 Distribution of GNP

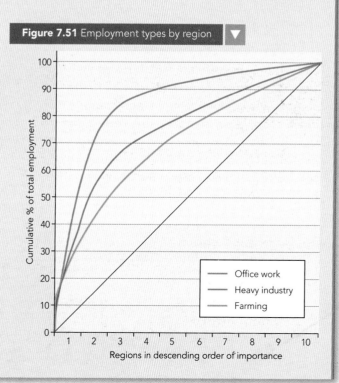

Figure 7.51 Employment types by region

Activities

1. What does the Lorenz curve that Alberto drew suggest about the distribution of wealth in that country? Why might this bring advantages as well as disadvantages to the country?
2. What does Figure 7.51 suggest about the distribution of employment for the three industries in the country? Suggest why this might have come about.

Limitations of the Lorenz curve include:

- it is sensitive to the size of the areal units used
- it is not very accurate statistically.

Testing for association, relationships and correlations

One of the most important uses of statistical techniques is to test data to see if there are any links – is there a relationship and/or correlation?

> **Key term**
>
> **Correlation:** to have a relationship (which may be positive or negative).

There are basically three types of relationship possible between two variables, as shown in Figure 7.52:

- positive or direct (as one value goes up so does the other)
- negative or inverse (as one value goes up the other goes down)
- no relationship.

An initial use of a scatter graph can help suggest the most likely outcome.

Tests not only indicate the type but also the strength of the relationship. Remember to work to the 95 per cent rule – to be confident of a significant relationship you must be 95 per cent, or more, certain that the null hypothesis (that there is no relationship – it came about by chance or accident) can be rejected. The simplest of these tests is the Spearman's Rank correlation test.

Figure 7.52 Three relationships between variables: a) positive, b) negative, c) no relationship.

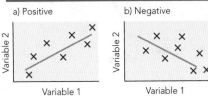
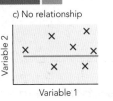

> **Key term**
>
> **Spearman's Rank:** statistical test for relationship using the ranks of each of the two data sets (maximum value is plus or minus one).

Spearman's Rank

This is used when both sets of data can be easily ranked, a quick and easy measure of correlation is needed and where exact values may be uncertain. This is a simple calculation that can indicate if two sets of data are related. It is best to start with a scatter graph as this will tell you if it looks like there may be a relationship and its likely direction – direct or inverse. It does not prove that one variable causes the other to vary, merely that they are related and the strength of that relationship.

Firstly, the data must be ranked. Think carefully what your null and alternative hypotheses are, and make sure you rank the most appropriate way either from largest to smallest or vice versa. What you shouldn't do is rank one variable one way and the other variable the other way.

The formula is:

$$R_s = 1 - \frac{6 \Sigma d^2}{n^3 - n}$$

Σ = sum of
d = the difference in rank order
n = number of pairs of data

The result will lie between +1 (a perfect direct or positive correlation) and –1 (a perfect negative correlation), with 0 representing no correlation. The exact significance of your result is checked by looking up the result in a table of significance, where the size of the result is compared to the number of pairs –1 (this is called the degrees of freedom as you are free to alter all the values but 1 in each set and still have the same totals).

> **Key term**
>
> **Degrees of freedom:** the number of values in a data set that you can change without changing the total (usually n–1).

Activities

Figure 7.53 shows two sets of data collected from the same sample points during a stream measuring exercise for river depth and speed of flow of the current. (Note – in this real example the candidate did not record the units of his measurement but because only the ranks of the values are used it doesn't matter in the calculation. It will matter in the interpretation and explanation.)

Figure 7.53 River depth and speed of flow of current

Point	Depth	Speed	Rank D	Rank S	Difference	Squared
1	15	10.1				
2	56	8.9				
3	43	9.1				
4	21	12.4				
5	19	12.7				
6	19	13.6				
7	21	15.0				
8	45	7.8				
9	42	8.4				
10	23	11.7				

1. Rank the two variables. Where there is a tied rank (depth at point 5 and 6) then you add the values they would have received if they had not been tied and divide that by the number of values that tied. For example, if the tied ranks would have been 7 and 8 then the tied ranks are 15 divided by 2 which gives a rank of 7.5.
2. Take the difference (in this case the rank of depth – rank of speed) and square it to remove the + or – signs.
3. Add up these squared values.
4. Put these figures into the following formula:

$$R_s = 1 - \frac{6 \sum d^2}{n^3 - n} \qquad n = 10$$

5. The value tells you if it is positive or negative, but to find how significant it is you again have to refer to significance tables or graphs as shown in Figure 7.54 below. In this case, the degrees of freedom is 9.
6. Look at Figure 7.54. Is the result significant? If so, at what level can you be confident that the null hypothesis has been disproved?
7. Now you must ask yourself why these two variables might have such a relationship.

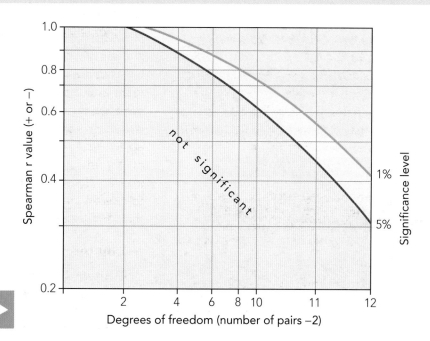

Figure 7.54 Significance graph for Spearman's Rank test

There are some limitations in the use of Spearman's Rank. It should not be used:

- when there are a lot (more than four) sets of tied ranks
- where there are a limited number of data sets – usually you need at least eight and preferably more
- where the two data sets are unequal in number.

Chi-squared technique

This is a more sophisticated version of the technique described on page 283. The Chi-squared is most useful when your data has been collected in categories or when you can group it into categories. It again tests if there is a difference between the observed pattern and that expected from random or chance events. For the test to be valid, the total number of observations should be more than twenty.

— Key term ——————————————————

Chi-squared: a statistical technique that compares the actual pattern with an expected one, to test if they are significantly different.

Investigating patterns of land use

Amos wanted to investigate if there was a pattern of land use in his local town centre. He sampled the occurrences of three main land uses within four concentric zones around the central market place. He placed his results into a table (Figure 7.55).

This shows one of the advantages of the Chi-squared technique in that it can cope with unequal sampling. In this case, 50 samples were taken but Amos took marginally more in the 1.01–1.5km zone. He then thought about his null and alternative hypotheses. As he was looking at land use with distance from the town centre, his null hypothesis was: 'There is no significant difference in land use with distance from the town centre', so his alternative hypothesis was: 'There is a significant difference in land use with distance from the town centre'.

Amos then had to calculate the even or equal distribution. In this use of the technique, the expected (if random) values are calculated using the observed values in the following formula:

$$\text{Expected value} = \frac{\text{Row total} \times \text{Column total}}{\text{Grand total}}$$

Amos's expected grid is shown in Figure 7.56. As the rows and column totals must remain unchanged, Amos didn't need to do the calculation for the final column. In the case of three of the rows, the end expected value must be 5.8 to give a row total of 12.

Figure 7.56 Amos's expected grid

Categories	Retail	Industry	Housing	Row total
0–0.5 km	$\frac{12 \times 20}{50}$	$\frac{12 \times 6}{50}$		12
0.51–1.0km	4.8	1.4		12
1.01–1.5km	5.6	1.8		14
Over 1.51 km	4.8	1.4		12
Column total	20	6	24	50

Activities

1. Finish Amos's expected grid.

By simply comparing the observed and expected data, Amos could easily identify whether the observed grid looked significantly different from the pattern expected if it was random, and which zones were most out of line with the expected. It was thus clear to him that retail land use was far lower than you would expect in the outer zone.

To continue the calculation, Amos had to take the expected value from the observed. He found it easier to set this out with the aid of columns, as shown in Figure 7.57.

Observed	Expected	O–E	(O–E)²	(O–E)²/E
12	4.8	7.7	59.29	12.4
0	1.4	–1.4	1.96	1.4

Figure 7.57 Amos's observed and expected data

Figure 7.55 Pattern of land use in a town centre

Categories	Retail	Industry	Housing	Row totals
0–0.5 km	12	0	0	12
0.51–1.0 km	6	2	4	12
1.01–1.5 km	2	4	8	14
Over 1.51 km	0	0	12	12
Column total	20	6	24	50

> **Activities**
>
> 1 In Figure 7.57, the lines for inner zone retail and industry have been done for you. Now complete the others.

These values can then be put into the Chi-squared formula:

$$\chi^2 = \sum \frac{(O - E)^2}{E}$$

Again, the higher the value of Chi-squared, the greater the difference between the data Amos observed and a random distribution. However, to know how significant the result is, Amos needed to compare the result with the degrees of freedom, which in this case is found by (number of rows −1) × (number of columns −1) = 6.

> **Activities**
>
> 1 Use Figure 7.58 below to see how significant Amos's result was. Then explain why his result was that significant.

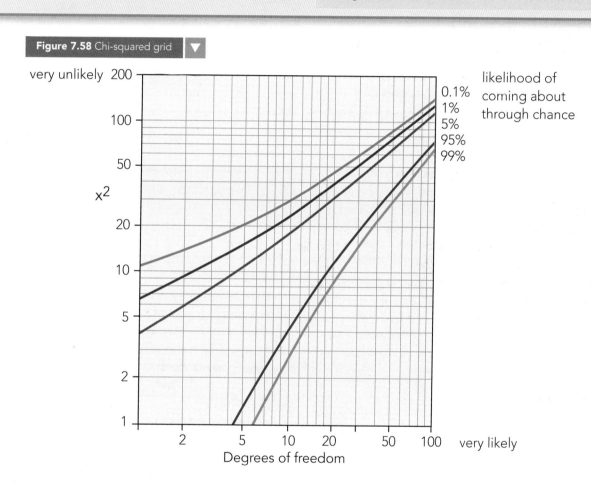

Figure 7.58 Chi-squared grid

Networks and movements

In some cases you need to analyse the pattern of movements and the routes they take. Such systems are referred to as networks. **Networks** are usually thought of in terms of transport such as roads or rail, but equally they could be less tangible such as emails, or conversations.

Key term

Network: a network consists of a number of nodes and the routes that link them.

Network analysis

Any network can be broken down into its main elements. Different terms can be used for the same thing which can cause some confusion.

- Centres in the network are called nodes or vertices (V).
- The routes are called routes or edges (E).
- Independent/unconnected parts are called sub-graphs (G).
- A completely linked set of nodes and routes is called a circuit.

Figure 7.59 shows a typical example of a network – the bus routes around a small market town

A number of measures have been devised to describe (and therefore compare) networks. The simplest is the Beta index, which is calculated using the following equation:

$$\beta = \frac{E}{V}$$

E = edges
V = vertices or nodes

This is a useful measure as a network with a complete circuit will give a score of 1. The maximum possible result is 3.

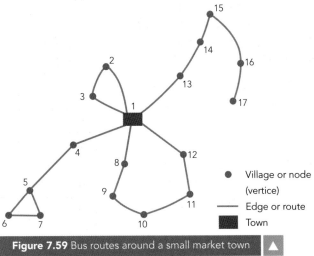

Figure 7.59 Bus routes around a small market town

The Beta index

Jodi used the Beta index to compare the motorway networks of a range of countries. She found that the Beta index increased as the level of GDP increased.

Activities

Suggest why this would happen?

A more complex measure is the Alpha index, which compares the actual number of circuits with the maximum possible within that network.

$$\alpha = \frac{E - V + G}{2V - 5}$$

E = edges
V = vertices or nodes
G = sub-graphs or independent networks

The answer is on a scale from 0 (no circuits) to 1.0, which is the maximum possible number of circuits (calculated by 2V–5).

could be based on how many edges that node has. In the example in Figure 7.59, the town has six but node 17 has one. Alternatively, an accessibility grid can be compiled. This shows how many edges are needed to connect one node to all the nodes in the network.

Key term

Centrality: a measure that indicates how central or accessible a place is in the network.

Key terms

Alpha index: a way of describing a network by comparing the actual number of circuits with the maximum possible.

Beta index: a way of describing a network. The total number of nodes is divided by the routes or vertices.

Another measure is centrality, which tells us how central or accessible a place is in the network. Centrality can be measured in a number of ways. It

Activities

1 Raja started an accessibility grid to assess the centrality of villages in Figure 7.59. Complete his grid opposite (Figure 7.60). (Remember, you only ever have to do half of it as one half duplicates the other.)

2 Add up the total edges for each node. The town (node 1) needs 35 edges to connect to all the other nodes in the network. What does node 17 score?

3 Which is the most central node? Which is second most central or accessible?

4 How could you improve the accessibility of node 17?

5 In what ways could you modify Figure 7.59 to better reflect reality?

These are just some of the types of ways of analysing networks but there are a number of limitations to these kinds of network analysis.

- They treat all routes as equally effective regardless of their width, surface quality, speed limit, etc.
- They disregard the actual distances of the routes and focus on their linkage to the nodes.
- They refer to planar networks, which means they are in the same plane so can't allow for routes going over each other, such as a motorway flyover.
- There are no allowances for the quality of the communications, there is only the route or edge. Hence you do not know what is using that route, its frequency, its speed or its capacity.

	1	2	3	4	5	6	7	8	9	10	11	12	13	14	15	16	17
1		1	1	1	2	3	3	1	2	3	2	1	1	2	3	4	5
2		0	1	2	3	4	4	2	3	4	3	2	2	3	4	5	6
3			0	2	3	4	4	2	3	4	3	2	2	3	4	5	6
4				0	1	2	2	2	3	4	3	2	2	3	4	5	6
5					0	1	1	3	4	5	4	3	3	4	5	6	7
6						0											
7							0										
8								0									
9									0								
10																	
11																	
12																	
13																	
14																	
15																	
16																	
17																	

Figure 7.60 Accessibility grid for Figure 7.59

Key term

Planar: on one plane or level. Used in network analysis to signify no flyovers, etc.

Gravity model

Another useful technique to aid investigations that involve looking at patterns of flows or movements is the gravity model. This is based on the idea that the attractiveness of one place to another is proportional to its size and the distance separating the two places.

Key term

Gravity model: formula used to predict the interaction between places based on their attractiveness and the distance separating them.

The gravity model can be used to predict the interaction between two or more places so is useful to compare against observed patterns. It assumes that the level of interaction between centres i (origin) and j (destination), for example traffic flow, shopping trips and migration, is directly proportional to the attractiveness of the destination (j) and indirectly proportional to the friction (distance, time, cost, etc.) separating them (D). This is shown as the following formula:

$$I_{ij} = f \frac{P_j}{D_{ij}}$$

I = interaction between i and j
P = attractiveness of that point
D = distance separating i and j
f = function of

Thus, the level of interaction with the destination j is a function of j's size (often shown as its population size) and how far away it is from i. If you want a two-way interaction then you add in Pi.

Usually two constants (a and b) are added to modify the friction or distance (b is often the square of the distance between the two places as distance is a big friction).

$$I_{ij} = a \frac{P_i P_j}{D_{ij}^b}$$

Activities

Imagine two towns 10 km apart. Town A has a population of 50 000 and town B has a population of 70 000.

1. Use the gravity model to forecast the level of interaction between the two towns.

$$I_{ab} = \frac{P_a P_b}{d_{ab}}$$

I = interaction
P = population
D = distance

2. Why do you think you need to add in the constants?
3. In what ways is the result unclear?
4. Do you think population is a good indicator of the attractiveness of each town? If not, what alternatives could you use?

they support it? It is not a failure if they do not, but you need to suggest why this is so.

It also includes possible explanations of any patterns or/and trends you found or anomalous results. This is where it is important to consider geographical theories and models that might explain your results, as well as possible cause and effect relationships. Which factor or combination of factors lead(s) to what pattern?

Don't forget that patterns may be random, clustered or regular, with variations such as linear and scattered (see Figure 7.61).

Key terms

Clustered: a pattern where points are close together.

Random: lacking a clear pattern.

Regular: recurring at fixed or uniform intervals.

Interpretation

This includes the interpretation of the results in relation to the original question/hypothesis that was posed – what did the results tell you about your hypothesis; did

It is also important that you suggest why anomalies (results or items of data that do not fit in to the overall pattern) occur. Sometimes they help prove a point.

Random	Clustered	Regular	Linear	Cuneiform
○ ○ ○ ○	○○○○○	○ ○ ○ ○ ○ ○	○ ○ ○ ○ ○	○ ○○○ ○

Figure 7.61 Patterns of results

Relationship between size of population and number of services

Jazar investigated the relationship between the sizes of population in villages and the number of services in an upland area. When he looked at his results it was clear that the village of Malham didn't fit into the pattern (Figure 7.62). Its population and services did not fit the trend. This meant either the village has more services than expected (due to high number of tourists) or a smaller population (due to outmigration from this relatively remote area).

Activities

Which do you think is the most likely explanation?

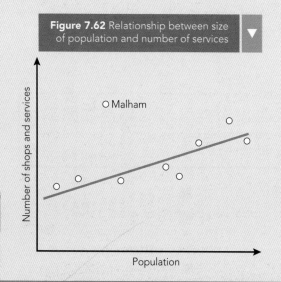

Figure 7.62 Relationship between size of population and number of services

Case study | Part of an investigation into whether soil conditions on sand dunes change with distance inland

Hypotheses

1. Soil temperature rises with distance inland.
2. Soil moisture increases with distance inland.
3. Soil pH falls with distance inland.

Data collected

Distance inland in ms	Soil temperature °C	Soil moisture %	Soil pH
3	10	20	8.0
6	11	22	7.8
9	12	24	7.3
12	12	37	7.1
15	13	33	6.5
18	9	76	5.2
21	11	47	6.7
24	12	56	6.3
27	10	54	6.2
30	9	79	5.5
33	9	78	5.3
36	10	65	5.6
39	12	75	5.4
42	10	74	5.2
45	9	76	5.4

Figure 7.63 Data collected during the investigation

Data analysis

Hypothesis 1   Soil temperature rises with distance inland

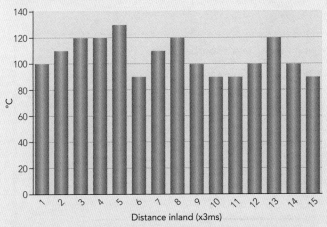

Figure 7.64 Soil temperature in relation to distance inland

Figure 7.64 shows little clear trend but does show an anomaly at 18 metres and again at 30 and 33 metres inland. To see if there is any trend, I used Spearman's Rank correlation (see appendix for workings). This uses Ranked data but the number of tied ranks reduced its effectiveness. The null hypothesis was that soil temperature increased with distance inland.

The result was:

$$Rs -.35 \text{ with 14 degrees of freedom}$$

This is outside the 95 per cent significance level but shows a negative trend suggesting temperatures fall with distance inland. This could reflect the greater shade, due to more plants, or moisture content which may keep soil temperature down. The anomalies coincided with slacks in the dunes which were shaded and wetter.

My hypothesis has not been proved.

Hypothesis 2   Soil moisture increases with distance inland

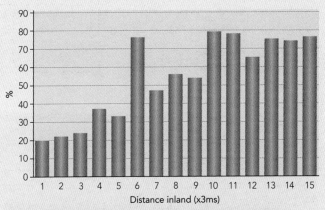

Figure 7.65 Soil moisture in relation to distance inland

Figure 7.65 shows a clear trend with soil moisture increasing inland but it does show a negative anomaly at 18 metres and again at 30 and 33 metres inland. To see what the strength of this trend is I again used Spearman's Rank correlation (see appendix for workings). The null hypothesis was that soil moisture increased with distance inland.

The result was:

$$Rs +.79 \text{ with 14 degrees of freedom}$$

This is well inside the 99 per cent significance level and shows that this is a strong positive correlation suggesting soil moisture does increase with distance inland. This could reflect the greater proportion of moisture retaining humus inland due to more plants or the higher temperatures and more sandy soil (which is easily drained) found nearer the sea. Again the anomalies coincided with slacks in the dunes which were shaded and wetter being lower, nearer the local watertable.

My hypothesis has been proved.

Hypothesis 3 Soil pH falls with distance inland

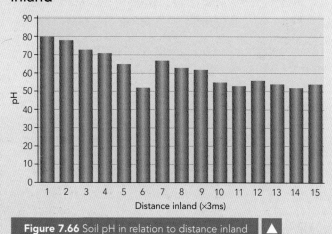

Figure 7.66 Soil pH in relation to distance inland

Figure 7.66 shows a clear trend with soil pH decreasing inland but again it does show a negative anomaly at 18 metres and again at 30 and 33 metres inland.

To see what the strength of this trend is, I again used Spearman's Rank correlation (see appendix for workings). The null hypothesis was that soil pH decreased with distance inland.

The result was:

$$Rs +.81 \text{ with 14 degrees of freedom}$$

This is well inside the 99 per cent significance level and shows that this is a strong positive correlation suggesting soil pH does decrease with distance inland. This could reflect the greater proportion of acidic humus inland due to more plants or the higher proportion of lime rich shell fragments and sea salt found nearer the sea. Again, the anomalies coincided with slacks in the dunes which had more decaying vegetation and being wetter probably had any lime fragments washed out.

My hypothesis has been proved.

Interpretation

The analysis suggests two of the hypotheses are valid and one is not – may even be reversed. Clearly the anomalies in the readings are consistent and suggest the slack – a lower wetter area between dune ridges – has its own micro-conditions. But why should there be a link between distance inland and these soil conditions?

Activities

Selina produced an interpretation section (see below) for her report on an investigation examining social differences between an inner urban and a suburban area in a large town.

1 Read Selina's interpretation and assess its effectiveness.
2 In what ways could you improve it?

> **An interpretation section examining social differences between inner and suburban areas**
>
> Marital status is another area with key differences. My bar graph shows more single (approx 15 per cent) people living in the Castle area than Walmsley and more married/re-married couples in Walmsley (approx 15 per cent). This could be due to married couples wanting to have families, which would suit the detached houses in the suburbs.
>
> Differences can be found in age between the two wards (see percentage bar graph); for instance, over 10 per cent more over-60s living in the Walmsley ward, likely due to retirement and wanting to spend their last days away from the city. This also explains the 7 per cent higher retirement figures in Walmsley (see enclosed pyramid graph).
>
> In conclusion, there are different physical characteristics in that the property condition is superior in the suburban ward and there are more housing types associated with more space. Socially and economically the above explains several differences, which again mostly swing in favour of the suburban ward being better.

7.6 Present a summary of the findings and an evaluation of the investigation

'A successful geographical investigation involves a clear summary of the findings and an evaluation of the success of the investigation in relation to wider geographical ideas and the methods employed and data collected.'

This means that you will need to:

- Use the evidence presented in your previous sections to provide a clear summary of what you found and so produce a conclusion which relates back specifically to the original question/hypothesis posed. Does your evidence support the hypothesis or not? If not, why not? It is not a disaster if the evidence from your analysis does not support your original hypothesis, provided you can explain why this has happened.

- Carry out an evaluation of the extent to which the study supports or otherwise the general geographical theories, ideas or concepts that underlie the topic being studied. Does the study fit the theory or model? If it does then this needs to be explained with the aid of the concepts that underpin the theory. If not then it is worthwhile considering the appropriateness of the model or theory or trying to offer an explanation of why your findings do not fit.

David's investigation

David found that the land use pattern he had mapped for Southend differed greatly from the Burgess model that he was comparing it with. So was the model wrong or was Southend's land use pattern based on some other model? David decided that it had elements of the Hoyt model along the main roads and railway lines, but the outer areas resembled the multi-nuclear model of land use.

An evaluation of the effectiveness and limitations of the study in terms of the methods used

It should be recognised that conclusions are often partial, tentative and incomplete – there are always limitations that prevent one from being totally confident in the effectiveness of the study.

Such limitations could include the aspects listed in Figure 7.67. Another set of limitations relate to the accuracy and reliability of the data collected. Some of this stems from the methodology but some may reflect the nature, type and source of that data. Suggested improvements or extensions should then be offered. These could include:

- repeating the investigation at different times of the day, week or year to see if results differed

- repeating the investigation in different locations, thereby comparing your results with a similar or different area

- taking a larger sample of data, usually by increasing the size of the sample or by changing the sampling unit, method or interval (see Katy's investigation on page 294)

- taking into account variables other than those looked at in the investigation; often you find there were factors or variables that were more important than you initially expected

- using a better methodology – this will usually focus on data collection methodology and should be a response to the limitations that had been identified earlier in this section; it might suggest other sources of primary and secondary data that could clarify a particular aspect

- improve the equipment used – again, this should be a practical response to previous limitations; it might be to improve existing equipment or improve its use or effectiveness, but equally it could be to suggest new types of equipment that might be better able to collect reliable and accurate data.

Time	There could be insufficient time allocated to complete the tasks or they could be carried out on the wrong time of day/year (see Tazir's investigation).
Place	The location of the investigation might have proved unsuitable due to limitations on access, its size or scale, risks of health and safety, and the ability to move around the area being limited by natural or human barriers such as fences.
Physical	This could include poor weather, difficult slope angles, soft or muddy ground, problems with rivers/drainage, and difficult vegetation, such as dense undergrowth causing problems or barriers (see Abdul's investigation).
Human error	There may be a lack of team work with some individuals underperforming or misunderstanding their allocated tasks, careless or inaccurate/unreliable measurements, and possibly different interpretations of the tasks or results by some individuals or groups.
Equipment	Equipment may be faulty, fail to do what was needed or provide inappropriate, unreliable or inaccurate data when used in the field.
Methodology	Sometimes the methodology is flawed from the start, often due to superficial planning of the investigation. Typically the size of sample or type of sampling proves inappropriate or inaccurate.

Figure 7.67 Potential limitations of the study

Abdul's investigation

Abdul found that his stream study was plagued by muddy ground following recent floods and the fact that it rained most of the day. Being thoroughly wet and miserable did not help the accuracy of his measurements.

Tazir's investigation

Tazir did her survey of tourist activity in Bournemouth during a week's fieldtrip in the summer to the area, but she did not survey at the weekends so her results might be unrepresentative of the full week.

Katy's investigation

Katy found that her dune survey produced confusing results on moisture content and pH of the soil. This was because her line transect and systematic sample showed conflicting results as areas of wetter slacks were included in her sample points. She realised that a stratified areal sample on the three ridges would have produced a more accurate picture of the changes in conditions on the dunes with distance from the sea.

It is important to be well focused and you should be wary of making vague statements such as: 'If the investigation was repeated I would use more scientific methods of collecting the data to reduce the likelihood of errors.'

An example of a candidate's summary

As a result of my investigation I conclude that Thumble Beck does fit the theories of stream channel changes with distance downstream. Although this fact was proven by my results, there were some anomalies which could have been prevented if different techniques had been used.

One way we could make the results more reliable would be to use a more systematic approach when choosing locations. The locations were very randomly spaced, some were 0.1 km apart and others were 1.7 km apart, meaning that the results were not as effective as if they had been equally spaced showing more clear changes. Any unreliable results due to this and other factors may have led to inaccurate conclusions.

In addition, often when we took results for averages at locations they were all measured very close together. For example, at location 4 (Edwin's farm) some of the measurements were slightly distorted because they were taken on an untypically narrow part of the meander. Therefore, with further spaced apart measurements at each location, the averages would show a more accurate representation of that particular stretch of river. In order to improve the results, I would have revisited this site to take more precise measurements.

Finally, we would have gained more accurate results if more measurements had been taken at each location, for example ten measurements of the depth instead of five. Consequently, more figures would have meant a more accurate average result would have been gained.

Activities

Look at the example of a candidate's summary of a research investigation above.
1. Is it clear what the research was about?
2. Is this an effective evaluation? Give reasons for your answer.
3. What would you do to improve its effectiveness?

Exam Café
Relax, refresh, result!

Relax and prepare

What I wish I had known at the start of the year...

Shahid

"I used to worry that I would have to do calculations in the exam and I always got the formula wrong. I now know that this won't happen as questions ask about the meaning of the end results and give the formula or table to help me interpret the result."

Mary

"The idea of a null hypothesis always confused me. Why were we testing that rather than what I wanted to test? It was explained to me that what we were testing for was the likelihood that the relationship had come about by chance, by accident or coincidence – the null hypothesis. If chance was unlikely then something must have caused that relationship. Then apparently it was up to me to use my logic to work out the most likely cause."

Student tips

Karjek

"I know that research expects a 95 per cent or better fit before it can be said that there can be confidence in the result. This helps me look for that level of significance in the various graphs or tables that are used to assess how meaningful the results are of various tests for difference or correlation."

Common mistakes – Iona

"It is all too easy to confuse some of the statistical tests especially as I don't like mathematics. Apparently I often use Spearman's Rank when it should be the Mann-Whitney U-test. The latter looks at two sets of ranked data and indicates if they are clearly different. The result is then looked up in tables and checked to see if it could come about by chance. That's why I confuse it with Spearman's Rank as you look that result up in tables of significance as well. Spearman's Rank tests for relationships (correlation) between two sets of ranked data. It seeks to show if there appears to be a link and its direction. It doesn't prove it's a causal link. So remember the key terms – testing for *difference* means Mann-Whitney U-test, whilst testing for *relationships* means Spearman's Rank."

Refresh your memory

7.1 Identify a suitable geographical question	
It should be:	Suitable in scale, size, area
	Capable of research – practical issues, e.g. time, area, data available
	Clearly geographical – strong sense of location, where, pattern, etc.
	Based on geographical theory, ideas, concepts, models
	Logical – it makes sense, especially cause-effect
	Clearly located – have an actual site where it can be carried out

7.2 Develop a plan and strategy	
Identify data needed	Sources of primary and secondary data, how much data, access to data sources
Strategy for collecting	Intended method, e.g. questionnaire, type of sampling strategy, timing, division of labour, equipment needed, timeline
Understand limitations	In terms of time, location, equipment, resources, access
Appreciate risks	Potential risks, their seriousness and likelihood, and possible ways of reducing these
Pilot	Need a pilot or reconnaissance to check viability, e.g. questionnaire

7.3 Collect and record data	
Primary data	Equipment and methods used – accuracy and reliability
	Questionnaire design – type of questions used (closed versus open)
	Sampling size and type – systematic, random, stratified, pragmatic
	Sampling unit – area, point, linear
	Map of sample sites, repeated sampling for accuracy
Secondary	Source, date, reliability, accuracy
Recording	Nature of tally sheet; any problems

7.4 Present the data in appropriate forms	
Presentation	Type – maps, diagrams, graphs, photos, charts; should fit type of data, e.g. continuous data = line graph. Remember spatial
	Use of colour, ease of comparability (e.g. same scale)
	Annotation – location, detail, etc.
Standards	Key, scale, title, north (if map)
	Logical organisation to help analysis, i.e. number figures

Refresh your memory

7.5 Analyse and interpret the data	
Analysis	Descriptive techniques, e.g. mean, mode, median, scatter graphs
	Central tendency – range, interquartile range, standard deviation
	Location of data, e.g. nearest neighbour
	Statistical test for difference, e.g. Mann-Whitney U-test, Chi-squared
	Statistical test for correlation (linkage), e.g. Spearman's Rank
	In tests, remember null hypothesis and degrees of freedom influence the result, and 95 per cent is the minimum expected accuracy
Interpretation	Look at results in context of original question(s)
	Try to offer explanation for patterns/links/trends and any anomalies
7.6 Present a summary and an evaluation	
Summary	Clear conclusion linked back to original question, proved or not?
	Does it support the model/concept – suggest why or why not
Evaluation	Limitations – time, equipment, method, personnel, location; explain effects of these on results
	Suggestions for improvement – other times/places, improved method/equipment

Top tips...

▷ You will never be asked to do a complete calculation of any statistical method in the examination. Instead you may be asked to interpret a result using tables of significance so you can say what a particular result means, or you will be asked to suggest appropriate statistical techniques for a particular aspect of data analysis such as identifying differences between sets of data. A2 is less about doing the technique and more about its appropriate use and its interpretation.

▷ It would still be advisable to attempt some of the techniques as you will learn the technique's limitations and requirements (such as the meaning of 'degrees of freedom').

▷ Do remember that a simple average is a statistical technique. So many candidates go for the complex rather than simple ones that are often more useful.

Get the result!

Example question

Outline the main uses of GIS in geographical investigations. [5 marks]

Student answer

Geographic Information Systems (GIS) have transformed the way data is portrayed as it ties data to a specific location. It replaces tedious mapping operations with interactive manipulation of large sets of digital maps and data to select, display and interpret spatial patterns and relationships. The use of overlays, vector map layers and 3D images help visualisation and allow researchers to experiment with cartographic techniques. They also allow access to up-to-date locational data via the Internet and can use data produced in the field such as weather data loggers.

Examiner says

This is a detailed and robust answer that sets out early on why it is important for geographical investigations. It then elaborates this with good use of terminology.

Examiner says

Again, this response is well linked to the question of use in research. It extends the use into its wider context, such as the ability to access the Internet and its application in the field.

Examiner's tips

Questions are usually marked on a level basis – level of knowledge, understanding, evaluation and the ability to draw conclusions. Higher levels can be reached by showing:

- detail with good understanding and knowledge
- ideas developed effectively, especially the relative merits of a particular strategy or choice of method of representation and analysis
- that examples, data and evidence are clearly integrated into the answer
- effective use of geographical research terminology
- an effective use of written communication often supported with maps or diagrams.

But above all for this paper, higher levels are achieved by the ability to justify the choice of:

- method, e.g. type of sampling strategy
- form of representing the data, e.g. pie chart
- analysis, e.g. use of Spearman's Rank correlation.

Exam Café

Index

Aberfan disaster (1966) 12
acid rain 54, 63, 116–17, 122
Africa 186, 188–9, 220, 237
agriculture 60, 62–3, 71, 81
   CAP (EU) 150, 153–6, 189
   hazards and 91, 99, 113, 121, 222
   UK 67, 68, 137
   water supplies 112, 125, 138, 139, 141, 142
   wheat supplies 142–3
aid 188, 195–8, 204, 217, 219, 228, 237
   development aid 33, 180, 195, 196, 197–8, 221
   emergency aid 31, 109–10, 111–12, 122, 195, 196
air masses 96–9, 99
air quality 102, 104, 118–19
anticyclones 99, 102–5, 118, 121
Arches National Park (US) 76–7
Asia, economic development 164–6
atmospheric systems 96–105

Bangladesh 128, 196, 197–8, 230
   demand for resources 146–7
   economic development 213, 214–15, 221–2, 223
blizzards 100, 101, 121
Bolivia 193–4, 199–201

California (USA) 139–42
CAP (Common Agricultural Policy) 150, 153–6, 189
Caribbean 87, 90–2, 108–10
central tendency 277–8
CFP (Common Fisheries Policy) 150–3
Chad, water development 111–12
Chi-squared test 281–2, 286–7
China 142, 146, 230
   and aid 220, 237
   economic development 18, 170, 213, 214, 215, 220, 223
   economic growth 124, 147–8, 165, 166, 176–7, 177, 212, 220
   and EU 176, 192, 193
   globalisation and 167, 168, 176–7
   industrialisation 147–9, 172, 175, 229
   inequalities 220, 224–5
   pollution 117, 176, 177, 220
   quality of life 213, 214, 215
   resource demand 146, 147–8, 176, 177
   trade 176, 191, 192–3, 220
   Yangtse River flooding 16–20
civil society 171, 174
classifying countries 209–12
climate change 19, 20, 22, 31, 105–7, 110
   see also global warming
climatic hazards 84–5, 105–8
   see also acid rain; anticyclones; drought; global dimming; global warming; hurricanes; smog; snowfalls; storms; tornadoes
coal 113, 134, 136, 147–8
coastal flooding 20–2, 24
cold spells 99–101, 102–3, 121
Colorado River (USA) 139–41
Common Agricultural Policy (CAP) 150, 153–6
Common Fisheries Policy (CFP) 150–3
conservation 67, 68, 72, 73–4, 78, 81, 203, 229
coral reefs 71–2, 73–4
core-periphery model 229–31
correlations 284–6
creep 10, 12–13, 43
cultural globalisation 170–1
cultural issues 14, 130, 158, 164

dams 16, 17, 18, 19, 112, 176
data 253–4
   analysis 263–4, 276–90, 298
   collection 253, 258–62, 262–3, 294, 297
   interpretation 263–4, 290–2, 298
   presentation 263, 265–76, 297
debt 180, 197, 237
deforestation 52, 70, 72, 110, 113, 114, 173, 177, 203
   and flooding 18, 92, 110
deindustrialisation 144, 145, 148, 174, 175
demographic change 129–30
demographic transition 126–8, 158, 159
dependency theory 231
depressions 86, 87, 88, 96–8, 99–101, 121
development 188–94, 206–7, 208, 209, 240–1, 242–3
   see also economic development
development gap 178–90, 225–31, 244

301

development stairway model   209–11, 217, 226, 229, 231, 240
differences, testing for   280–2, 296
dispersion, measures of   278
drought   104–5, 106, 110–12, 121, 142, 143
dry deposition (acids)   116–17

earth hazards   6–7, 12, 13, 14–15, 26, 30, 31–3, 34–41
  see also earthquakes; flooding; mass movement; volcanic hazards
earthquakes   8, 24–6, 30, 31, 41
  impacts 6, 26–7, 29–30, 34–5, 36, 37, 41
  management   38, 39–40
ecological footprints   145–6
economic development   144, 206–7, 208, 209, 216–17, 229–31, 240–1, 242–3
  Asia   164–6
  Bangladesh   213, 214–15, 221–2, 223
  China   213, 214, 215, 220, 223
  and inequalities   235–6
  Japan   213, 214–15, 218–19, 223
  measures of   210–11
  and pollution   177
  and quality of life   212–16, 225–6
  and resource consumption   146, 147
economic globalisation   167–70
economic growth   112, 130, 133, 166–7, 212, 217, 240–1
  China   124, 147–8, 212, 220
ecosystems   46–8, 52–5, 81, 106–7
  stores and flows   48–52
  see also acid rain; salt-marsh ecosystems; woodlands
education   130, 159, 175, 199, 203, 214, 215
Eldfell (Iceland)   36, 41
employment   144, 176, 203
energy flows, ecosystems   48–51
enhanced greenhouse effect   113–14
environment   29, 206
  globalisation and   162, 164, 172–3, 175, 176, 177, 203
  inequalities and   232, 234–5, 244
  see also ecosystems; physical environments; pollution
Epping Forest   64, 65–6, 69
Essex, salt-marshes   62, 68
EU (European Union)   114, 164, 175, 190, 191, 238
  CAP   150, 153–6, 189
  CFP   150–3
  and China   176, 192, 193
  trade   189, 191, 192, 193
Europe, drought (2003)   104–5
evaluation   293–4, 298

FDI (foreign direct investment)   165, 166, 176, 180, 181, 198–9, 201, 236
  UK and   174, 175

financial markets   167, 173–4, 174, 175, 203
flash floods   107–8, 121
flood management   19–20, 21–2, 23–4, 38, 39
flooding   15, 16, 42, 43, 86, 221
  coastal flooding   20–2, 24
  flash floods   107–8, 121
  hurricanes and   89, 92
  impacts   16–17, 18–19, 20, 43, 107–8
  The Netherlands   21–4
  river flooding   16–20, 22–4
  Yangtse River (China)   16–20
flow lines   272–4
food   124, 125, 131, 137, 142–3, 150, 175, 214
  EU dumping of surpluses   156, 189
food webs   49–51, 56
footpath trampling   63, 65–6, 67, 75
forest fires   105–7, 121
fossil fuels   54, 113, 116, 117, 122, 134, 177
frequency diagrams   279
frontal weather   96–8

GDP (gross domestic product)   170
  per capita   210–11, 233, 243
geographical investigations   248–9
  analysis   263–4, 276–90, 298
  data collection   253, 258–62, 262–3, 294, 297
  evaluation   293–4, 298
  interpretation   263–4, 290–2, 298
  planning and strategy   253–8, 297
  presentation of data   263, 265–76, 297
  summary   293, 295, 298
geographical questions   248–9, 250–2, 297
Gini coefficient   214, 215, 216
GIS (Geographic Information Systems)   262–4
global dimming   115, 122
global shift   164–6
global warming   74, 113–14, 122, 203
globalisation   162–71, 202, 203–4, 217
  and development gap   178–90
  and environment   162, 164, 172–3, 175, 176, 177, 203
  impact of   145, 174–7, 188, 221, 228–9, 231, 235–6
  issues   164, 172–80, 203
  managing impact of   198–201
  see also aid; TNCs; trade
globalised countries   168–9
GNI (gross national income) per capita   210–11, 211–12, 212, 213–15, 225, 226, 228, 240
Goma (Democratic Republic of Congo)   31, 35
GPS   33, 263
graphs   269, 277
gravity model   289–90
Great Barrier Reef (Australia)   73–4
greenhouse effect   113–14
greenhouse gas emissions   20, 72, 113, 114
Grenada, Hurricane Ivan   90–1

Haiti   92, 109, 110, 111
hazards see climatic hazards; earth hazards
HDI (Human Development Index)   178, 215, 226, 232, 243
health   118–19, 121, 178, 214, 215, 235
heatwaves   102, 104–5
Heimaey (Iceland)   36, 41
high pressure   99, 102–5, 118, 121
Holbeck Hall Hotel (Scarborough)   11, 12, 13
human resources   133, 158, 169, 218, 221
hurricanes   86–92, 94, 108–10, 121
hypotheses   248–9, 250–2, 253, 277, 290

ICs (industrialising countries)   210, 226, 231, 234
IMF (International Monetary Fund)   173, 189, 195, 198, 237
India   142, 167, 170, 212, 229
    economic growth   165, 177
    industrialisation   172, 212, 229
Indonesia   70–2, 105–7
industrialisation   18, 124, 125, 128, 144–5
    China   147–9, 172, 175
    India   172, 212, 229
industrialising countries see ICs
inequalities   174, 175, 206–7, 208–9, 223–5, 242, 244
    China   220, 224–5
    Gini coefficient   214, 215, 216
    issues   232–5, 244
    reducing   235–40, 244
    spatial   224–5, 233, 238–9, 242, 243
International Monetary Fund see IMF
Internet   171, 202, 203, 215, 254, 275
investigations   248–9

Japan   114, 165, 180–1, 230
    earthquake observation   6, 36, 38
    economic development   164, 166, 213, 214–15, 218–19, 223
    resources   133, 219
    trade   186, 187, 192, 219

lahars   30, 31, 35–6, 36
landslides   8, 9, 11, 12, 18, 27, 31, 38, 121
LDCs (least developed countries)   179–80, 210, 211
LEDCs (less economically developed countries)   78, 114, 136, 145, 209, 210, 211
    and globalisation   164, 228–9
    impact of hazards   6, 14–15, 26, 31–3, 34, 39, 105–8, 120
    and physical environments   70–2
    population   127, 128, 129, 161
    trade   188–9, 193–4, 204
living standards   142, 144, 206
Lorenz curve   283–4
Los Angeles (USA)   6, 26, 118, 139, 140, 141

low pressure systems see depressions; hurricanes

Malthus, Rev. Thomas (1766–1834)   131, 132, 160
Mann-Whitney U-test   280–1, 296
maps   270–2
mass movement   8–15, 38, 43
MEDCs (more economically developed countries)   114, 129, 144, 145, 209, 210, 211
    and aid   195, 237
    demographic transition   127, 128
    flooding 20, 21–4
    and globalisation   164, 228–9
    impact of hazards   6, 12, 13, 14–15, 26, 30, 34, 36, 39, 105
    and physical environments   67–9, 234
    resource demand   136, 144–5, 146, 148, 149
    trade   188–9, 190–1, 193–4, 202, 204
Mediterranean, pollution   69
Mexico   234
Mexico City, pollution   234–5
migration   159, 164, 169, 170, 203, 227, 244
    international   173
    net migration   130, 159
    rural-urban   127, 176, 224–5
    UK   174, 175
Montserrat   31–3, 35–6, 37
Mount Pinatubo (Philippines)   35, 36, 37, 38
Mount St Helens (US)   30, 35, 37
mudflows   10, 11, 27, 30, 31, 35–6, 36, 121
mudslides   14–15, 18, 86, 107–8

nation states   164, 165, 171, 198
national parks   67, 68, 69, 72, 75, 76–7, 82
nationalisation   200
natural gas   113, 136, 137, 147, 190, 200
natural hazards   38
    see also climatic hazards; earth hazards
natural resources   133, 217, 219, 220
    see also resources
nearest neighbour   282–3
The Netherlands   21–4, 155, 237
network analysis   287–9
NGOs (Non-Governmental Organisations)   179, 188–9, 190, 195, 197
NICs (newly industrialised countries)   124, 127, 164–5, 167, 172, 181, 204, 210, 211
    industrialisation   145, 148
    trade   176, 192–3
Nike   182, 183–4
North Sea   20–2, 136, 137, 190
North-South Divide   209, 225, 226, 228–9
Northridge, Los Angeles (US)   6, 26
null hypothesis   277, 280, 296
nutrient cycles   51–2, 56

oil   113, 125, 134, 135, 201, 217
    consumption   136, 146, 147, 177
    North Sea   136, 137, 190
OPEC (Organization of the Petroleum Exporting Countries)   210, 211
optimum population   131
overfishing   125, 133, 150–2
ozone   104, 118–19, 235
ozone layer   114

Philippines, mudslides   107–8
photochemical smog   102, 118–19
photographs   259, 275–6
physical environments
    human impact on   62–72, 81–2
    managing   72–8, 81, 82
    see also ecosystems
planning controls   75–7
plate tectonics   24–6
pollution   60, 63, 81, 121, 203, 234, 244
    China   117, 176, 177, 220
    and economic development   175, 177
    Mediterranean   69
    Mexico City   234–5
    see also acid rain; smog
population   126, 129, 131
    and resources   124–5, 131–2, 160, 223
population density   34, 35, 144, 146
population growth   18, 70, 112, 126, 129, 131–2, 146, 159, 175, 206
    demographic transition   126–8, 158, 159
poverty   178–9, 214–15, 224, 227, 240, 244
primary data   253–4, 254, 258, 297
privatisation   199, 200–1
proportional symbols   267–8

qualitative data   265, 276–7
quality of life   135, 206, 212–16, 223, 225–6, 232–3
quantitative data   265–6, 277

recycling   137, 141, 149–50, 158
renewable energy   133–4, 137
resource management   133–4, 141–2
resources   124–5, 133–7, 158, 159
    demand   138, 144–9, 160, 172–3, 176, 177
    and development   146, 147, 206
    managing supply and demand   149–56, 160
    population and   124–5, 131–2, 160
    supply and use   137–43, 159
Rhine, River, floods (1995)   22–3
risk assessment   32, 33, 37–40, 122
    for investigations   256
river flooding   16–20, 22–4
rural-urban migration   127, 176, 224–5

salt-marsh ecosystems   55–62, 68
sampling   253, 254–5, 260, 294
satellites   89–90, 122
sea level rise   22, 113
secondary data   254, 258, 297
sector model   144
service sector   144, 145, 212, 213–14, 218, 220
    globalisation and   165, 174, 176
services   186, 187, 191, 202
significance, statistical   280, 284
sketches   259, 274
slumps   11, 12, 13, 38, 43
smog   102, 105–7, 118–19, 122, 235
snowfalls   99–101, 121
social issues   130, 159, 164, 175, 176, 203
    inequalities   174, 175, 232–3, 235, 244
The Solent   60, 61
solifluction   8, 9
Soufrière Hills volcano (Montserrat)   31–3, 35, 37
Soviet Union   167, 171, 209, 210
spatial inequalities   224–5, 233, 238–9, 242, 243
spatial patterns   270–4, 282–4
Spearman's Rank   284–6, 296
stages of growth model   231
standard deviation   278, 279
storm surges   20–2, 24, 89, 91
storms   98, 113
successions   52–5, 58–9
Sumatra (Indonesia)   105–7
sustainability   72–8, 82, 124, 149–50, 158, 160
sustainable development   78, 133, 199
systems approach   48, 81

technological advances   131, 136–7, 149, 150, 151, 159
tectonic processes   24–6
tertiary sector   144, 145, 165, 174, 176, 212, 213–14, 218, 220
Three Gorges Dam (China)   16, 17, 18, 19, 176
TNCs (transnational corporations)   70, 164, 165, 168, 171, 175, 180–6, 198, 201, 217, 236
    Japan-based   180–1, 219
tornadoes   93–5
tourism   69, 73, 74, 91, 107, 135, 165, 176
    sustainable   72, 78
    UK   104, 175
Toyota   184–5
trade   166–7, 170, 202, 204, 217, 228, 236
    Bangladesh   221–2
    Bolivia   193–4, 201–2
    China   176, 191, 192–3, 220
    and development   188–94
    Japan   186, 187, 192, 219
    LEDCs   188–9, 193–4, 204
    MEDCs   188–9, 190–1, 193–4, 202, 204
    NICs   176, 192–3

    TNCs and 180
    UK 174, 175, 186, 187, 190–1
    Zimbabwe 227
trade blocs 164, 190, 193
tragedy of the commons 151
transnational corporations see TNCs
transport 137, 171, 173, 175, 203, 219
    climatic hazards 99, 101, 121
tropical rainforests 70, 78, 113, 133
tropical storms 86–92, 121
tsunamis 6, 20, 27–9, 31, 39–41

UK (United Kingdom)
    agriculture 67, 68, 137, 155
    aid 196, 197–8, 237
    air masses 96
    climatic hazards 93, 98, 101, 102–3, 104–5, 118
    deindustrialisation 174, 175
    economic growth 177
    and environment 67–8, 69, 175, 234
    and EU 155, 175
    FDI 174, 175, 180
    globalisation and 174–5
    greenhouse gas reductions 114
    migration 174, 175
    North Sea oil and gas 190
    reducing inequalities 238–9
    services 174, 186, 187, 191
    technological advances 136–7
    tourism 104, 175
    trade 174, 175, 186, 187, 190–1
    and Zimbabwe 226

UN (United Nations) 164, 217, 237
urbanisation 18, 128, 144, 239
USA (United States of America) 69, 148, 176, 185, 186, 193, 201
    climatic hazards 87, 88, 93, 93–4, 95, 101
    ecological footprint 146
    economy 170, 180
    greenhouse gas reductions 114
    industrialisation 144–5
    population 128, 132, 148
    resource consumption 136, 144–5, 145, 148, 149
    and South America 201
    trade 186, 187, 189, 191, 192, 193
    water supply problems 138–42

Venezuela, mudslides 14–15
Vietnam 165, 166, 185, 186
volcanic activity, distribution 24–5
volcanic hazards 30–3, 35–8, 41

water development, Chad 111–12
water supplies 110–12, 125, 173, 178, 214, 234, 235
    western USA 138–42
wheat, global supplies 142–3
woodlands 62–8
World Bank 110, 189, 195, 199, 211, 217, 237
WTO (World Trade Organization) 156, 164, 167, 179, 189, 193, 217, 236

Yangtse River (China) 16–20

Zimbabwe 212, 226–8

Your A2 Geography CD-ROM

Opposite you will find the A2 Geography CD-ROM. Open up the CD, explore its contents and develop your geographical knowledge and skills further.

LiveText

On the CD you will find an electronic version of the Student Book, powered by LiveText. As well as the student book and the LiveText tools there are additional case studies, extension activities and weblinks. Within the electronic version of the Student Book, you will also find the interactive Exam Café.

Immerse yourself in our contemporary interactive Exam Café environment! With a click of your mouse you can visit 3 separate areas in the café to **Relax, Refresh your memory** or **Get the result**. You'll find a wealth of material including:

- Revision tips from students, Common mistakes and Examiner's hints
- Language of the exam (an interactive activity)
- Revision checklists
- Example questions (which you can try) with student answers and examiner comments.

Minimum system requirements

- Windows 2000, XP Pro or Vista
- Internet Explorer 6 or Firefox 2.0
- Flash Player 8 or higher plug-in
- Pentium III 900 MHZ with 256 Mb RAM (512 Mb for Vista)
- Adobe Reader 6® or higher

To run your Exam Café CD, insert it into the CD drive of your computer. It should start automatically; if not, please go to My Computer (Computer on Vista), click on the CD drive and double-click on 'start.html'.

If you have difficulties running the CD, or if your copy is not there, please contact the helpdesk number given below.

Software support

For further software support between the hours of 8.30–5.00 (Mon–Fri), please contact:

Tel: 01865 888108

Fax: 01865 314091

Email: software.enquiries@pearson.com